Selected Titles in This Series

Volume

11 **Tibor Krisztin, Hans-Otto Walther, and Jianhong Wu**
Shape, smoothness and invariant stratification of an attracting set for delayed monotone positive feedback
1999

10 **Jiří Patera, Editor**
Quasicrystals and discrete geometry
1998

9 **Paul Selick**
Introduction to homotopy theory
1997

8 **Terry A. Loring**
Lifting solutions to perturbing problems in C^*-algebras
1997

7 **S. O. Kochman**
Bordism, stable homotopy and Adams spectral sequences
1996

6 **Kenneth R. Davidson**
C*-Algebras by example
1996

5 **A. Weiss**
Multiplicative Galois module structure
1996

4 **Gérard Besson, Joachim Lohkamp, Pierre Pansu, and Peter Petersen**
Miroslav Lovric, Maung Min-Oo, and McKenzie Y.-K. Wang, Editors
Riemannian geometry
1996

3 **Albrecht Böttcher, Aad Dijksma and Heinz Langer, Michael A. Dritschel and James Rovnyak, and M. A. Kaashoek**
Peter Lancaster, Editor
Lectures on operator theory and its applications
1996

2 **Victor P. Snaith**
Galois module structure
1994

1 **Stephen Wiggins**
Global dynamics, phase space transport, orbits homoclinic to resonances, and applications
1993

Fields Institute Monographs

The Fields Institute for Research in Mathematical Sciences

Shape, Smoothness and Invariant Stratification of an Attracting Set for Delayed Monotone Positive Feedback

Tibor Krisztin
Hans-Otto Walther
Jianhong Wu

American Mathematical Society
Providence, Rhode Island

The Fields Institute for Research in Mathematical Sciences

The Fields Institute is named in honour of the Canadian mathematician John Charles Fields (1863–1932). Fields was a visionary who received many honours for his scientific work, including election to the Royal Society of Canada in 1909 and to the Royal Society of London in 1913. Among other accomplishments in the service of the international mathematics community, Fields was responsible for establishing the world's most prestigious prize for mathematics research—the Fields Medal.

The Fields Institute for Research in Mathematical Sciences is supported by grants from the Ontario Ministry of Education and Training and the Natural Sciences and Engineering Research Council of Canada. The Institute is sponsored by McMaster University, the University of Toronto, the University of Waterloo, and York University and has affiliated universities in Ontario and across Canada.

1991 *Mathematics Subject Classification.* Primary 34K15;
Secondary 58F12, 58F22, 34C30.

Library of Congress Cataloging-in-Publication Data

Krisztin, Tibor, 1956–
Shape, smoothness, and invariant stratification of an attracting set for delayed monotone positive feedback / Tibor Krisztin, Hans-Otto Walther, Jianhong Wu.
p. cm. — (Fields institute monographs, ISSN 1069-5273 ; 11)
Includes bibliographical references and index.
ISBN 0-8218-1074-X (alk. paper)
1. Delay differential equations—Numerical solutions. 2. Initial value problems. 3. Attractors (Mathematics) I. Walther, Hans-Otto. II. Wu, Jianhong. III. Title. IV. Series.
QA371.K735 1999
515′.35—dc21 98-44070
CIP

∞ The paper used in this book is acid-free and falls within the guidelines
established to ensure permanence and durability.
This publication was prepared by The Fields Institute.
Visit the AMS home page at URL: `http://www.ams.org/`

10 9 8 7 6 5 4 3 2 1 04 03 02 01 00 99

Contents

Preface

This book contains recent results about the global dynamics defined by a class of delay differential equations which model basic feedback mechanisms and arise in a variety of applications such as neural networks. We describe in detail the geometric structure of a fundamental invariant set, which in special cases is the global attractor, and the asymptotic behaviour of solution curves on it.

Our approach makes use of advanced tools which in recent years have been developed for the investigation of infinite-dimensional dynamical systems: Local invariant manifolds and inclination lemma for noninvertible maps, Floquet theory for delay differential equations, a-priori estimates controlling the growth and decay of solutions with prescribed oscillation frequency, a discrete Lyapunov functional counting zeros, methods to represent invariant sets as graphs, Poincaré–Bendixson techniques for classes of delay differential systems.

Several appendices provide the general results needed in our case study, so that the presentation is self-contained. Some of these general results seem not to be available elsewhere in the literature. We mention in particular Appendix II on smooth infinite-dimensional center-stable manifolds for maps. We believe that the results in the appendices will be useful also for future studies of more complicated attractors of delay and partial differential equations.

A brief description of a part of our main result is that for the delay differential equation $\dot{x}(t) = -\mu x(t) + f(x(t-1))$ with $\mu > 0$ and increasing bounded C^1-function $f : \mathbb{R} \to \mathbb{R}$ with $f(0) = 0$, under natural and mild additional conditions, the leading 3-dimensional local unstable manifold at the stationary point 0 extends in forward time to a smooth solid spindle with singularities at its tips, which are further stationary points, both stable and attractive; an invariant smooth disk of solution curves winding from 0 towards a bordering unstable periodic orbit splits the spindle into invariant halves each of which is attracted to one of its tips.

The major part of this work was carried out while T. Krisztin and J. Wu were supported by the Alexander von Humboldt Foundation during their guest stay with Humboldt fellowships at the University of Giessen in 1996–97. Both are indebted to the Foundation for its generous support and to the Institute of Mathematics at the University of Giessen for the kind hospitality during their visit. We are grateful to Professor J. Kincses for calling our attention to some topological results used in Chapter 16.

We would like to acknowledge the support from the Hungarian National Foundation for Scientific Research (T. Krisztin), and from the Natural Sciences and Engineering Research Council of Canada (J. Wu).

T. Krisztin, H. O. Walther and J. Wu
June 1998

CHAPTER 1

Introduction

The differential equation

$$\dot{x}(t) = -\mu x(t) + f(x(t-1)) \tag{1.1}$$

with a function $f : \mathbb{R} \to \mathbb{R}$ and a constant $\mu \geq 0$ models a system with instantaneous friction (in case where $\mu > 0$) and delayed feedback. In this monograph, we consider continuously differentiable feedback functions $f : \mathbb{R} \to \mathbb{R}$ which satisfy

$$f(0) = 0, \quad f'(\xi) > 0 \quad \text{for } \xi \in \mathbb{R},$$

and thereby represent monotone positive feedback with respect to the rest state given by $\xi = 0$. Equation (1.1) with such a function f is related to models for single neurons and for synchronized activities in neural networks. For example, the behaviour of a single, self-excitatory neuron with graded delayed response is modelled by equation (1.1) with $f = f_{\alpha\beta}$, where $f_{\alpha\beta}(x) = \alpha \tanh(\beta x)$ with $\alpha > 0$ and $\beta > 0$. See Herz **[32]** and Pakdaman, Malta, Grotta-Ragazzo and Vibert **[50]**. Also, a system of equations

$$C\dot{x}_i(t) = -\frac{1}{R}x_i(t) + \sum_{j=1}^{n} T_{ij} f_{\alpha\beta}(x_j(t-\tau)), \qquad i \in \{1, \dots, n\}$$

was used as a model for a network of n identical saturating amplifiers (or neurons) with delayed outputs which are coupled by a resistive interconnection matrix $T = (T_{ij})$. Here, x_i represents the voltage input at the ith neuron which is characterized by input capacitance C, parallel resistance R and activation function $f_{\alpha\beta}$ as in the previous model for single neurons. If the coefficients T_{ij} of the interconnection matrix satisfy

$$\sum_{j=1}^{n} T_{ij} = \sum_{j=1}^{n} T_{kj} \qquad \text{for all } i, k \text{ in } \{1, \dots, n\},$$

then the system has synchronized solutions. These are solutions satisfying $x_1(t) = x_2(t) = \ldots = x_n(t)$ for all t. Clearly, synchronized solutions are completely characterized by the scalar equation (1.1) with certain μ and f. The system above without delay ($\tau = 0$) was proposed by Hopfield **[35, 36]**, a similar model is also due to Cohen and Grossberg **[20]**. Marcus and Westervelt **[46]** incorporated the time delay in order to account for the finite switching speed of the amplifiers. For studies of the dynamics of such a system, see the work of Bélair **[8]**, Bélair, Campbell and van den Driessche **[9]**, Bélair and Dufour **[10]**, Herz **[32]**, Olien and Bélair **[49]**, Wu **[72]** and references therein. The investigation of equation (1.1) may also help to understand the dynamics of networks of neurons with internal dynamic threshold, which are described by a system of $2n$ equations (see, for example, Arndt **[7]**)

$$\begin{cases} \dot{x}_i(t) = -\mu_i x_i(t) + \sum_{j=1}^{n} T_{ij} f_{\alpha\beta}(x_j(t-\tau_{ij}) - s_j(t-\tau_{ij})), \\ \dot{s}_i(t) = -\mu_s s_i(t) + T f_{\alpha\beta}(x_i(t-\tau) - s_i(t-\tau)). \end{cases}$$

Here, s_i is the threshold in the ith neuron, and it is assumed that the input at the ith neuron and its rate of change do not depend on the states $x_j(t-\tau_{ij})$ but on their distances from the thresholds $s_j(t-\tau_{ij})$. Again, if one looks for a single neuron or if one looks for the synchronized activities and if $\sum_{j=1}^{n} T_{ij} > T$, $\mu_s = \mu_i$, and $\tau = \tau_{ij}$ for all i and j, then $x_i - s_i$ satisfies equation (1.1) with positive feedback. The technique developed for equation (1.1) may also be useful in the study of the dynamics of other models with delayed positive feedback, see, for example, **[5, 6, 26, 39, 40, 54, 55]**.

Every element ϕ of the Banach space C of continuous real functions on the initial interval $[-1,0]$ determines a solution $x^\phi : [-1,\infty) \to \mathbb{R}$ of equation (1.1), i.e., a continuous function which is differentiable on $(0,\infty)$ and satisfies equation (1.1) for all $t > 0$. The relations

$$F(t,\phi) = x_t,\ x = x^\phi,\ x_t(s) = x(t+s),\ s \in [-1,0]$$

define a continuous semiflow $F : \mathbb{R}^+ \times C \to C$ such that all maps $F(t,\cdot)$, $t \geq 0$, are injective and continuously differentiable, and F is monotone with respect to the pointwise ordering on C. We shall see that the positively invariant set

$$S = \{\phi \in C : (x^\phi)^{-1}(0) \text{ is unbounded}\}$$

separates the domain of absorption into the interior of the positively invariant cone

$$K = \{\phi \in C : \phi(s) \geq 0 \text{ for all } s \in [-1,0]\}$$

from the domain of absorption into the interior of $-K$, and that the separatrix S is a Lipschitz graph over a closed hyperplane in C. So, we can speak of the parts of C above and below S.

We are interested in a detailed description of the long term behaviour of solutions of equation (1.1). A natural object to study would be the global attractor of the semiflow, i.e., a compact set $A \subset C$ which is invariant in the sense that $F(t,A) = A$ for all $t \geq 0$ and which attracts every bounded set $B \subset C$ in the sense that for every open set U containing A there exists $T \geq 0$ with $F([T,\infty) \times B) \subset U$ (see Hale **[28]**). Unfortunately, in general, there is no such global attractor for our semiflow F. For example, in case $\mu = 0$ every solution on $[-1,\infty)$ of equation (1.1) with an initial datum in the interior of $K \cup (-K)$ is unbounded, and thus a global attractor does not exist. So we have to look for substitutes. These are the closure of the unstable set

$$\begin{aligned} W^u = \{\phi \in C :\ & \text{There is a solution } x : \mathbb{R} \to \mathbb{R} \text{ of equation (1.1)} \\ & \text{with } x_0 = \phi \text{ and } x_t \to 0 \text{ as } t \to -\infty\} \end{aligned}$$

of the stationary point $0 \in C$, and subsets thereof. Notice that in the presence of a global attractor A, necessarily $\overline{W^u} \subset A$ since for every ball B centered at 0 and for every $t \geq 0$, $W^u \subset F([t,\infty) \times (W^u \cap B))$, which implies that W^u is contained in every neighbourhood of the compact set A, yielding $W^u \subset \overline{A} = A$ and $\overline{W^u} \subset A$.

In order to introduce our further hypotheses on μ and f and the subsets of $\overline{W^u}$ which we shall study, we linearize equation (1.1) at the stationary point $0 \in C$. The derivatives $D_2F(t,0)$, $t \geq 0$, form a strongly continuous semigroup, and

$$D_2F(t,0)\phi = y_t^\phi,$$

where y^ϕ is the solution of

$$\dot{y}(t) = -\mu y(t) + f'(0)y(t-1) \tag{1.2}$$

with $y_0^\phi = \phi$. The spectrum of the generator of the semigroup consists of simple eigenvalues which coincide with the zeros of the characteristic function

$$\mathbb{C} \ni \lambda \mapsto \lambda + \mu - f'(0)e^{-\lambda} \in \mathbb{C}.$$

There is one real eigenvalue λ_0, and the others form a sequence of complex conjugate pairs $(\lambda_j, \overline{\lambda_j})$ with

$$\operatorname{Re}\lambda_{j+1} < \operatorname{Re}\lambda_j < \lambda_0 \quad \text{and} \quad (2j-1)\pi < \operatorname{Im}\lambda_j < 2j\pi$$

for all integers $j \geq 1$, and $\operatorname{Re}\lambda_j \to -\infty$ as $j \to \infty$. The number of eigenvalues in the open right halfplane depends on μ and $f'(0)$. Any odd integer $2j+1$ can be achieved.

In the simplest case where $\lambda_0 < 0$, $W^u = \{0\}$. If $\operatorname{Re}\lambda_1 < 0 < \lambda_0$, then one can show that W^u consists of 0 and the segments of two solutions on $\mathbb{R}$, one being positive and the other negative. Interesting structures of W^u appear only when W^u is higher-dimensional. We are going to make an assumption on μ and $f'(0)$ which is equivalent to

$$0 < \operatorname{Re}\lambda_1.$$

Then the topological dimension of the set W^u is at least 3, and there is a 3-dimensional C^1-submanifold $W_{loc} \subset W^u$ of C which is tangent at 0 to the realified eigenspace of the spectral set $\{\lambda_0, \lambda_1, \overline{\lambda_1}\}$ and has the property that for every $\phi \in W_{loc}$ there exist a solution $x : \mathbb{R} \to \mathbb{R}$ and $T \in \mathbb{R}$ such that $x_0 = \phi$ and $x_s \in W_{loc}$ for all $s \leq T$. The forward extension

$$W = F\left(\mathbb{R}^+ \times W_{loc}\right)$$

of W_{loc} is an invariant subset of W^u. If $\operatorname{Re}\lambda_2 < 0$ then $W = W^u$. Moreover, in some special cases the set $\overline{W}$ should be the global attractor. We suspect that this is true, for example, if $\mu > 0$, $\operatorname{Re}\lambda_2 < 0$, and if f is bounded with f' being strictly increasing on $(-\infty, 0)$ and strictly decreasing on $(0, \infty)$. Compare Walther **[66]** where a similar situation is studied for equation (1.1) with negative feedback.

We shall investigate the structure of $\overline{W}$ under mild additional assumptions on μ and f which are always satisfied by the above models for neural networks. We require that either

$$\mu = 0 \text{ and } -\infty < \inf f \text{ or } \sup f < \infty$$

or

$$\mu > 0 \text{ and } \frac{f(\xi)}{\xi} < \mu \quad \text{for } \xi \text{ outside a bounded neighbourhood of } 0.$$

The last property and the inequality $\mu < f'(0)$, which follows from $\operatorname{Re}\lambda_1 > 0$, together imply that in case $\mu > 0$ there are a largest negative zero ξ^- of $f - \mu\,\mathrm{id}$ and a smallest positive zero ξ^+ of $f - \mu\,\mathrm{id}$, and

$$f'(\xi^-) \le \mu, \quad f'(\xi^+) \le \mu.$$

The final assumption is that in case $\mu > 0$

$$f'(\xi^-) < \mu \quad \text{and} \quad f'(\xi^+) < \mu.$$

Our main results concerning $\overline{W}$ begin with the relatively easy facts that $\overline{W}$ is invariant, and that the semiflow defines a continuous flow $F_W : \mathbb{R} \times \overline{W} \to \overline{W}$. For $\overline{W}$ and for the part $\overline{W} \cap S$ of the separatrix S in $\overline{W}$, we then obtain graph representations as follows:

There exist subspaces $G_2 \subset G_3$ of C of dimensions 2 and 3, respectively, a complementary space G_1 of G_2 in G_3, a closed complementary space E of G_3 in C, a compact set $D_S \subset G_2$ and a closed set $D_W \subset G_3$, and continuous mappings $w : D_W \to E$ and $w_S : D_S \to G_1 \oplus E$ such that

$$\overline{W} = \{\chi + w(\chi) : \chi \in D_W\}, \quad \overline{W} \cap S = \{\chi + w_S(\chi) : \chi \in D_S\}.$$

We have

$$D_W = \partial D_W \cup \overset{\circ}{D}_W, \quad W = \{\chi + w(\chi) : \chi \in \overset{\circ}{D}_W\},$$

and the restriction of w to $\overset{\circ}{D}_W$ is C^1-smooth, where ∂D_W and $\overset{\circ}{D}_W$ stand for the topological boundary and interior of D_W, respectively. The restriction of F_W to $\mathbb{R} \times W$ is C^1-smooth. The domain D_S is homeomorphic to the closed unit disk in $\mathbb{R}^2$, and consists of the trace of a simple closed C^1-curve and its interior. The map w_S is C^1-smooth in the sense that $w_S|_{\overset{\circ}{D}_S}$ is C^1-smooth, and for each $\chi \in \partial D_S$ there is an open neighbourhood N of χ in G_2 so that $w_S|_{N \cap D_S}$ extends to a C^1-map on N.

Concerning the dynamics we have that

the set

$$\mathcal{O} = \overline{W} \cap S \setminus (W \cap S) = \{\chi + w_S(\chi) : \chi \in \partial D_S\}$$

is a periodic orbit, and there is no other periodic orbit in $\overline{W}$. The open annulus $(W \cap S) \setminus \{0\}$ consists of heteroclinic connections from 0 to $\mathcal{O}$. For every $\phi \in W$, $F_W(t, \phi) \to 0$ as $t \to -\infty$.

The further properties of $\overline{W}$ are different in the cases $\mu > 0$ and $\mu = 0$.

For $\mu > 0$, $\overline{W}$ is compact and contains the stationary points ξ_- and ξ_+ in C given by the values ξ^- and ξ^+, respectively. For every $\phi \in \overline{W}$,

$$\xi_- \le \phi \le \xi_+.$$

There exist homeomorphisms from $\overline{W}$ and D_W onto the closed unit ball in $\mathbb{R}^3$ which send

$$\operatorname{bd} W = \overline{W} \setminus W = \{\chi + w(\chi) : \chi \in \partial D_W\} \quad \textit{and} \quad \partial D_W$$

onto the unit sphere $S^2 \subset \mathbb{R}^3$, respectively. If we define χ_- and χ_+ by $\xi_- = \chi_- + w(\chi_-)$ and $\xi_+ = \chi_+ + w(\chi_+)$, respectively, then the set $\partial D_W \setminus \{\chi_-, \chi_+\}$ is a 2-dimensional C^1-submanifold of G_3, and the restriction of w to $D_W \setminus \{\chi_-, \chi_+\}$ is C^1-smooth (in the sense explained above for w_S). The points $\phi \in \overline{W} \setminus \overline{W} \cap S$

above the separatrix S form a connected set and satisfy $F_W(t,\phi) \to \xi_+$ as $t \to \infty$, and all $\phi \in \overline{W} \setminus \overline{W} \cap S$ below the separatrix S form a connected set and satisfy $F_W(t,\phi) \to \xi_-$ as $t \to \infty$. Finally, for every ϕ in the set bd W *and different from ξ_- and ξ_+, $F_W(t,\phi) \to \mathcal{O}$ as $t \to -\infty$.*

Combining all the results stated so far, one may visualize the invariant set $\overline{W}$ in case $\mu > 0$ as a smooth solid spindle which is split by an invariant disk into the basins of attraction towards the tips ξ_- and ξ_+.

For $\mu = 0$, the sets $\overline{W}$ and D_W are unbounded. There exist homeomorphisms from $\overline{W}$ and D_W onto the solid cylinder $\{z \in \mathbb{R}^3 : z_1^2 + z_2^2 \leq 1\}$ which send

$$\operatorname{bd} W = \{\chi + w(\chi) : \chi \in \partial D_W\} \quad \textit{and} \quad \partial D_W$$

onto the cylinder $S^1 \times \mathbb{R}$, respectively. The boundary ∂D_W is a 2-dimensional C^1-submanifold of G_3, and w is C^1-smooth. The points $\phi \in \overline{W} \setminus \overline{W} \cap S$ above the separatrix S form a connected set and satisfy $x^\phi(t) \to \infty$ as $t \to \infty$, and the points $\phi \in \overline{W} \setminus \overline{W} \cap S$ below S form a connected set and satisfy $x^\phi(t) \to -\infty$ as $t \to \infty$. Finally, for every $\phi \in \operatorname{bd} W$, $F_W(t,\phi) \to \mathcal{O}$ as $t \to -\infty$.

Precise theorems containing all our results on $\overline{W}$ are stated in the final chapter, Chapter 17, where we also discuss open problems and directions for further research.

Key passages on the way to the description of $\overline{W}$ are the existence of the periodic orbit $\mathcal{O} = \overline{W \cap S} \setminus (W \cap S)$ in Chapter 7, the minimal linear instability of $\mathcal{O}$ in Chapter 8, the smoothness of $W \cap S \setminus \{0\}$ as a consequence of transversal intersection of stable and center-stable manifolds with a piece of W in a hyperplane in Chapters 9 and 10, the identification of bd W with the closure of the unstable set of $\mathcal{O}$ in Chapters 11 and 12, the smoothness of the intersection of $\overline{W}$ with a hyperplane in Chapter 13, and a uniform local connectedness property of W in Chapter 16.

Our investigation requires results on local invariant manifolds for maps in Banach spaces and a-priori results on the dynamics of certain delay differential systems which are more general than equation (1.1). All of these tools are developed in a series of appendices. Appendix I collects results about leading stable and unstable manifolds, asymptotic phases, equivalent norms, Poincaré maps and a lemma on C^1-submanifolds. Appendices II and III deal with existence and smoothness of center-stable and generalized center-unstable manifolds close to fixed points of C^1-maps on open sets in Banach spaces. The general versions of the smoothness results we present seem not to be available elsewhere in the literature. Appendix IV studies invariant cones close to neutrally stable fixed points of C^1-maps with one-dimensional linear center space. In case there is a trajectory converging to the fixed point we construct positively invariant open sets close to the fixed point. Appendix V discusses for a given periodic orbit of a continuous semiflow the connection of the unstable set of the period map and the unstable set of the Poincaré map associated with a transversal hyperplane. In Appendix VI, a version of Mallet-Paret's discrete Lyapunov functional counting sign changes of elements $\phi \in C$ is introduced. Also important are growth estimates for solutions, which go back to Mallet-Paret **[42]** and Walther **[61, 62]**. In Appendix VII, we give a complete proof of a-priori results on Floquet multipliers of periodic solutions, which alternatively can be deduced from a result of Mallet-Paret and Sell **[43, 44]**. The final appendix, Appendix VIII recalls the Schoenfliess Theorem and a theorem of Bing from the topology and geometry of 3-dimensional manifolds which we use in Chapter 16 to construct homeomorphisms from $\overline{W}$ onto the closed unit ball and onto the solid cylinder.

We mention that results on unstable sets and attractors for delay differential equations related to ours are also contained in the doctoral dissertation of Ammar **[2]** who showed, among others, that in case $\mu = 0$ the set W is a smooth graph over an open set in the realified generalized eigenspace of the leading eigenvalues $\{\lambda_0, \lambda_1, \overline{\lambda_1}\}$, and in the work of Walther **[67, 69, 70]**, Walther and Yebdri **[71]**, and Mallet-Paret and Walther **[45]** on equations with negative feedback. In preparation is the work of Arino and Krisztin **[4]** on a 2-dimensional attractor for a negative feedback equation with a state-dependent time lag.

In the sequel, we introduce some notations and terminologies to be used throughout this monograph.

The symbols $\mathbb{N}$ and $\mathbb{R}^+$ denote the nonnegative integers and reals, respectively. $\mathbb{R}$ and $\mathbb{Z}$ stand for the set of all reals and all integers, respectively.

An upper index tr denotes the transpose of a matrix or of a vector in $\mathbb{R}^n$ or $\mathbb{C}^n$, $1 \leq n \in \mathbb{N}$. $(\mathbb{R}^n)^{tr}$ and $(\mathbb{C}^n)^{tr}$ are spaces of row vectors.

A trajectory of a map $g : M \to N$, $M \subset N$, is a finite or infinite sequence $(x_j)_{j \in I \cap \mathbb{Z}}$, $I \subset \mathbb{R}$ an interval, in M with $x_{j+1} = g(x_j)$ for all $j \in I \cap \mathbb{Z}$ with $j + 1 \in I \cap \mathbb{Z}$.

If M is a subset of a topological space, then $\overset{\circ}{M}$, $\overline{M}$ and ∂M stand for the interior, closure, and topological boundary of M, respectively. The boundary $\operatorname{bd} W = \overline{M} \setminus M$ is, in general, different from ∂M, of course.

S^n is the unit sphere with respect to the Euclidean norm in $\mathbb{R}^{n+1}$, $n \in \mathbb{N}$. D^n is the closed unit ball with respect to the Euclidean norm in $\mathbb{R}^n$, $1 \leq n \in \mathbb{N}$. $S^1_{\mathbb{C}}$ is the unit circle in the complex plane $\mathbb{C}$.

Let c be a simple closed curve, i.e., a continuous and injective map from S^1 into $\mathbb{R}^2$. The Jordan Curve Theorem guarantees that $\mathbb{R}^2 \setminus c(S^1)$ consists of two nonempty open connected sets, one bounded and one unbounded, and that $c(S^1)$ is the boundary of each of these components. We denote the bounded component by $\operatorname{int}(c)$ and the unbounded one by $\operatorname{ext}(c)$.

If X is a Banach space over $\mathbb{R}$ with a norm $||\cdot||$ and $r > 0$, then

$$X_r = \{x \in X : ||x|| < r\}.$$

For a subset $M \subset X$ and a point $x \in M$, the tangent vectors v of M at x are the tangent vectors $v = Dc(0)1 = c'(0)$ of differentiable curves $c : (-1,1) \to X$ with $c(0) = x$ and range, or trace, $|c| = c((-1,1))$ in M. The set $T_x M$ of all such tangent vectors is always nonempty since $0 \in T_x M$, but not necessarily a linear subspace of X. If $g : U \to Y$ is a differentiable map from an open set $U \subset X$ into a Banach space Y and if $x \in M \subset U$, $N \subset Y$ and $g(M) \subset N$, then $Dg(x)T_x M \subset T_{g(x)} N$ by the chain rule.

C and $C_{\mathbb{C}}$ denote the Banach spaces of continuous real and complex functions on $[-1,0]$, over the fields $\mathbb{R}$ and $\mathbb{C}$, respectively, with the norms given by

$$||\phi|| = \max_{t \in [-1,0]} |\phi(t)|.$$

C^1 denotes the Banach space of all continuously differentiable functions $\phi \in C$, with the norm given by

$$||\phi||_{C^1} = ||\phi|| + ||\dot{\phi}||.$$

For Banach spaces E and F over $\mathbb{R}$ or $\mathbb{C}$, the Banach space of all continuous linear operators $L : E \to F$ is denoted by $L(E, F)$.

Spectra of linear operators in X are defined as the spectra of complexifications. If a decomposition $X = Y \oplus Z$ into closed linear subspaces is given, then Pr_Y denotes the associated projection along Z onto Y, considered as a map from X into X.

Let $\Phi : \mathbb{R}^+ \times X \to X$ be a semiflow. A set $M \subset X$ is called positively invariant if $\Phi(\mathbb{R}^+ \times M) \subset M$. It is called invariant (resp. negatively invariant) if for every $x \in M$ there is a complete phase curve, i.e., a map $\gamma : \mathbb{R} \to X$ with $\gamma(t+s) = \Phi(t, \gamma(s))$ for all $s \in \mathbb{R}$ and $t \geq 0$, which satisfies $\gamma(0) = x$ and $\gamma(\mathbb{R}) \subset M$ (resp. $\gamma((-\infty, 0]) \subset M$). For a given $x \in X$, the ω-limit set of x is defined as

$$\omega(x) = \{y \in X : \text{There is a sequence } (t_n)_0^\infty \subset \mathbb{R}^+ \text{ such that } t_n \to \infty \text{ and } \Phi(t_n, x) \to y \text{ as } n \to \infty\}.$$

If Φ is continuous and $\Phi(\mathbb{R}^+ \times \{x\})$ is relatively compact, then $\omega(x)$ is nonempty, compact, connected and invariant. Similarly, for a given complete phase curve $\gamma : \mathbb{R} \to X$, the α-limit set of γ is defined as

$$\alpha(\gamma) = \{y \in X : \text{There is a sequence } (t_n)_0^\infty \subset (-\infty, 0] \text{ such that } t_n \to -\infty \text{ and } \gamma(t_n) \to y \text{ as } n \to \infty\}.$$

If Φ is continuous and $\gamma((-\infty, 0])$ is relatively compact, then $\alpha(\gamma)$ is nonempty, compact, connected and invariant. In case all maps $\Phi(t, \cdot)$, $t \geq 0$, are injective, γ is uniquely determined by its value $\gamma(0)$, and we set $\alpha(x) = \alpha(\gamma)$ for every $x \in X$ so that there exists a complete phase curve $\gamma : \mathbb{R} \to X$ with $\gamma(0) = x$.

Let g be a real function defined either on a set $[t_0, \infty) \times \mathbb{R}^2$ with $t_0 \in \mathbb{R}$ or on $\mathbb{R}^3$. In the first case, a solution of the delay differential equation

$$\dot{x}(t) = g(t, x(t), x(t-1)) \tag{1.3}$$

is a continuous real function x on an interval $[t_1 - 1, \infty)$ with $t_1 \geq t_0$ which is differentiable and satisfies (1.3) on (t_1, ∞). In the second case, solutions are either continuous functions on intervals $[t_1 - 1, \infty)$ with $t_1 \in \mathbb{R}$, which are differentiable and satisfy (1.3) on (t_1, ∞), or differentiable functions $x : \mathbb{R} \to \mathbb{R}$ which satisfy (1.3) for all $t \in \mathbb{R}$. Solutions of autonomous equations and complex-valued solutions of linear equations are defined analogously.

If $x : I \subset \mathbb{R} \to M$ is a continuous map and if $[t-1, t] \subset I$, then the segment $x_t : [-1, 0] \to M$ is defined by $x_t(s) = x(t+s)$ for $s \in [-1, 0]$.

We recommend as references Dieudonné **[23]** for calculus in Banach spaces, Abraham and Robbin **[1]** for submanifolds and transversality, Diekmann, van Gils, Verduyn Lunel and Walther **[22]** and Hale and Verduyn Lunel **[30]** for basic properties of delay differential equations and their semiflows.

CHAPTER 2

The delay differential equation and the hypotheses

Let $f : \mathbb{R} \to \mathbb{R}$ be continuously differentiable with

(H1) $f(0) = 0$ and $f'(\xi) > 0$ for all $\xi \in \mathbb{R}$.

We study solutions of the delay differential equation

$$\dot{x}(t) = -\mu x(t) + f(x(t-1)) \tag{1.1}$$

with

$$\mu \geq 0.$$

Obviously, the zero 0 of f defines a constant solution on $\mathbb{R}$. Other solutions are easily obtained from the variation-of-constants formula for ordinary differential equations on successive intervals of length one,

$$x(t) = e^{-\mu(t-n)}x(n) + \int_n^t e^{-\mu(t-s)} f(x(s-1))ds \tag{2.1}$$

with $n \in \mathbb{N}$, $n \leq t \leq n+1$. One sees that each $\phi \in C$ uniquely determines a solution $x^\phi : [-1, \infty) \to \mathbb{R}$, with $x_0^\phi = \phi$. Also any two solutions x, y on a common domain which coincide on an interval of length one must be equal. We have continuous dependence on initial data in the sense that given $\phi \in C$, $t \geq 0$, $\epsilon > 0$ there exists $\delta > 0$ with $|x^\psi(s) - x^\phi(s)| < \epsilon$ for all $\psi \in \phi + C_\delta$ and for all $s \in [-1, t]$.

A constant real function on $\mathbb{R}$ or on $[-1, \infty)$ is a solution if and only if its value is a zero of $f - \mu\,\mathrm{id}$.

The map $F : \mathbb{R}^+ \times C \ni (t, \phi) \mapsto x_t^\phi \in C$ is a continuous semiflow with stationary point 0. An element $\phi \in C$ is a stationary point of F if and only if ϕ is constant and $f(\phi(0)) = \mu\phi(0)$. All maps $F(t, \cdot) : C \to C$, $t \geq 0$, are injective. It follows that for every $\phi \in C$ there is at most one solution $x : \mathbb{R} \to \mathbb{R}$ of equation (1.1) which satisfies $x_0 = \phi$. We shall also use the symbol x^ϕ to denote such a solution on $\mathbb{R}$ whenever it exists for a given $\phi \in C$. This should not cause confusion.

Using the differential equation and the Arzèla–Ascoli Theorem, we conclude that for every $\phi \in C$ so that $x^\phi : [-1, \infty) \to \mathbb{R}$ is bounded, and for every solution $x : \mathbb{R} \to \mathbb{R}$ with $x((-\infty, 0])$ bounded, the sets $F(\mathbb{R}^+ \times \{\phi\})$ and $\{x_t : t \leq 0\}$, respectively, are relatively compact. Consequently, the limit sets $\omega(\phi)$ and $\alpha(x) = \alpha(x_0)$ are nonempty, compact, connected and invariant.

Each map $F(t, \cdot)$, $t \geq 0$, is continuously differentiable, and for every $t \geq 0$ and $\phi \in C$,

$$D_2 F(t, \phi)\psi = v_t$$

with the solution $v : [-1,\infty) \to \mathbb{R}$ of the variational equation

$$\dot{v}(t) = -\mu v(t) + f'(x^\phi(t-1))v(t-1) \tag{2.2}$$

along x^ϕ given by $v_0 = \psi$. All derivatives $D_2F(t,\phi)$ are injective. For every $\phi \in C$, the restriction of the flowline $F(\cdot,\phi)$ to the interval $(1,\infty)$ is continuously differentiable, with

$$D_1F(t,\phi)1 = \dot{x}_t^\phi = (\dot{x}^\phi)_t \qquad \text{for all } t > 1.$$

The restriction of F to the open set $(1,\infty) \times C$ is continuously differentiable.

For $r > 0$, set

$$f_r = \max_{-r\le\xi\le r} |f(\xi)| = \max\{f(r), -f(-r)\}$$

and

$$f'_r = \max_{-r\le\xi\le r} f'(\xi).$$

Proposition 2.1 *(Growth Bound, Compactness)*

(i) For every $r > 0$ and all $t \in [0,1]$, $F(t, C_r) \subset C_{r+f_r}$; and for every $\phi \in C$ and $t \in [0,1]$, $||D_2F(t,\phi)|| \le 1 + f'_{||\phi||}$. For every bounded set $B \subset C$ and for every $t \ge 0$ there is a constant $c > 0$ so that

$$||F(s,\phi) - F(s,\psi)|| \le c||\phi - \psi|| \qquad \textit{for all } \phi \in B,\ \psi \in B,\ s \in [0,t].$$

(ii) $F(t,\cdot)$ and $D_2F(t,\phi)$, $t \ge 1$ and $\phi \in C$, map bounded sets into relatively compact sets.

Proof 1. The first part of assertion (i) is obvious from the variation-of-constants formulae representing x^ϕ and solutions v of the variational equation along x^ϕ on $[0,1]$.

Let a bounded set $B \subset C$ and $t \ge 0$ be given. Choose an integer $n \ge t+1$. The first part of assertion (i) implies that for some $b \ge 0$,

$$F([0,n] \times B) \subset C_b.$$

Set $\tilde{c} = \max_{[-b,b]} f'$. Let ϕ, ψ in B, an integer $j \in \{1,\dots,n\}$ and $s \in [j-1,j]$ be given. The function $y = x^\phi - x^\psi$ satisfies

$$\dot{y}(u) = -\mu y(u) + \int_0^1 f'(x^\psi(u-1) + \theta y(u-1))\, d\theta\, y(u-1)$$

for $0 < u$. Hence

$$y(s) = e^{-\mu(s-(j-1))}y(j-1) + \int_{j-1}^s e^{\mu(s-u)} \int_0^1 f'(x^\psi(u-1)+\theta y(u-1))\, d\theta\, y(u-1)\, du,$$

and thus

$$|y(s)| \le ||y_{j-1}|| + \tilde{c}||y_{j-1}|| = (1+\tilde{c})||y_{j-1}||.$$

It follows that for all $s \in [0,n]$,

$$||F(s,\phi) - F(s,\psi)|| = ||y_s|| \le (1+\tilde{c})^n ||y_0|| = (1+\tilde{c})^n ||\phi - \psi||.$$

2. In order to deduce the first part of assertion (ii), we note first that due to $F(t,\cdot) = F(t-1,\cdot) \circ F(1,\cdot)$ it is sufficient to show that $F(1,\cdot)$ maps bounded sets into relatively compact sets. Assertion (i) and equation (1.1) show that for

every bounded set $B \subset C$ there is a bound for all $x_1^\phi(s)$ and $\dot{x}_1^\phi(s)$ with $\phi \in B$ and $s \in [-1, 0]$. Using the Arzèla–Ascoli Theorem, we then obtain the first part of (ii). The proof of the last assertion is similar. □

For ϕ, ψ in C, we define

$$\phi \leq \psi \text{ if and only if } \phi(s) \leq \psi(s) \text{ for all } s \in [-1,0];$$
$$\phi < \psi \text{ if and only if } \phi \leq \psi \text{ and } \phi \neq \psi;$$
$$\phi \ll \psi \text{ if and only if } \phi(s) < \psi(s) \text{ for all } s \in [-1,0].$$

The relations $\phi \geq \psi$, $\phi > \psi$ and $\phi \gg \psi$ are defined analogously.

The set

$$K = \{\phi : 0 \leq \phi\}$$

is closed and convex. We have $(0,\infty)K \subset K$, $K \cap (-K) = \{0\}$ and

$$\overset{\circ}{K} = \{\phi \in C : 0 \ll \phi\}.$$

Proposition 2.2 *(Invariance.) Let $\phi \in C$, $\psi \in C$, $t \geq 0$.*

(i) $F(t, \overset{\circ}{K}) \subset \overset{\circ}{K}$ and $F(t, -\overset{\circ}{K}) \subset -\overset{\circ}{K}$. If $\psi - \phi \in K$ $(\in \overset{\circ}{K})$ then $F(t,\psi) - F(t,\phi) \in K$ $(\in \overset{\circ}{K})$. If $0 \neq \psi - \phi \in K$ and $t \geq 2$ then $F(t,\psi) - F(t,\phi) \in \overset{\circ}{K}$.

(ii) $D_2F(t,\phi)K \subset K$, $D_2F(t,\phi)\overset{\circ}{K} \subset \overset{\circ}{K}$, and in case $t \geq 2$, $D_2F(t,\phi)(K \setminus \{0\}) \subset \overset{\circ}{K}$.

Proof The assertions in (i) follow by means of formula (2.1) and hypothesis (H1). To obtain (ii), we use the variation-of-constants formulae

$$v(t) = e^{-\mu(t-n)}v(n) + \int_n^t e^{-\mu(t-s)} f'(x^\phi(s-1))v(s-1)\,ds$$

with $t \in [n, n+1]$, $n \in \mathbb{N}$, for solutions of equation (2.2). □

One observation which is occasionally useful is that for every solution x of equation (1.1) with $\mu > 0$ the function $y : t \mapsto e^{\mu t}x(t)$ is a solution of the equation

$$\dot{y}(t) = e^{\mu t} f(e^{-\mu(t-1)}y(t-1))$$

which is nonautonomous but has no decay term on the right-hand side.

The second hypothesis concerns the instability of the stationary point 0. The operators $D_2F(t,0), t \geq 0$, form a strongly continuous semigroup on C, which is given by solutions of the linear equation

$$\dot{y}(t) = -\mu y(t) + f'(0)y(t-1). \tag{1.2}$$

The spectrum of the generator coincides with the zeros of the characteristic function

$$\mathbb{C} \ni \lambda \mapsto \lambda + \mu - f'(0)e^{-\lambda} \in \mathbb{C}$$

obtained from the Ansatz $x(t) = e^{\lambda t}$ for a solution of equation (1.2). These zeros are all simple. There is one real zero λ_0; the others form a sequence of complex conjugate pairs $(\lambda_k, \overline{\lambda_k})$, $1 \leq k \in \mathbb{N}$, with $(2k-1)\pi < \operatorname{Im}\lambda_k < 2k\pi$ and $\lambda_0 > \operatorname{Re}\lambda_k > \operatorname{Re}\lambda_{k+1}$ for $k \geq 1$, and $\operatorname{Re}\lambda_k \to -\infty$ as $k \to \infty$.

We consider equation (1.1) under the condition

(H2) $\mu = 0$ and $f'(0) > 3\pi/2$, or
$\mu > 0$ and $f'(0) > \mu/cos\theta_\mu$ for $\theta_\mu \in (3\pi/2, 2\pi)$ with $\theta_\mu = -\mu \tan \theta_\mu$.

(H2) is equivalent to

$$\operatorname{Re} \lambda_1 > 0$$

which means that the linear unstable space of the generator of the semigroup of equation (1.2) is at least 3-dimensional.

We note that the realified generalized eigenspace P of the generator associated with λ_0 is 1-dimensional and given by 0 and the segments of the solution $\mathbb{R} \ni t \mapsto e^{\lambda_0 t} \in \mathbb{R}$ of equation (1.2). For $\chi_0 : [-1, 0] \ni t \mapsto e^{\lambda_0 t} \in \mathbb{R}$, $P = \mathbb{R}\chi_0$. The realified generalized eigenspaces L_k of the generator associated with the pairs $(\lambda_k, \overline{\lambda_k})$, $1 \leq k \in \mathbb{N}$, are 2-dimensional and consist of the segments of all solutions

$$\mathbb{R} \ni t \mapsto e^{\operatorname{Re} \lambda_k t}(a \cos(\operatorname{Im} \lambda_k t) + b \sin(\operatorname{Im} \lambda_k t)) \in \mathbb{R}, \qquad a \in \mathbb{R},\ b \in \mathbb{R},$$

of equation (1.2). We set $L = L_1$. Let Q denote the realified generalized eigenspace of the generator which is given by the spectral set of all λ_k and $\overline{\lambda_k}$ with $k \geq 2$.

The next hypothesis requires bounds and growth bounds. We assume

(H3) $\mu = 0$ and f is bounded from above or from below, or
$\mu > 0$ and $f(\xi)/\xi < \mu$ outside a bounded neighbourhood of 0.

It follows that in case $\mu > 0$ there are $\xi^- < 0$ and $\xi^+ > 0$ so that

$$f(\xi^-) = \mu\xi^-,\ f(\xi) < \mu\xi \text{ on } (\xi^-, 0),\ \text{ and } f(\xi^+) = \mu\xi^+,\ \mu\xi < f(\xi) \text{ on } (0, \xi^+).$$

ξ^- and ξ^+ are values of constant solutions, and the elements ξ_- and ξ_+ of C with ranges $\{\xi^-\}$ and $\{\xi^+\}$, respectively, are stationary points of F. Of course, further stationary points given by values in $(-\infty, \xi^-) \cup (\xi^+, \infty)$ are not excluded.

Proposition 2.3 *(Zeros, Bounds)*

(i) If $z \geq 0$ is a zero of a solution $x : [-1, \infty) \to \mathbb{R}$ of equation (1.1), then for every $t \in [0, z]$ there is a zero of x in $[t-1, t]$.

(ii) Let $\mu = 0$. There exists $B > 0$ so that for every solution $x : [-1, \infty) \to \mathbb{R}$ of equation (1.1) and for every zero $z > 2$ of x, $|x(t)| \leq B$ on $[z-1, z]$.

(iii) Let $\mu > 0$. Then every solution $x : [-1, \infty) \to \mathbb{R}$ of equation (1.1) is bounded.

Proof Assertion (i) follows from Proposition 2.2(i).

Let $\mu = 0$ and suppose f is bounded from above. Set $B = \sup f$. Consider a solution $x : [-1, \infty) \to \mathbb{R}$ of equation (1.1) and a zero $z > 2$ of x. Let $t \in [z-1, z]$. In case $x(t) > 0$, consider the zero $z' \geq 0$ in $[t-1, t]$ with $x > 0$ in $(z', t]$. Hence,

$$x(t) = 0 + \int_{z'}^{t} \dot{x}(s)\, ds = \int_{z'}^{t} f(x(s-1))\, ds \leq B.$$

In case $x(t) < 0$ there is a zero $z' \in (t, t+1] \cap (t, z]$ with $x < 0$ in (t, z'), and

$$-x(t) = \int_{t}^{z'} \dot{x}(s)\, ds = \int_{t}^{z'} f(x(s-1))\, ds \leq B.$$

In case $\mu = 0$ and f bounded from below, one can argue analogously. This proves (ii).

To prove (iii), choose $M > 0$ with $|f(\xi)| < \mu|\xi|$ for $|\xi| \geq M$. Let $x : [-1, \infty) \to \mathbb{R}$ be a solution of equation (1.1). We claim

$$|x(t)| \leq \max\{M, ||x_0||\} \qquad \text{for all } t \geq -1.$$

Assume that there exist $\epsilon > 0$ and $s \in (0, \infty)$ so that $|x(s)| > \epsilon + \max\{M, ||x_0||\}$. Then there is a minimal $t \in (0, \infty)$ with $|x(t)| = \epsilon + \max\{M, ||x_0||\}$. Clearly, $\dot{x}(t) \geq 0$ if $x(t) = \epsilon + \max\{M, ||x_0||\}$, and $\dot{x}(t) \leq 0$ if $x(t) = -\epsilon - \max\{M, ||x_0||\}$. From the minimal property of t it follows that $|x(t-1)| < \epsilon + \max\{M, ||x_0||\}$. This fact, the monotonicity of f, the definition of M and equation (1.1) combined imply $\dot{x}(t) < 0$ in case $x(t) = \epsilon + \max\{M, ||x_0||\}$, and $\dot{x}(t) > 0$ in case $x(t) = -\epsilon - \max\{M, ||x_0||\}$, a contradiction. □

Our last hypothesis, which will not be used before Chapter 16, is

(H4) $\mu = 0$, or
$\mu > 0$ and $f'(\xi^-) < \mu$ and $f'(\xi^+) < \mu$.

This implies that in case $\mu > 0$ the leading real eigenvalues λ_0^- and λ_0^+ of the generators of the semigroups $(D_2F(t, \xi_-))_{t \geq 0}$ and $(D_2F(t, \xi_+))_{t \geq 0}$ are negative. So the stationary points ξ_- and ξ_+ of F are stable and locally attractive.

CHAPTER 3

The separatrix

The set

$$S = \{\phi \in C : (x^{\phi})^{-1}(0) \text{ is not bounded from above}\}$$

is positively invariant. By Proposition 2.2,

$$C \setminus S = \bigcup_{t \geq 0} F(t, \cdot)^{-1}(\overset{\circ}{K} \cup (-\overset{\circ}{K}));$$

it follows that S is closed. The set S is a separatrix for the semiflow F in the sense that its complement $C \setminus S$ splits into the set of initial data for solutions which are positive on some unbounded interval, and into the set of initial data for solutions which are negative on some unbounded interval. Proposition 2.3 guarantees that, for every $\phi \in S$, $x^{\phi}[-1, \infty) \to \mathbb{R}$ is a bounded solution of equation (1.1).

Proposition 3.1 *(Nonordering of S) For all ϕ, ψ in C with $\phi < \psi$, either $\phi \in C \setminus S$ or $\psi \in C \setminus S$.*

Proof Assume that S contains elements $\tilde{\phi}, \tilde{\psi}$ with $\tilde{\phi} < \tilde{\psi}$. Then the positive invariance of S and Proposition 2.2(i) yield $\phi = x_2^{\tilde{\phi}} \ll \psi = x_2^{\tilde{\psi}}$ with ϕ, ψ in S.

In case $\mu = 0$, $y = x^{\psi} - x^{\phi}$ satisfies $0 \ll y_0$ and

$$\dot{y}(t) = \int_0^1 f'((1-s)x^{\phi}(t-1) + s x^{\psi}(t-1))\, ds\, y(t-1) \qquad \text{for all } t > 0.$$

Therefore, y is monotonically increasing on $\mathbb{R}^+$. y is also bounded as both x^{ϕ} and x^{ψ} are bounded. Let $c = \lim_{t \to \infty} y(t)$. Then $0 < c < \infty$. Choose $T > 0$ so that $y(t) \geq c/2$ for $t \geq T - 1$ and let

$$d = \inf_{t \geq T} \int_0^1 f'((1-s)x^{\phi}(t-1) + s x^{\psi}(t-1))\, ds.$$

Note that $d > 0$ since x^{ϕ} and x^{ψ} are bounded and $f'(x) > 0$ for all $x \in \mathbb{R}$. For $t \geq T$ we get $\dot{y}(t) \geq cd/2$ which yields $y(t) \geq y(T) + \frac{cd}{2}(t - T) \to \infty$ as $t \to \infty$, a contradiction to the boundedness of y.

Assume now $\mu > 0$. Let E denote the set of all stationary points of F. Using Theorem 4.3 in Smith **[56]** (see also Hirsch **[33]** and Smith and Thieme **[57]**), we find $(\phi^*, \psi^*) \in C \times C$ such that

$$\phi \ll \phi^* \ll \psi^* \ll \psi,$$

$$x_t^{\phi^*} \to E \text{ and } x_t^{\psi^*} \to E \qquad \text{as } t \to \infty.$$

Consider $b : \mathbb{R}^+ \ni t \mapsto \int_0^1 f'((1-s)x^{\psi^*}(t-1)+sx^{\phi^*}(t-1))\,ds \in \mathbb{R}$. $y = x^{\phi^*} - x^{\psi^*}$ satisfies $0 \ll y_t$ for all $t \geq 0$ (see Proposition 2.2(i)) and

$$\dot{y}(t) = -\mu y(t) + b(t)y(t-1) \qquad \text{for all } t > 0.$$

Suppose both $x_t^{\phi^*}$ and $x_t^{\psi^*}$ converge to 0 as $t \to \infty$. Then $0 < b(t) \to f'(0)$ as $t \to \infty$. The number

$$\epsilon = \frac{1}{2}[f'(0) - \mu]e^{-\mu}$$

is positive since $f'(0) > \mu$. It follows that there exists $T \geq 0$ so that

$$b(t) \geq f'(0) - \epsilon \qquad \text{for all } t \geq T.$$

Notice that $0 \ll y_t$ for all $t \geq 0$ implies

$$\frac{d}{dt}\left(e^{\mu t}y(t)\right) = e^{\mu t}b(t)y(t-1) > 0 \qquad \text{for all } t > 0.$$

Therefore

$$e^{\mu(t-1)}y(t-1) \leq e^{\mu t}y(t)$$

for $t \geq 2$, and hence for all $t \geq T+2$ we have

$$\begin{aligned} \dot{y}(t) &\geq -\mu y(t) + [f'(0) - \epsilon]y(t-1) \\ &\geq -(\mu + \epsilon e^{\mu})y(t) + f'(0)y(t-1). \end{aligned}$$

The choice of ϵ ensures the existence of a real number $\lambda^* > 0$ such that

$$\lambda^* = -(\mu + \epsilon e^{\mu}) + f'(0)e^{-\lambda^*}.$$

Choose $\delta > 0$ so that

$$y(t) > \delta e^{\lambda^* t} \qquad \text{for all } t \in [T+1, T+2].$$

The function $z : \mathbb{R} \ni t \mapsto \delta e^{\lambda^* t} \in \mathbb{R}$ satisfies $\dot{z}(t) = -(\mu + \epsilon e^{\mu})z(t) + f'(0)z(t-1)$ for all $t \in \mathbb{R}$. Define $u : [-1, \infty) \ni t \mapsto y(t) - z(t) \in \mathbb{R}$. Then $0 \ll u_{T+2}$ and

$$\dot{u}(t) \geq -(\mu + \epsilon e^{\mu})u(t) + f'(0)u(t-1) \qquad \text{for all } t \geq T+2.$$

If the assertion that $u(t) > 0$ for all $t \geq T+1$ is not true, then there exists $s > T+2$ such that $u(s) = 0$ and $u(t) > 0$ for all $t \in [T+1, s)$. Clearly $\dot{u}(s) \leq 0$. On the other hand, the facts $u(s) = 0$ and $u(s-1) > 0$, and the differential inequality for u combined imply $\dot{u}(s) > 0$, a contradiction. Therefore $u(t) > 0$ for all $t \geq T+1$. This yields

$$y(t) > z(t) = \delta e^{\lambda^* t} \qquad \text{for all } t \geq T+1,$$

which is in contradiction to the boundedness of y.

So at least one of the ω-limit sets of ϕ^* and ψ^* contains a non-zero stationary point. Assume $\omega(\psi^*)$ contains $p \in E \setminus \{0\}$. As the stationary points of F are constant functions, either $0 \ll p$ or $p \ll 0$. If $0 \ll p$ then there exists $t^* > 0$ such that $0 \ll x_{t^*}^{\psi^*}$. Due to Proposition 2.2(i), we have $0 \ll x_{t^*}^{\psi^*} \ll x_{t^*}^{\psi}$. Using Proposition 2.2(i) again, $x_t^{\psi} \in \overset{\circ}{K}$ follows for all $t \geq t^*$, a contradiction to $\psi \in S$. If $p \ll 0$ then there exists $t^* > 0$ such that $x_{t^*}^{\psi^*} \ll 0$. By Proposition 2.2(i), we obtain $x_{t^*}^{\phi} \ll x_{t^*}^{\psi^*}$, and $x_t^{\phi} \in -\overset{\circ}{K}$ follows for all $t \geq t^*$, a contradiction to $\phi \in S$. Analogously, the assumption $p \in \omega(\phi^*) \cap (E \setminus \{0\})$ leads to a contradiction. $\square$

Recall that we have a decomposition

$$C = Q \oplus L \oplus P$$

into realified generalized eigenspaces of the generator of the semigroup given by the linear equation (1.2).

Proposition 3.2 *(Graph Representation of S) There exists a map*

$$\mathrm{Sep} : Q \oplus L \to P$$

with

$$S = \{\chi + \mathrm{Sep}(\chi) : \chi \in Q \oplus L\}$$

and

$$||\,\mathrm{Sep}(\chi) - \mathrm{Sep}(\tilde{\chi})|| \leq e^{\lambda_0}||\chi - \tilde{\chi}|| \qquad \text{for all } \chi, \tilde{\chi} \text{ in } Q \oplus L.$$

Proof 1. In order to show that $\mathrm{Pr}_{Q\oplus L}|_S$ is injective, consider $\phi \in S$, $\psi \in S$ with $\mathrm{Pr}_{Q\oplus L}\,\phi = \mathrm{Pr}_{Q\oplus L}\,\psi$. Then $\phi - \psi \in P = \mathbb{R}\chi_0$. So either $\phi = \psi$, or $\psi \ll \phi$, or $\phi \ll \psi$. The last two possibilities are excluded by Proposition 3.1.

It follows that there is a map $\mathrm{Sep} : \mathrm{Pr}_{Q\oplus L}\, S \to P$ with $S = \{\chi + \mathrm{Sep}(\chi) : \chi \in \mathrm{Pr}_{Q\oplus L}\, S\}$.

2. Proof of $\mathrm{Pr}_{Q\oplus L}\, S \supset Q \oplus L$. Fix $\chi \in Q \oplus L$. There are $s_+ > 0$ and $s_- < 0$ with $\chi + s_+\chi_0 \in \overset{\circ}{K}$ and $\chi + s_-\chi_0 \in -\overset{\circ}{K}$. The sets

$$A_+ = \{s \in \mathbb{R} : F(t, \chi + s\chi_0) \in \overset{\circ}{K} \text{ for some } t \geq 0\}$$

and

$$A_- = \{s \in \mathbb{R} : F(t, \chi + s\chi_0) \in -\overset{\circ}{K} \text{ for some } t \geq 0\}$$

are open, nonempty and disjoint. It follows from the connectedness of $\mathbb{R}$ that there exists $s \in \mathbb{R} \setminus (A_- \cup A_+)$. So, for all $t \geq 0$, $F(t, \chi + s\chi_0) \in C \setminus (\overset{\circ}{K} \cup (-\overset{\circ}{K}))$. That is, $\chi + s\chi_0 \in S$ and consequently,

$$\chi = \mathrm{Pr}_{Q\oplus L}(\chi + s\chi_0) \in \mathrm{Pr}_{Q\oplus L}\, S.$$

3. For every $\chi \in Q \oplus L$, $\mathrm{Sep}(\chi) = \mathrm{Sep}(\chi)(0)\chi_0$. For $\chi, \tilde{\chi}$ in $Q \oplus L$ we get

$$||\,\mathrm{Sep}(\chi) - \mathrm{Sep}(\tilde{\chi})|| = |\,\mathrm{Sep}(\chi)(0) - \mathrm{Sep}(\tilde{\chi})(0)|\, ||\chi_0|| = |\,\mathrm{Sep}(\chi)(0) - \mathrm{Sep}(\tilde{\chi})(0)|.$$

Suppose $|\,\mathrm{Sep}(\chi)(0) - \mathrm{Sep}(\tilde{\chi})(0)| > e^{\lambda_0}||\chi - \tilde{\chi}||$. Then the elements $\phi = \chi + \mathrm{Sep}(\chi)$ and $\tilde{\phi} = \tilde{\chi} + \mathrm{Sep}(\tilde{\chi})$ of S satisfy, for $s \in [-1, 0]$, the inequalities

$$\begin{aligned} |\phi(s) - \tilde{\phi}(s)| &\geq |\,\mathrm{Sep}(\chi)(s) - \mathrm{Sep}(\tilde{\chi})(s)| - |\chi(s) - \tilde{\chi}(s)| \\ &\geq e^{\lambda_0 s}|\,\mathrm{Sep}(\chi)(0) - \mathrm{Sep}(\tilde{\chi})(0)| - ||\chi - \tilde{\chi}|| \\ &> e^{-\lambda_0}e^{\lambda_0}||\chi - \tilde{\chi}|| - ||\chi - \tilde{\chi}|| = 0. \end{aligned}$$

So, $\phi - \tilde{\phi}$ has no zero. Hence, $\phi \ll \tilde{\phi}$ or $\tilde{\phi} \ll \phi$, is a contradiction to Proposition 3.1. □

For each $\phi \in C \setminus S$, we have $\mathrm{Pr}_P\,\phi \neq \mathrm{Sep}(\mathrm{Pr}_{Q\oplus L}\,\phi)$. Using $P = \mathbb{R}\chi_0$ and $0 \ll \chi_0$, we obtain

$$\mathrm{Pr}_P\,\phi \ll \mathrm{Sep}(\mathrm{Pr}_{Q\oplus L}\,\phi) \quad \text{or} \quad \mathrm{Sep}(\mathrm{Pr}_{Q\oplus L}\,\phi) \ll \mathrm{Pr}_P\,\phi.$$

It is convenient to say in the first case that ϕ is below S, and in the second case that ϕ is above S.

Proposition 3.3 *Assume* $\mu = 0$.
(i) If ϕ *is above (below)* S*, then* $x^\phi(t) \to +\infty$ $(\to -\infty)$ *as* $t \to \infty$.
(ii) $K \setminus \{0\}$ *is above* S*, and* $(-K) \setminus \{0\}$ *is below* S.

Proof 1. Let $\phi \in C$ with $\mathrm{Sep}(\mathrm{Pr}_{Q\oplus L}\,\phi) \ll \mathrm{Pr}_P\,\phi$ be given. Set $\psi = \mathrm{Pr}_{Q\oplus L}\,\phi + \mathrm{Sep}(\mathrm{Pr}_{Q\oplus L}\,\phi) \in S$. Then $\psi \ll \phi$, and the function $y = x^\phi - x^\psi$ satisfies $0 \ll y_0$, and

$$\dot{y}(t) = \int_0^1 f'(x^\psi(t-1) + sy(t-1))\,ds\,y(t-1) \qquad \text{for all } t > 0.$$

It follows that $y(t) > 0$ for all $t \geq -1$ and $\dot{y}(t) > 0$ for all $t > 0$. Suppose y is bounded on $[-1, \infty)$. Then y converges to some $c > 0$ as $t \to \infty$. Set $d = \sup_{-1 \leq t} |x^\psi(t)|$. Then

$$\dot{y}(t) \geq \frac{c}{2} \min_{-d \leq \xi \leq d+2c} f'(\xi) > 0$$

for t sufficiently large, which leads to a contradiction to the assumption that y is bounded on $[-1, \infty)$. We infer $y(t) \to +\infty$ as $t \to \infty$. Using the boundedness of x^ψ, we obtain $x^\phi(t) \to +\infty$ as $t \to \infty$. The argument for ϕ below S is analogous.

2. If $\phi \in K \setminus \{0\}$ then Proposition 2.2 and the above argument yield $x^\phi(t) \to \infty$ as $t \to \infty$. Thus, $\phi \notin S$ by Proposition 2.3, and the conclusion in (i) also implies that ϕ is above S. The other case can be dealt with similarly. □

CHAPTER 4

The leading unstable set of the origin

Observe that $L \oplus P$ is the realified generalized eigenspace of $D_2F(1,0)$ given by the spectral set $\{e^{\lambda_0}, e^{\lambda_1}, e^{\overline{\lambda_1}}\}$ of this operator while Q is also the realified generalized eigenspace of $D_2F(1,0)$ for the complementary spectral set $\{e^{\lambda_k} : k \geq 2\} \cup \{e^{\overline{\lambda_k}} : k \geq 2\} \cup \{0\}$.

Choose $\beta > 1$ with $e^{\operatorname{Re}\lambda_2} < \beta < e^{\operatorname{Re}\lambda_1}$. According to Theorem I.4 there exist convex bounded open neighbourhoods N_Q, N_L, N_P of 0 in Q, L, P, respectively, and a C^1-map $\omega_{loc} : N_L + N_P \to Q$ with range in N_Q and $w_{loc}(0) = 0$, $Dw_{loc}(0) = 0$, so that the β-unstable set of the fixed point 0 of $F(1,\cdot)$ in $N_{loc} = N_Q + N_P + N_P$, namely

$$\begin{aligned} W_\beta(N_{loc}) = \{\phi \in N_{loc} : \text{ There is a trajectory } (\phi_n)_{-\infty}^0 \text{ of } F(1,\cdot) \\ \text{with } \phi_0 = \phi,\ \phi_n\beta^{-n} \in N_{loc} \text{ for all integers } n \leq 0, \\ \text{and } \phi_n\beta^{-n} \to 0 \text{ as } n \to -\infty\}, \end{aligned}$$

coincides with the graph

$$W_{loc} = \{\chi + w_{loc}(\chi) : \chi \in N_L + N_P\}$$

with tangent space $L \oplus P$ at 0.

One easily shows that every ϕ in the 3-dimensional submanifold W_{loc} uniquely determines a solution $x^\phi : \mathbb{R} \to \mathbb{R}$ of equation (1.1). Using the growth bound of Proposition 2.1(i), we obtain that this solution satisfies $x^\phi(t) \to 0$ as $t \to -\infty$, there exists $t \in \mathbb{R}$ with $x^\phi_s \in W_{loc}$ for all $s \leq t$. We call the forward extension

$$W = F(\mathbb{R}^+ \times W_{loc})$$

the leading unstable set of the origin. Our goal is to describe the closure $\overline{W}$ of W and the solutions with segments in $\overline{W}$.

Proposition 4.1 *(Invariance, Heteroclinic Connections and Compactness for $\mu > 0$, Unboundedness for $\mu = 0$)*

(i) For every $\phi \in W$ there exists one and only one solution $x^\phi : \mathbb{R} \to \mathbb{R}$ of equation (1.1) with $x^\phi_0 = \phi$. For all $t \in \mathbb{R}$, $x^\phi_t \in W$, and $x^\phi_t \to 0$ as $t \to -\infty$.

(ii) In case $\mu > 0$, $\xi_- \ll \phi \ll \xi_+$ for all $\phi \in W$, and there exist $\chi \in W$, $\eta \in W$ so that x^χ is negative, strictly decreasing, and $x^\chi(t) \to \xi^-$ as $t \to \infty$, while x^η is positive, strictly increasing, and $x^\eta(t) \to \xi^+$ as $t \to \infty$.

(iii) In case $\mu > 0$, $\overline{W}$ is compact, invariant, and contains ξ_- and ξ_+. For all $\phi \in \overline{W}$, $\xi_- \leq \phi \leq \xi_+$.

(iv) In case $\mu = 0$, there exist $\chi \in W$, $\eta \in W$ so that x^χ is negative, strictly decreasing, and $x^\chi(t) \to -\infty$ as $t \to \infty$, while x^η is positive, strictly increasing, and $x^\eta(t) \to \infty$ as $t \to \infty$.

Proof 1. The definition of W and the properties of W_{loc} imply assertion (i).

2. Proof of (ii). Let $\mu > 0$. We have $\xi_- \ll 0 \ll \xi_+$. For every $\phi \in W$ there exists $t < -2$ with $\xi_- \ll x_t^\phi \ll \xi_+$. Proposition 2.2 yields $\xi_- \ll \phi \ll \xi_+$.

Choose $\tilde{\beta} \in (e^{\operatorname{Re}\lambda_1}, e^{\lambda_0})$. Theorem I.4 implies that for $F(1,\cdot)$ and $\tilde{\beta}$ there is a one-dimensional analogue of W_{loc} with tangent space P at 0. In particular, there exist $\epsilon_0 > 0$ and a C^1-map $\tilde{w} : (-\epsilon_0, \epsilon_0)\chi_0 \to Q \oplus L$ with $\tilde{w}(0) = 0$ and $D\tilde{w}(0) = 0$ so that for every $\epsilon \in (-\epsilon_0, \epsilon_0)$ there is a trajectory $(\phi_n)_{-\infty}^0$ of $F(1,\cdot)$ with $\phi_0 = \tilde{w}(\epsilon\chi_0) + \epsilon\chi_0$, $\phi_n\tilde{\beta}^{-n} \in N_{loc}$ for all integers $n \le 0$, and $\phi_n\tilde{\beta}^{-n} \to 0$ as $n \to -\infty$. It follows that each $\tilde{w}(\epsilon\chi_0) + \epsilon\chi_0$ with $|\epsilon| < \epsilon_0$ belongs to $W_{loc} \subset W$. Also, for every $\epsilon \in (-\epsilon_0, \epsilon_0)$ there exists $t_\epsilon \in \mathbb{R}$ so that all segments x_t, $t \le t_\epsilon$, of the solution $x : \mathbb{R} \to \mathbb{R}$ with $x_0 = \tilde{w}(\epsilon\chi_0) + \epsilon\chi_0$ belong to

$$\widetilde{W} = \{\tilde{w}(\delta\chi_0) + \delta\chi_0 : |\delta| < \epsilon_0\}.$$

Choose $\epsilon_1 \in (0, \epsilon_0)$ so small that $||D\tilde{w}(\epsilon\chi_0)|| < e^{-\lambda_0}$ for $|\epsilon| < \epsilon_1$. Fix $\epsilon \in (0, \epsilon_1)$ and set

$$\begin{aligned}\eta &= \tilde{w}(\epsilon\chi_0) + \epsilon\chi_0 \\ &= \epsilon\chi_0 + \int_0^1 D\tilde{w}(\theta\epsilon\chi_0)\epsilon\chi_0\, d\theta.\end{aligned}$$

It follows that $0 \ll \eta \in W$. Consider the solution $x : \mathbb{R} \to \mathbb{R}$ with $x_0 = \eta$.

We have $x_s \ne x_t$ for $s \ne t$, for otherwise x is periodic with $x_0 \ne 0$ which leads to a contradiction to $x_u \to 0$ as $u \to -\infty$.

Proof of that $x_t - x_s$ has no zero for $s \ne t$: Let reals $s \ne t$ be given. We first note that there exists $j \in \mathbb{N}$ such that $x_{t-2k} \in \widetilde{W}$, $x_{s-2k} \in \widetilde{W}$ for all integers $k \ge j$. We have $\operatorname{Pr}_P x_u \to 0$ as $u \to -\infty$, since $x_0 \in W$. It follows that there is an integer $k \ge j$ such that for each integer $m \ge k$,

$$\begin{aligned}&x_{t-2m} - x_{s-2m} \\ &= \operatorname{Pr}_P x_{t-2m} - \operatorname{Pr}_P x_{s-2m} + \tilde{w}(\operatorname{Pr}_P x_{t-2m}) - \tilde{w}(\operatorname{Pr}_P x_{s-2m}) \\ &= \operatorname{Pr}_P x_{t-2m} - \operatorname{Pr}_P x_{s-2m} \\ &\quad + \int_0^1 D\tilde{w}(\operatorname{Pr}_P x_{s-2m} + \theta(\operatorname{Pr}_P x_{t-2m} - \operatorname{Pr}_P x_{s-2m}))[\operatorname{Pr}_P x_{t-2m} - \operatorname{Pr}_P x_{s-2m}]\, d\theta\end{aligned}$$

and

$$\begin{aligned}&\left\| \int_0^1 D\tilde{w}(\operatorname{Pr}_P x_{s-2m} + \theta(\operatorname{Pr}_P x_{t-2m} - \operatorname{Pr}_P x_{s-2m}))[\operatorname{Pr}_P x_{t-2m} - \operatorname{Pr}_P x_{s-2m}]\, d\theta \right\| \\ &\qquad \le \frac{1}{2} e^{-\lambda_0} || \operatorname{Pr}_P x_{t-2m} - \operatorname{Pr}_P x_{s-2m}||.\end{aligned}$$

A remark above and the injectivity of Pr_P on $\widetilde{W}$ give

$$\operatorname{Pr}_P x_{t-2m} - \operatorname{Pr}_P x_{s-2m} = c_m\chi_0 \qquad \text{with } c_m \ne 0$$

for all integers $m \geq k$. Now it follows that

$$|x_{t-2m}(u) - x_{s-2m}(u)| \geq |c_m|\chi_0(u) - \frac{e^{-\lambda_0}}{2}|c_m|\,||\chi_0|| > 0$$

for all integers $m \geq k$ and for all $u \in [-1, 0]$. Hence Proposition 2.2 yields that $x_t - x_s$ has no zero.

Proof of that x is strictly increasing: x is injective since otherwise there exist $s \neq t$ with $x(s) = x(t)$, and $x_s - x_t$ has a zero at 0, a contradiction to the previous result. It follows that x is strictly monotone. If x is strictly decreasing, then $0 < \eta(0) \leq x(t)$ for all $t \leq 0$, contradicting $x_t \to 0$ as $t \to -\infty$. So, x is strictly increasing.

Proof of $x(t) \to \xi^+$ for $t \to \infty$: Assertion (i) and the first part of assertion (ii) give $x(t) \leq \xi^+$ for all $t \in \mathbb{R}$. It follows that $x(t)$ converges to some $\xi \in (0, \xi^+]$ as $t \to \infty$. By equation (1.1), $\dot{x}(t) \to -\mu\xi + f(\xi)$ as $t \to \infty$. In case $-\mu\xi + f(\xi) \neq 0$, $\dot{x}$ is bounded away from 0 on some unbounded interval in $\mathbb{R}^+$, and we obtain a contradiction to $\lim_{t\to\infty} x(t) = \xi$. Thus, $\mu\xi = f(\xi)$ from which it follows that $\xi = \xi^+$.

The existence of $\chi \in W$ with $x_t^\chi \to \xi_-$ as $t \to \infty$ can be proved in the same way.

3. Proof of (iii). Assume $\mu > 0$. From (ii) it is obvious that ξ_- and ξ_+ belong to $\overline{W}$ and $\xi_- \leq \phi \leq \xi_+$ for all $\phi \in \overline{W}$. It also follows from (i) that each $\phi \in W$ is continuously differentiable. Using equation (1.1), we obtain a bound for all $\dot{\phi}$, $\phi \in W$. Therefore, the Arzèla–Ascoli Theorem shows that $\overline{W}$ is compact. In order to show the invariance of $\overline{W}$, let $\phi \in \overline{W}$ and choose a sequence of points ϕ_n, $n \in \mathbb{N}$, in W such that $\phi_n \to \phi$ as $n \to \infty$. For every $t \geq 0$,

$$F(t, \phi) = F(t, \lim_{n\to\infty} \phi_n) = \lim_{n\to\infty} F(t, \phi_n),$$

and $F(t, \phi_n) \in W$ for all $n \in \mathbb{N}$. Hence $F(t, \phi) \in \overline{W}$. This shows the positive invariance of $\overline{W}$. We now consider an integer $k < 0$. The compactness of $\overline{W}$ implies that every subsequence of $(x_k^{\phi_n})_{n=0}^\infty$ has a convergent subsequence which is convergent to some $\psi \in \overline{W}$. Using the continuity of $F(-k, \cdot)$, we get $F(-k, \psi) = \phi$. By the injectivity of $F(-k, \cdot)$, we infer that the limits of all convergent subsequences are equal, so that the sequence $(x_k^{\phi_n})_{n=0}^\infty$ itself converges to some $\psi_k \in \overline{W}$ with $F(-k, \psi_k) = \phi$. We have $\psi_k = F(1, \psi_{k-1})$ for all integers $k < 0$. It follows that there is a solution $x : \mathbb{R} \to \mathbb{R}$ of equation (1.1) with $x_k = \psi_k$ for all integers $k < 0$, $x_0 = \phi$, and $x_t \in \overline{W}$ for all $t \in \mathbb{R}$.

4. Proof of (iv). Let $\mu = 0$. As in the proof of (ii) we find $\eta \in W$ so that $0 \ll \eta$ and x^η is strictly increasing. Then equation (1.1) with $\mu = 0$ yields $\dot{x}(t) \geq f(\eta(-1)) > 0$ for all $t \geq 0$, and hence $x(t) \to \infty$ as $t \to \infty$. Similarly, we get χ as required. □

Proposition 4.2 *Let $\mu = 0$. Then $\overline{W}$ is invariant, and there exists $B_0 > 0$ so that for every $b > B_0$ the set $\overline{W} \cap \overline{C_b}$ is negatively invariant and compact.*

Proof 1. Claim: There exists $B_0 > 0$ so that for every $b > B_0$ and for every solution $x : \mathbb{R} \to \mathbb{R}$ of equation (1.1) with $|x(0)| = b$, $\dot{x}(0) \neq 0$ and $|x(t)| < b$ for all $t < 0$.

Proof. Set $B_0 = B + f_B$ with B as in Proposition 2.3(ii). Let $b > B_0$ and consider a solution $x : \mathbb{R} \to \mathbb{R}$ of equation (1.1) with $|x(0)| = b$. If x has no zero in $(-\infty, 0]$, then equation (1.1) shows that $|x|$ is strictly increasing and $\dot{x}$ has no zero, and thus the claim holds. If x has zero in $(-\infty, 0]$, then there is the largest zero z of x in $(-\infty, 0]$. Propositions 2.3 and 2.1 combined yield $|x(t)| \le B_0$ on $(-\infty, z+1]$. Consequently, $z + 1 < 0$, and x has no zero on $(z, 0]$. Equation (1.1) implies that $\dot{x}$ has no zero on $(z+1, 0]$ and that $|x|$ is strictly increasing on $(z+1, 0]$.

2. It follows that for every $b > B_0$ and for every solution $x : \mathbb{R} \to \mathbb{R}$ of equation (1.1) with $x_0 \in \overline{C_b}$, we have $x_t \in \overline{C_b}$ for all $t \le 0$.

3. Proof of that $\overline{W}$ is invariant. Let $\phi \in \overline{W}$. As in part 3 of the proof of Proposition 4.1 we obtain $F(t, \phi) \in \overline{W}$ for all $t \ge 0$. Choose $b > B_0$ such that $||\phi|| < b$. Then there is a sequence $(\phi_n)_0^\infty$ in $W \cap C_b$ with $\phi_n \to \phi$ as $n \to \infty$. Using part 2 and equation (1.1) we infer that there is a bound for all $x_k^{\phi_n}$ and $\dot{x}_k^{\phi_n}$, $k \in -\mathbb{N}$ and $n \in \mathbb{N}$. Now we can construct a solution $x : \mathbb{R} \to \mathbb{R}$ of equation (1.1) with $x_0 = \phi$ and all segments x_t in $\overline{W}$ as in part 3 of the proof of Proposition 4.1.

4. Proof of that $\overline{W} \cap \overline{C_b}$ is negatively invariant for $b > B_0$. Let $\phi \in \overline{W} \cap \overline{C_b}$. Choose $r > b$. There is a sequence $(\phi_n)_0^\infty$ in $W \cap C_r$ with $\phi_n \to \phi$ as $n \to \infty$. Let $t < 0$ and choose an integer $k < t$. As in part 3, $x_k^{\phi_n} \to x_k^\phi$ as $n \to \infty$. By the continuity of $F(t-k, \cdot)$, $x_t^{\phi_n} \to x_t^\phi$ as $n \to \infty$. Part 2 yields $x_t^{\phi_n} \in \overline{C_r}$ for all $n \in \mathbb{N}$. Hence $x_t^\phi \in \overline{C_r}$, and $||x_t^\phi|| \le \inf_{r > b} r = b$.

5. Let $b > B_0$. Compactness of the bounded closed set $\overline{W} \cap \overline{C_b}$ follows from the Arzèla–Ascoli Theorem, since part 4 and equation (1.1) combined yield a bound for the derivatives $\dot{\phi}$, $\phi \in \overline{W} \cap \overline{C_b}$. □

Note that all three sets W, $\overline{W}$ and

$$\operatorname{bd} W = \overline{W} \setminus W$$

are invariant.

Proposition 4.3 *The map $F_W : \mathbb{R} \times \overline{W} \ni (t, \phi) \mapsto x_t^\phi \in \overline{W}$ is a continuous flow. Each $\phi \in \overline{W}$ is continuously differentiable, and the map*

$$\overline{W} \ni \phi \mapsto \dot{\phi} \in C$$

is continuous.

Proof It is easy to see that $F_W(t+s, \phi) = F_W(t, F_W(s, \phi))$ and $F_W(0, \phi) = \phi$ for all $t \in \mathbb{R}$, $s \in \mathbb{R}$ and for all $\phi \in \overline{W}$. To obtain the continuity, let $\phi \in C$ and $t \in \mathbb{R}$ be given. Choose an integer $k < t$. For every $s > k$ and $\psi \in C$,

$$F_W(s, \psi) = F(s - k, F_W(k, \psi)).$$

Using the continuity of F we see that it is sufficient to show that the map $F_W(k, \cdot) : \overline{W} \to \overline{W}$ is continuous. In case $k \ge 0$, this is obvious from $F_W(k, \psi) = F(k, \psi)$ on $\overline{W}$. Let $k < 0$. As in the proofs of Propositions 4.1 and 4.2, we see that for every sequence $(\psi_n)_0^\infty$ in $\overline{W}$ with $\psi_n \to \psi$ as $n \to \infty$, $F_W(k, \psi_n) = x_k^{\psi_n} \to x_k^\psi = F_W(k, \psi)$ as $n \to \infty$.

The invariance of $\overline{W}$ shows that each $\phi \in \overline{W}$ is continuously differentiable, with

$$\dot{\phi} = -\mu\phi + f \circ F_W(-1, \phi)$$

from which the continuity of $\overline{W} \ni \phi \mapsto \dot{\phi} \in C$ follows. □

The remaining results of this chapter concern the part of the separatrix S in the invariant set $\overline{W}$.

Proposition 4.4 *$\overline{W} \cap S$ and $\overline{W \cap S}$ are compact and invariant.*

Proof 1. Recall that S is closed. In case $\mu > 0$, $\overline{W}$ is compact, and it follows that $\overline{W} \cap S$ and $\overline{W \cap S} \subset \overline{W} \cap \overline{S} = \overline{W} \cap S$ are compact. In case $\mu = 0$, Proposition 2.3(i) and (ii) yield $\overline{W} \cap S \subset \overline{C_B}$, and the closed sets $\overline{W \cap S} \subset \overline{W} \cap S$ are contained in a compact set $\overline{W} \cap \overline{C_b}$ with $b > B_0$ (see Proposition 4.2).

2. To see the invariance of $\overline{W} \cap S$, let $\phi \in \overline{W} \cap S$. As $\overline{W}$ is invariant there is a solution $x^{\phi} : \mathbb{R} \to \mathbb{R}$ of (1.1) such that $x_0^{\phi} = \phi$ and $x_t^{\phi} \in \overline{W}$ for all $t \in \mathbb{R}$. For every $t \in \mathbb{R}$ the zeroset of the solution $y : \mathbb{R} \to \mathbb{R}$ with $y_0 = x_t^{\phi}$ is not bounded from above since $y = x^{\phi}(t + \cdot)$; it follows that $x_t^{\phi} = y_0 \in S$. The invariance of $W \cap S$ can be shown in the same way since W is invariant. To prove the invariance of $\overline{W \cap S}$, let $\phi \in \overline{W \cap S}$ and $\phi_n \in W \cap S$, $n \in \mathbb{N}$, be given with $\phi_n \to \phi$ for $n \to \infty$. By $\overline{W \cap S} \subset \overline{W}$ and the invariance of $\overline{W}$, there exists a solution $x^{\phi} : \mathbb{R} \to \mathbb{R}$ of (1.1) with $x_0^{\phi} = \phi$ and $x_t^{\phi} \in \overline{W}$ for all $t \in \mathbb{R}$. Let $t \in \mathbb{R}$. As F_W is a continuous flow, $F_W(t, \phi_n) \to F_W(t, \phi)$ as $n \to \infty$. Because of the invariance of $W \cap S$, $F_W(t, \phi_n) \in W \cap S$ for all $n \in \mathbb{N}$. Consequently, $x_t^{\phi} = F_W(t, \phi) \in \overline{W \cap S}$. □

Proposition 3.3 shows that in case $\mu = 0$, for each $\phi \in \overline{W}$ above (below) S, the solution x^{ϕ} converges to $+\infty$ $(-\infty)$ as $t \to \infty$. The following is an analogue in the case $\mu > 0$.

Proposition 4.5 *Let $\mu > 0$.*
(i) For each $\phi \in \overline{W}$ above (below) S, $x^{\phi}(t) \to \xi^+$ (ξ^-) as $t \to \infty$.
(ii) $(K \cap \overline{W}) \setminus \{0\}$ is above S, and $((-K) \cap \overline{W}) \setminus \{0\}$ is below S.

Proof 1. For $\phi \in \overline{W}$ above S, $x_t^{\phi} \in \overset{\circ}{K} \cup (-\overset{\circ}{K})$ for some $t \geq 0$. Suppose $x_t^{\phi} \in -\overset{\circ}{K}$. By Proposition 2.2,

$$F(t, \mathrm{Pr}_{Q \oplus L}\, \phi + \mathrm{Sep}(\mathrm{Pr}_{Q \oplus L}(\phi))) \leq x_t^{\phi} \ll 0,$$

which is a contradiction to the fact that $F(t, \mathrm{Pr}_{Q \oplus L}\, \phi + \mathrm{Sep}(\mathrm{Pr}_{Q \oplus L}\, \phi)) \in S$ has a zero. It follows that $0 \ll x_t^{\phi}$.

Recall the solution x^{η} from Proposition 4.1(ii). For some $s \in \mathbb{R}$, we have $0 \ll x_s^{\eta} \ll x_t^{\phi}$. Since $x_t^{\phi} \leq \xi_+$ and $x^{\eta}(s+u) \to \xi^+$ as $u \to \infty$, Proposition 2.2 yields $x^{\phi}(u) \to \xi^+$ as $u \to \infty$. The proof for the case where ϕ is below S is analogous.

2. Using the same argument as above, we can show that if $\phi \in (K \cap \overline{W}) \setminus \{0\}$ then $x_t^{\phi} \to \xi_+$ as $t \to \infty$. This implies $\phi \notin S$. Assertion (i) excludes the case where ϕ is below S. Therefore, ϕ is above S. The other case can be dealt with similarly. □

The behaviour of the solutions $x^{\phi} : \mathbb{R} \to \mathbb{R}$ with $\phi \in \overline{W} \cap S$ will be discussed in later chapters.

We need the following local graph representation of $W_{loc} \cap S$. See Figure 4.1 for illustration.

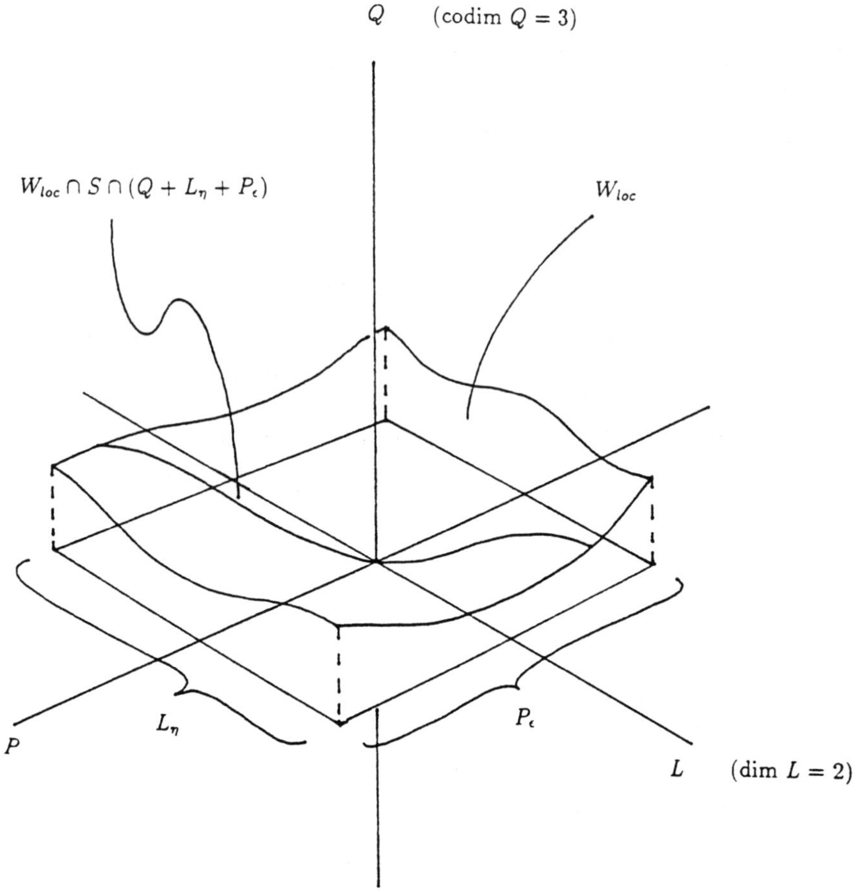

Figure 4.1 The local representation of $W_{loc} \cap S$

Proposition 4.6 *There exist $\eta > 0$, $\epsilon > 0$, and a Lipschitz continuous map*

$$\mathrm{Sep}_\eta : L_\eta \to Q \oplus P$$

with $\mathrm{Sep}_\eta(0) = 0$, $\mathrm{Sep}_\eta(L_\eta) \subset Q + P_\epsilon$ *and*

$$(W_{loc} \cap S) \cap (Q + L_\eta + P_\epsilon) = \{\chi + \mathrm{Sep}_\eta(\chi) : \chi \in L_\eta\}.$$

Proof 1. Choose $\epsilon > 0$ with $\overline{L_\epsilon} \subset N_L$, $\overline{P_\epsilon} \subset N_P$ so that

$$c = \sup_{\chi \in L_\epsilon + P_\epsilon} ||Dw_{loc}(\chi)||$$

satisfies $ce^{\lambda_0} < \frac{1}{2}$. Obviously,

$$W_{loc} \cap (Q + L_\epsilon + P_\epsilon) = \{\chi + w_{loc}(\chi) : \chi \in L_\epsilon + P_\epsilon\}$$

and

$$\begin{aligned}
&W_{loc} \cap S \cap (Q + L_\epsilon + P_\epsilon) \\
&\qquad = \{\phi \in C : \Pr_L \phi + \Pr_P \phi \in L_\epsilon + P_\epsilon, \\
&\qquad\qquad \Pr_Q \phi = w_{loc}(\Pr_L \phi + \Pr_P \phi),\ \Pr_P \phi = \operatorname{Sep}(\Pr_Q \phi + \Pr_L \phi)\} \\
&\qquad \subset \{\phi \in Q + L_\epsilon + P_\epsilon : \operatorname{Sep}(\Pr_Q \phi + \Pr_L \phi) \in P_\epsilon, \\
&\qquad\qquad \Pr_Q \phi = w_{loc}(\Pr_L \phi + \operatorname{Sep}(\Pr_Q \phi + \Pr_L \phi)), \\
&\qquad\qquad \Pr_P \phi = \operatorname{Sep}(w_{loc}(\Pr_L \phi + \Pr_P \phi) + \Pr_L \phi)\}.
\end{aligned}$$

Denote the last set by W^*. For any $\phi, \hat{\phi}$ in W^*, we have from Proposition 3.2 the inequalities

$$||\Pr_Q \phi - \Pr_Q \hat{\phi}|| \leq c[||\Pr_L \phi - \Pr_L \hat{\phi}|| + e^{\lambda_0}(||\Pr_Q \phi - \Pr_Q \hat{\phi}|| + ||\Pr_L \phi - \Pr_L \hat{\phi}||)]$$

and

$$||\Pr_P \phi - \Pr_P \hat{\phi}|| \leq e^{\lambda_0}[||\Pr_L \phi - \Pr_L \hat{\phi}|| + c(||\Pr_L \phi - \Pr_L \hat{\phi}|| + ||\Pr_P \phi - \Pr_P \hat{\phi}||)].$$

Thus,

$$||\Pr_{Q\oplus P} \phi - \Pr_{Q\oplus P} \hat{\phi}|| \leq \frac{c(1+e^{\lambda_0}) + e^{\lambda_0}(1+c)}{1 - ce^{\lambda_0}} ||\Pr_L \phi - \Pr_L \hat{\phi}|| \text{ for all } \phi, \hat{\phi} \text{ in } W^*.$$

It follows that the restriction of $\Pr_L$ to W^* is injective, and there is a map

$$\operatorname{Sep}^* : \Pr_L(W_{loc} \cap S \cap (Q + L_\epsilon + P_\epsilon)) \to Q \oplus P$$

which is Lipschitz continuous with Lipschitz constant

$$\frac{c(1+e^{\lambda_0}) + e^{\lambda_0}(1+c)}{1 - e^{\lambda_0}c}$$

so that

$$W_{loc} \cap S \cap (Q + L_\epsilon + P_\epsilon) = \{\chi + \operatorname{Sep}^*(\chi) : \chi \in \Pr_L(W_{loc} \cap S \cap (Q + L_\epsilon + P_\epsilon))\}.$$

Note $\operatorname{Sep}^*(0) = 0$.

2. We show that for $\delta = \frac{\epsilon}{4(c+1)} e^{-\lambda_0}$, $L_\delta \subset \Pr_L(W_{loc} \cap S \cap (Q + L_\epsilon + P_\epsilon))$.

Let $\chi \in L_\delta$ be given. We want to find $\phi \in Q$, $\psi \in P$ such that $\phi + \chi + \psi \in W_{loc} \cap S \cap (Q + L_\epsilon + P_\epsilon)$. For this purpose, consider the map

$$T : \overline{P_\epsilon} \ni \psi \mapsto \operatorname{Sep}(w_{loc}(\chi + \psi) + \chi) \in P.$$

For all $\psi \in \overline{P_\epsilon}$, we have

$$||T(\psi)|| \leq e^{\lambda_0}[||\chi|| + c(||\psi|| + ||\chi||)] \leq e^{\lambda_0}(c(\epsilon + \delta) + \delta) < \epsilon$$

and hence $T(\overline{P_\epsilon}) \subset P_\epsilon$. Also, for all $\psi, \hat{\psi}$ in $\overline{P_\epsilon}$, we have

$$||T(\psi) - T(\hat{\psi}|| \leq ce^{\lambda_0}||\psi - \hat{\psi}||.$$

So there is a fixed point ψ of the contraction T in P_ϵ,

$$\psi = T(\psi) = \operatorname{Sep}(w_{loc}(\chi + \psi) + \chi).$$

Set $\phi = w_{loc}(\chi + \psi)$. Then

$$\psi = \operatorname{Sep}(w_{loc}(\chi + \psi) + \chi) = \operatorname{Sep}(\phi + \chi).$$

Hence,

$$\phi + \chi + \psi \in W_{loc} \cap S \cap (Q + L_\delta + P_\epsilon) \subset W_{loc} \cap S \cap (Q + L_\epsilon + P_\epsilon),$$

and

$$\chi \in \mathrm{Pr}_L(W_{loc} \cap S \cap (Q + L_\epsilon + P_\epsilon)).$$

3. Choose $\eta \in (0, \delta)$ with $\mathrm{Pr}_P \circ \mathrm{Sep}^*(L_\eta) \subset P_\epsilon$, and set $\mathrm{Sep}_\eta = \mathrm{Sep}^*|_{L_\eta}$. Then

$$L_\eta \subset L_\delta \subset \mathrm{Pr}_L(W_{loc} \cap S \cap (Q + L_\epsilon + P_\epsilon)),$$

and Sep_η is Lipschitz continuous, and

$$\{\chi + \mathrm{Sep}_\eta(\chi) : \chi \in L_\eta\} \subset [W_{loc} \cap S \cap (Q + L_\epsilon + P_\epsilon)] \cap [Q + L_\eta + P] = W_{loc} \cap S \cap (Q + L_\eta + P_\epsilon).$$

Conversely, if $\phi \in W_{loc} \cap S \cap (Q + L_\eta + P_\epsilon)$ then

$$\mathrm{Pr}_L\, \phi \in L_\eta \subset L_\epsilon, \quad \text{or} \quad \phi \in Q + L_\epsilon + P_\epsilon.$$

Using also $\phi \in W_{loc} \cap S$, we infer

$$\phi = \chi + \mathrm{Sep}^*(\chi) \qquad \text{for some } \chi \in L_\epsilon.$$

Then

$$\chi = \mathrm{Pr}_L(\chi + \mathrm{Sep}^*(\chi)) = \mathrm{Pr}_L\, \phi \in L_\eta.$$

Hence

$$\phi \in \{\chi + \mathrm{Sep}^*(\chi) : \chi \in L_\eta\}.$$

This shows that $W_{loc} \cap S \cap (Q + L_\eta + P_\epsilon) \subset \{\chi + \mathrm{Sep}_\eta(\chi) : \chi \in L_\eta\}$. □

Proposition 4.7 *Let a C^1-curve $c : (-1, 1) \to C$ and a sequence $(s_n)_0^\infty$ in $(-1, 1)$ be given so that $c(0)$ and all $c(s_n)$, $n \in \mathbb{N}$, belong to $(W_{loc} \cap S) \cap (Q + L_\eta + P_\epsilon)$, and $0 \neq s_n \to 0$ as $n \to \infty$ and $c'(0) \neq 0$. Then $\mathrm{Pr}_L\, c'(0) \neq 0$.*

Proof Let λ be a Lipschitz constant for Sep_η. By Proposition 4.6,

$$||\,\mathrm{Pr}_{Q\oplus P}(c(s_n) - c(0))|| \leq \lambda ||\,\mathrm{Pr}_L(c(s_n) - c(0))||$$

for all $n \in \mathbb{N}$. Dividing both sides by $|s_n|$ and taking the limit as $n \to \infty$, we get

$$||\,\mathrm{Pr}_{Q\oplus P}\, c'(0)|| \leq \lambda ||\,\mathrm{Pr}_L\, c'(0)||,$$

from which the assertion follows. □

To conclude this chapter, we recall the explicit formula

$$\mathrm{Pr}_P\, \phi = \frac{1}{1 + f'(0)e^{-\lambda_0}} \left(\phi(0) + f'(0) \int_{-1}^{0} e^{-\lambda_0(s+1)} \phi(s)\, ds \right) \chi_0$$

for all $\phi \in C$ (see Hale **[27]**), and introduce the continuous linear functional

$$c_P : C \ni \phi \mapsto \phi(0) + f'(0) \int_{-1}^{0} e^{-\lambda_0(s+1)} \phi(s)\, ds \in \mathbb{R}.$$

Clearly, $c_P^{-1}(0) = Q \oplus L$.

CHAPTER 5

Oscillation frequencies

We first recall a discrete (integer-valued) Lyapunov functional, and then use this to obtain various qualitative results for solutions of (1.1) in terms of oscillation frequencies.

For any $\phi \in C \setminus (K \cup (-K))$, define the number of sign changes

$$\mathrm{sc}(\phi) = \sup\{k \in \mathbb{N} \setminus \{0\} : \text{ There exists a strictly increasing finite sequence } (s^i)_0^k \text{ in } [-1,0] \text{ with } \phi(s^{i-1})\phi(s^i) < 0 \text{ for all } i \in \{1,\dots,k\}\}.$$

Certainly, $\mathrm{sc}(\phi) = \infty$ is possible. Set $\mathrm{sc}(\phi) = 0$ if $0 \neq \phi \in K \cup (-K)$. Define the discrete Lyapunov functional $V : C \setminus \{0\} \to \{0, 2, 4, \dots, \infty\}$ by

$$V(\phi) = \begin{cases} \mathrm{sc}(\phi) & \text{if } \mathrm{sc}(\phi) \text{ is even or } \infty, \\ \mathrm{sc}(\phi) + 1 & \text{if } \mathrm{sc}(\phi) \text{ is odd.} \end{cases}$$

V is lower semicontinuous, and the restriction of V to the set

$$\begin{aligned} R = \{\phi \in C^1 : \; & \phi(0) \neq 0 \text{ or } \dot{\phi}(0)\phi(-1) > 0, \\ & \phi(-1) \neq 0 \text{ or } \dot{\phi}(-1)\phi(0) < 0, \\ & \text{all zeros of } \phi \text{ in } (-1,0) \text{ are simple}\} \end{aligned}$$

is continuous with respect to the C^1-topology, see Lemma VI.1. The set R is open and dense in C^1.

Assume that $I \subseteq \mathbb{R}$ is an interval, $\mu \geq 0$, $b : I \to \mathbb{R}$ is a positive continuous function, $z : I + [-1, 0] \to \mathbb{R}$ is a continuous not identically zero function, $z|_I$ is differentiable, and

$$\dot{z}(t) = -\mu z(t) + b(t) z(t-1) \tag{5.1}$$

holds for all $t > \inf I$ in I. Then Lemma VI.2(i) implies that the function $I \ni t \mapsto V(z_t) \in \{0, 2, 4, \dots, \infty\}$ is monotone nonincreasing. For the proof of this and other properties of V we refer to Appendix VI.

Various functions related to equation (1.1) satisfy (5.1). For example, if $x : [-1, \infty) \to \mathbb{R}$ is a solution of (1.1) then

$$\dot{x}(t) = -\mu x(t) + \left(\int_0^1 f'(s x(t-1))\, ds \right) x(t-1)$$

for all $t > 0$. If $\hat{x} : [-1, \infty) \to \mathbb{R}$ is another solution of (1.1) and $c > 0$, then the weighted difference $y = \frac{1}{c}(x - \hat{x})$ satisfies

$$\dot{y}(t) = -\mu y(t) + \left(\int_0^1 f'((1-s)\hat{x}(t-1) + s x(t-1))\, ds \right) y(t-1)$$

for all $t > 0$ which is again of the form (5.1). The linear variational equation along any solution of (1.1) is also of the form (5.1) with some positive b.

The aim of this chapter is to characterize the sets W, $\overline{W}$, S, $W \cap S$, $\overline{W} \cap S$ and $\overline{W \cap S}$ in terms of oscillation frequencies, or equivalently, the values of the Lyapunov functional V. First we need a preparatory result.

Proposition 5.1 *If $y : \mathbb{R} \to \mathbb{R}$ is a solution of equation (1.2) with $y_0 \neq 0$ then*

$$y_0 \in L \oplus P \quad \textit{if and only if} \quad V(y_t) \leq 2 \ \textit{ for all}\, t \leq 0.$$

Let $\mu > 0$. If $y : \mathbb{R} \to \mathbb{R}$ is a solution of the equation

$$\dot{y}(t) = -\mu y(t) + f'(\xi^+)y(t-1) \quad \left[\dot{y}(t) = -\mu y(t) + f'(\xi^-)y(t-1)\right]$$

with $y_0 \neq 0$ then there exists $c \in \mathbb{R}$ with $y(t) = ce^{\lambda_0^+ t}$ $\left[y(t) = ce^{\lambda_0^- t}\right]$ for all $t \in \mathbb{R}$ if and only if $V(y_t) = 0$ for all $t \leq 0$.

Proof 1. Proof of the statement concerning solutions of equation (1.2).

Necessity: If $y_0 \in L \oplus P$ then there exist constants a_0, a_1, b_1 such that for all $t \in \mathbb{R}$,

$$\begin{aligned} y(t) &= a_0 e^{\lambda_0 t} + a_1 e^{\operatorname{Re}\lambda_1 t}\cos(\operatorname{Im}\lambda_1(t+b_1)) \\ &= e^{\operatorname{Re}\lambda_1 t}[a_0 e^{(\lambda_0 - \operatorname{Re}\lambda_1)t} + a_1\cos(\operatorname{Im}\lambda_1(t+b_1))]. \end{aligned}$$

As $\lambda_0 > \operatorname{Re}\lambda_1$ and $\operatorname{Im}\lambda_1 \in (\pi, 2\pi)$, it easily follows that $V(y_t) \leq 2$ for large negative t. By Lemma VI.2(i), $\mathbb{R} \ni t \mapsto V(y_t) \in \mathbb{R}$ is monotone nonincreasing. Thus, we find that $V(y_t) \leq 2$ for all $t \leq 0$.

Sufficiency: Suppose $V(y_t) \leq 2$ for all $t \leq 0$. By Lemma VI.5, there exists $c > 0$ such that

$$||y_t|| \leq ||y_0||ce^{-ct} \qquad \text{for all } \ t \leq 0.$$

Choose $\alpha < -c$ with $\operatorname{Re}\lambda_{k+1} < \alpha < \operatorname{Re}\lambda_k$ for some integer $k \geq 1$. Let $\hat{P}_k$ be the realified generalized eigenspace of the generator of the semigroup $(D_2F(t,0))_{t\geq 0}$ associated with the eigenvalues $\lambda_0, \lambda_1, \overline{\lambda_1}, \ldots, \lambda_k, \overline{\lambda_k}$, and let $\hat{Q}_k$ be the generalized eigenspace associated with the remaining spectrum. Then $C = \hat{P}_k \oplus \hat{Q}_k$, and there exists a constant $M_k > 0$ such that

$$||D_2F(t,0)\phi|| \leq M_k e^{\alpha t}||\phi|| \qquad \text{for all } t \geq 0 \text{ and } \phi \in \hat{Q}_k.$$

Let $t \leq 0$. For every $\sigma \leq t$,

$$\begin{aligned} ||\operatorname{Pr}_{\hat{Q}_k} y_t|| &= ||D_2F(t-\sigma, 0)\operatorname{Pr}_{\hat{Q}_k} y_\sigma|| \\ &\leq M_k e^{\alpha(t-\sigma)}||\operatorname{Pr}_{\hat{Q}_k} y_\sigma|| \\ &\leq M_k||\operatorname{Pr}_{\hat{Q}_k}||e^{\alpha(t-\sigma)}||y_\sigma|| \\ &\leq M_k||\operatorname{Pr}_{\hat{Q}_k}||e^{\alpha(t-\sigma)}||y_0||ce^{-c\sigma} \\ &= M_k||\operatorname{Pr}_{\hat{Q}_k}||||y_0||ce^{\alpha t}e^{-(\alpha+c)\sigma}. \end{aligned}$$

It follows that $\operatorname{Pr}_{\hat{Q}_k} y_t = 0$. Consequently,

$$y_t = \operatorname{Pr}_{\hat{P}_k} y_t \in \hat{P}_k \qquad \text{for all } \ t \leq 0.$$

Thus, there exist an integer N and reals $a_0, a_j, b_j, j \in \{1, \dots, N\}$, with $0 \le N \le k$, $a_N \neq 0$ and

$$y(t) = a_0 e^{\lambda_0 t} + \sum_{j=1}^{N} a_j e^{\operatorname{Re}\lambda_j t} \cos(\operatorname{Im}\lambda_j (t + b_j)) \qquad \text{for all } t \le 0.$$

If $N > 1$ then from $\operatorname{Re}\lambda_N < \operatorname{Re}\lambda_{N-1} < \dots < \operatorname{Re}\lambda_1 < \lambda_0$ it follows that

$$y(t) = e^{\operatorname{Re}\lambda_N t}[a_N \cos(\operatorname{Im}\lambda_N (t + b_N)) + o(1)]$$

as $t \to -\infty$. Hence and from $\operatorname{Im}\lambda_N > (2N-1)\pi \ge 3\pi$ we obtain $V(y_t) > 2$ for sufficiently large negative t, a contradiction. Consequently, $N \le 1$, and thus $y_0 \in L \oplus P$.

2. The assertions in case $\mu > 0$ about solutions of the linear variational equations along the solutions $t \mapsto \xi^+$, $t \mapsto \xi^-$ are proved analogously.

□

Proposition 5.2 *If $\phi \in \overline{W}$, $\psi \in \overline{W}$ and $\phi \neq \psi$ then $V(\phi - \psi) \le 2$.*

Proof By the lower semicontinuity of V (Lemma VI.1(i)), it is sufficient to prove that $V(\phi - \psi) \le 2$ provided $\phi \in W$, $\psi \in W$ and $\phi \neq \psi$.

Let $\phi \in W$, $\psi \in W$ with $\phi \neq \psi$. Proposition 4.1 implies that the unique solutions $x^\phi : \mathbb{R} \to \mathbb{R}$, $x^\psi : \mathbb{R} \to \mathbb{R}$ of (1.1) with $x_0^\phi = \phi$, $x_0^\psi = \psi$ satisfy $x_t^\phi \in W$, $x_t^\psi \in W$ for all $t \in \mathbb{R}$ and $x_t^\phi \to 0$, $x_t^\psi \to 0$ as $t \to -\infty$. Clearly, $x_t^\phi \neq x_t^\psi$ for all $t \in \mathbb{R}$. Since $x_t^\phi \to 0$, $x_t^\psi \to 0$ as $t \to -\infty$, there exists a sequence $(t_n)_0^\infty$ with $t_n \to -\infty$ as $n \to \infty$ and

$$|x^\phi(t_n) - x^\psi(t_n)| = \sup_{s \le 0} |x^\phi(t_n + s) - x^\psi(t_n + s)| \qquad \text{for all } n \in \mathbb{N}.$$

The functions $z^n : \mathbb{R} \to \mathbb{R}$, $n \in \mathbb{N}$, given by

$$z^n(t) = \frac{x^\phi(t_n + t) - x^\psi(t_n + t)}{|x^\phi(t_n) - x^\psi(t_n)|}$$

satisfy

$$|z^n(t)| \le |z^n(0)| = 1 \qquad \text{for all } t \le 0,$$

and

$$\dot{z}^n(t) = -\mu z^n(t) + b^n(t) z^n(t-1) \qquad \text{for all } t \in \mathbb{R},$$

with $b^n : \mathbb{R} \to \mathbb{R}$ given by

$$b^n(t) = \int_0^1 f'(s x^\phi(t_n + t - 1) + (1-s) x^\psi(t_n + t - 1))\, ds.$$

Clearly,

$$b^n(t) \to f'(0) \qquad \text{as } n \to \infty \text{ uniformly on } (-\infty, 0].$$

Then the conditions of Lemma VI.4 are satisfied with $I_0 = (-\infty, 0]$. Thus there exist a C^1-function $z : (-\infty, 0] \to \mathbb{R}$ and a subsequence $(z^{n_k})_{k=0}^\infty$ of $(z^n)_0^\infty$ such that

$$z^{n_k}(t) \to z(t),\ \dot{z}^{n_k}(t) \to \dot{z}(t) \qquad \text{as } k \to \infty$$

uniformly on compact subsets of $(-\infty, 0]$, and z satisfies

$$\dot{z}(t) = -\mu z(t) + f'(0) z(t-1) \qquad \text{for all } t \le 0.$$

It also follows that

$$|z(t)| \leq |z(0)| = 1 \qquad \text{for all } t \leq 0.$$

We claim that $z_0 \in L \oplus P \setminus \{0\}$. Recall from Chapter 4 that for sufficiently large n we have

$$x^{\phi}_{t_n} = \chi_n + w_{loc}(\chi_n) \qquad \text{with } \chi_n \in N_L + N_P,$$
$$x^{\psi}_{t_n} = \eta_n + w_{loc}(\eta_n) \qquad \text{with } \eta_n \in N_L + N_P,$$

and

$$\chi_n \to 0,\ \eta_n \to 0 \qquad \text{as } n \to \infty.$$

From $Dw_{loc}(0) = 0$ it follows that

$$\frac{||\operatorname{Pr}_Q z_0^{n_k}||}{||\operatorname{Pr}_{L\oplus P} z_0^{n_k}||} = \frac{||w_{loc}(\chi_{n_k}) - w_{loc}(\eta_{n_k})||}{||\chi_{n_k} - \eta_{n_k}||} \to 0$$

as $k \to \infty$, and we obtain from $\lim_{k\to\infty} z_0^{n_k} = z_0$ that $\operatorname{Pr}_Q z_0^{n_k} \to 0$ as $k \to \infty$. Consequently, $\operatorname{Pr}_Q z_0 = 0$, or

$$z_0 \in L \oplus P.$$

Proposition 5.1 implies $V(z_t) \leq 2$ for all $t \leq 0$.

The monotonicity of V and the fact that $V(z_t) \in \{0, 2\}$ imply that there exists $T \leq 0$ such that $V(z_t)$ is constant for $t \in (-\infty, T]$. Thus $z_T \in R$ by Lemma VI.2(iii). Therefore, using Lemma VI.1(ii), and that $z_T^{n_k} \to z_T$, $\dot{z}_T^{n_k} \to \dot{z}_T$ as $k \to \infty$, we find

$$2 \geq V(z_T) = \lim_{k\to\infty} V(z_T^{n_k}).$$

This implies

$$V(z_T^{n_k}) = V(x^{\phi}_{t_{n_k}+T} - x^{\psi}_{t_{n_k}+T}) \leq 2$$

for all sufficiently large integers k. The monotonicity of V implies $V(\phi - \psi) \leq 2$. □

We can now characterize the set W in terms of sign changes.

Proposition 5.3 $W \setminus \{0\} = \{\phi \in C \setminus \{0\}$: *There is a solution* $x : \mathbb{R} \to \mathbb{R}$ *of* (1.1) *with* $x_0 = \phi$, $V(x_t) \leq 2$ *for all* $t \in \mathbb{R}$, *and* $x_t \to 0$ *as* $t \to -\infty\}$.

Proof 1. The fact that $W \setminus \{0\}$ is contained in the set on the right-hand side is a direct consequence of Proposition 5.2, the invariance of W and that $0 \in W$.

2. To prove the remaining inclusion, we let ϕ be an element of the set on the right-hand side of the equality in the proposition. Consider the associated solution $x : \mathbb{R} \to \mathbb{R}$ with $x_0 = \phi$, $V(x_t) \leq 2$ on $\mathbb{R}$, and $x_t \to 0$ as $t \to -\infty$. It suffices to show that for some $t \leq 0$, $x_t \in W_{loc}$. Recall from Chapter 4

$$W_{loc} = W_\beta(N_{loc}),\ \beta > 1,\ e^{\operatorname{Re}\lambda_2} < \beta < e^{\operatorname{Re}\lambda_1}.$$

Choose $\hat{\beta} \in (\beta, e^{\operatorname{Re}\lambda_1})$. Theorem I.1 permits to find a norm $|\cdot|$ on C equivalent to $||\cdot||$ so that

$$|D_2F(1,0)\psi| \geq \hat{\beta}\psi \qquad \text{on } L \oplus P.$$

Claim:

$$\gamma = \limsup_{t\to-\infty} \frac{|x_{t-1}|}{|x_t|} < \frac{1}{\beta}.$$

Proof: Lemma VI.3 can be used to get $\gamma < \infty$. There is a sequence $(t_n)_0^\infty$ with $t_n \to -\infty$ and

$$\frac{|x_{t_n-1}|}{|x_{t_n}|} \to \gamma \qquad \text{as } n \to \infty.$$

Define the functions $z^n : \mathbb{R} \to \mathbb{R}$, $n \in \mathbb{N}$, by

$$z^n(t) = \frac{x(t_n + t)}{|x_{t_n}|}.$$

Then

$$|z_0^n| = 1, \quad V(z_t^n) \le 2 \quad \text{for all } n \in \mathbb{N} \text{ and } t \le 0,$$

and each z^n satisfies

$$\dot{z}^n(t) = -\mu z^n(t) + b^n(t) z^n(t-1) \quad \text{for all } t \le 0,$$

with

$$b^n(t) = \int_0^1 f'(sx(t + t_n - 1))\, ds \to f'(0) \qquad \text{as } n \to \infty \text{ uniformly in } (-\infty, 0],$$

since $x_t \to 0$ as $t \to -\infty$. Clearly there are $b_1 > b_0 > 0$ such that for all $t \le 0$ and for all $n \in \mathbb{N}$, $b_0 \le b^n(t) \le b_1$. By Lemma VI.5 there exists $c > 0$ such that

$$|z^n(t)| \le ce^{-ct} \qquad \text{for all } t \le 0 \text{ and all } n \in \mathbb{N}.$$

Then Lemma VI.4 with $I_0 = (-\infty, 0]$ can be applied to find a subsequence $(z^{n_k})_{k=0}^\infty$ of $(z^n)_{n=0}^\infty$ and a continuously differentiable function $z : (-\infty, 0] \to \mathbb{R}$ with

$$z^{n_k}(t) \to z(t), \quad \dot{z}^{n_k}(t) \to \dot{z}(t)$$

as $k \to \infty$ uniformly on compact subsets of $(-\infty, 0]$, and

$$\dot{z}(t) = -\mu z(t) + f'(0) z(t-1) \qquad \text{for all } t \le 0.$$

Moreover

$$|z_0| = 1, \quad |z_{-1}| = \gamma.$$

By Lemma VI.1(i),

$$V(z_t) \le \liminf_{k\to\infty} V(z_t^{n_k}) \le 2 \qquad \text{for all } t \le 0.$$

Then we can apply Proposition 5.1 to get $z_0 \in L \oplus P \setminus \{0\}$. Thus,

$$1 = |z_0| = |D_2F(1,0) z_{-1}| \ge \hat{\beta} |z_{-1}| = \hat{\beta}\gamma,$$

and therefore $\gamma \le \frac{1}{\hat{\beta}} < \frac{1}{\beta}$.

3. Choose $\epsilon > 0$ so that for all $\psi \in C$ with $|\psi| < \epsilon$, $\psi \in N_{loc}$. Fix $t < 0$ with $|x_t| < \epsilon$ and

$$\frac{|x_{s-1}|}{|x_s|} < \frac{1}{\beta} \qquad \text{for all } s \le t.$$

Then

$$|x_{t-k}|\beta^k \le |x_t| < \epsilon \qquad \text{for all } k \in \mathbb{N},$$

and thereby

$$x_{t-k}\beta^k \in N_{loc} \qquad \text{for all } k \in \mathbb{N}.$$

Theorem I.4(iii) yields $x_{t-k}\beta^k \to 0$ as $k \to \infty$. It follows that

$$x_t \in W_\beta(N_{loc}) = W_{loc}.$$

$\square$

Proposition 5.4 *For all ϕ, ψ in S with $\phi \neq \psi$, $V(\phi - \psi) \geq 2$.*

Proof This follows from the nonordering property of S in Proposition 3.1 and the definition of V, since $V(\phi-\psi) < 2$ for ϕ, ψ in S with $\phi \neq \psi$ implies $V(\phi-\psi) = 0$ so that ϕ and ψ are order related. $\square$

The following corollaries are immediate from the definition of S and from Propositions 5.2, 5.3 and 5.4.

Corollary 5.5 $S \setminus \{0\} = \{\phi \in C \setminus \{0\}$: *The solution* $x^\phi : [-1, \infty) \to \mathbb{R}$ *of (1.1) satisfies* $V(x_t^\phi) \geq 2$ *for all* $t \geq 0\}$.

Corollary 5.6 $(W \cap S) \setminus \{0\} = \{\phi \in C \setminus \{0\}$: *There is a solution* $x : \mathbb{R} \to \mathbb{R}$ *of (1.1) with* $x_0 = \phi$, $x_t \to 0$ *as* $t \to -\infty$, *and* $V(x_t) = 2$ *for all* $t \in \mathbb{R}\}$.

Proposition 5.7 *For all ϕ, ψ in $\overline{W} \cap S$ with $\phi \neq \psi$, $V(\phi - \psi) = 2$ and $\phi - \psi \in R$.*

Proof Consider ϕ, ψ in $\overline{W} \cap S$ with $\phi \neq \psi$. Propositions 5.2 and 5.4 imply $V(\phi - \psi) = 2$. By the invariance of $\overline{W} \cap S$ in Proposition 4.4, there exist solutions $x^\phi : \mathbb{R} \to \mathbb{R}$ and $x^\psi : \mathbb{R} \to \mathbb{R}$ such that $x_t^\phi \in \overline{W} \cap S$ and $x_t^\psi \in \overline{W} \cap S$ for all $t \in \mathbb{R}$. It follows that $V(x_t^\phi - x_t^\psi) = 2$ for all $t \in \mathbb{R}$. Lemma VI.2(iii) implies $x_t^\phi - x_t^\psi \in R$ for all $t \in \mathbb{R}$. In particular, $\phi - \psi \in R$. $\square$

Corollary 5.8 *If $\phi \in \overline{W} \cap S \setminus \{0\}$ and $x = x^\phi$, then $V(\phi) = 2$, $V(\dot{\phi}) \geq 2$, and there exists a sequence $(t^n)_{-\infty}^{\infty}$ such that for all $n \in \mathbb{Z}$ we have*

$$\begin{aligned} &t^{n+1} - t^n < 1, \ t^{n+2} - t^n > 1, \\ &x(t^n) = 0, \ \dot{x}(t^{2n}) > 0, \ \dot{x}(t^{2n+1}) < 0, \\ &x(t) > 0 \qquad \textit{for all } t \in (t^{2n}, t^{2n+1}), \\ &x(t) < 0 \qquad \textit{for all } t \in (t^{2n-1}, t^{2n}). \end{aligned}$$

Proof Let $\phi \in \overline{W} \cap S \setminus \{0\}$. By the invariance of $\overline{W} \cap S$, the solution $x : \mathbb{R} \to \mathbb{R}$ of (1.1) with $x_0 = \phi$ satisfies $x_t \in \overline{W} \cap S$ for all $t \in \mathbb{R}$. Note $x_t \neq 0$ for all $t \in \mathbb{R}$. As $0 \in \overline{W} \cap S$, Proposition 5.7 gives $V(x_t) = 2$ and $x_t \in R$ for all $t \in \mathbb{R}$. The definition of R implies that all zeros of x are simple. Consequently, there exists an increasing sequence $(t^n)_{-\infty}^{\infty}$ so that for all $n \in \mathbb{Z}$, $x(t^n) = 0$, $\dot{x}(t^{2n}) > 0$, $\dot{x}(t^{2n+1}) < 0$, and $x(t) > 0$ for all $t \in (t^{2n}, t^{2n+1})$, and $x(t) < 0$ for all $t \in (t^{2n-1}, t^{2n})$. Clearly, for every $t \in \mathbb{R}$, $\mathrm{sc}(x_t)$ is given by the number of zeros of x in the interval $(t-1, t)$, and $\mathrm{sc}(x_t) \in \{1, 2\}$. Hence $t^{n+1} - t^n < 1$ and $t^{n+2} - t^n \geq 1$ for all $n \in \mathbb{Z}$. For every $n \in \mathbb{Z}$, $t^{n+2} - t^n \neq 1$ since $x_{t^{n+2}} \in R$. Thus $t^{n+2} - t^n > 1$ for all $n \in \mathbb{Z}$. Now it is obvious that $V(\dot{x}_t) \geq \mathrm{sc}(\dot{x}_t) \geq 1$ for all $t \in \mathbb{R}$. In particular $V(\dot{\phi}) \geq 2$. $\square$

Remark 5.9 As S is closed, it follows that $\overline{W \cap S} \subseteq \overline{W} \cap S$. Therefore, Proposition 5.7 and Corollary 5.8 remain true if $\overline{W} \cap S$ is replaced with $\overline{W \cap S}$.

CHAPTER 6

Graph representations

Ammar **[2]** showed, among others, that in case $\mu = 0$ the set W is given by a C^1-map from an open subset of $L \oplus P$ into Q. Proposition 4.6 says that close to 0 the subset $W_{loc} \cap S$ is given by a map from a subset of L into $Q \oplus P$. This may suggest that the full part $W \cap S$ of the separatrix in W can be written as a map from a set in L into $Q \oplus P$. Unfortunately, we do not know whether this is true or not.

So we abandon the decomposition of C into P, L, Q in this chapter and construct a 2-dimensional space $G_2 \subset C$ contained in a 3-dimensional space $G_3 \subset C$, a complementary space G_1 of G_2 in G_3 and a closed complementary space E of G_3 in C so that $\overline{W}$ and $\overline{W} \cap S$ are given by maps from subsets of G_3 into E and from G_2 into $G_1 \oplus E$, respectively. Furthermore, we derive first results on smoothness of $\overline{W}$, $\overline{W} \cap S$, W, and the restriction of the flow F_W to $\mathbb{R} \times W$.

We begin with the linear continuous maps

$$\Pi_2 : C \ni \phi \mapsto (\phi(0), \phi(-1))^{tr} \in \mathbb{R}^2$$

and

$$\Pi_3 : C \ni \phi \mapsto (\phi(0), \phi(-1), c_P(\phi))^{tr} \in \mathbb{R}^3,$$

which can be regarded as certain evaluations or coordinates of states in C.

Proposition 6.1 *The restrictions of* Π_2 *to* $\overline{W} \cap S$*, of* Π_3 *to* $\overline{W}$*, and of* Π_3 *to* $L \oplus P$ *are injective.*

Proof 1. Consider ϕ, ψ in $\overline{W} \cap S$ with $\phi \neq \psi$. Proposition 5.7 implies that $\phi - \psi \in R$ and hence $\Pi_2(\phi - \psi) \neq 0$. This shows that $\Pi_2|_{\overline{W} \cap S}$ is injective.

2. Consider ϕ, ψ in $\overline{W}$ with $\phi \neq \psi$. Assume $\Pi_3 \phi = \Pi_3 \psi$. Then $\phi(-1) - \psi(-1) = \phi(0) - \psi(0)$. Hence, $\phi - \psi \notin R$. By Lemma VI.2(iii) and Proposition 5.2, $V(\phi - \psi) < V(F_W(-3, \phi) - F_W(-3, \psi)) \leq 2$. It follows that $V(\phi - \psi) = 0$ and consequently, $\phi \leq \psi$ or $\psi \leq \phi$. The assumption $\phi \neq \psi$, the last inequalities and $c_P(\phi) = c_P(\psi)$ combined yield a contradiction.

3. By Proposition 5.1 and Lemma VI.2(i), for every solution $x : \mathbb{R} \to \mathbb{R}$ of equation (1.2) with $0 \neq x_0 \in L \oplus P$, $V(x_t) \leq 2$ for all $t \in \mathbb{R}$. The injectivity of $\Pi_3|_{L \oplus P}$ follows by arguments as in part 2. □

Let $\Pi_2^{-1} : \Pi_2(\overline{W} \cap S) \to C$ and $\Pi_3^{-1} : \Pi_3 \overline{W} \to C$ be the maps given by the inverses of

$$\overline{W} \cap S \ni \phi \mapsto \Pi_2 \phi \in \Pi_2(\overline{W} \cap S), \quad \overline{W} \ni \phi \mapsto \Pi_3 \phi \in \Pi_3 \overline{W},$$

respectively. Recall the property of the constant B_0 from Proposition 4.2.

Proposition 6.2

(i) Π_2^{-1} *is Lipschitz continuous.*

(ii) In case $\mu = 0$, *all restrictions of* Π_3^{-1} *to the sets* $\Pi_3(\overline{W} \cap \overline{C_b})$ *with* $b > B_0$ *are Lipschitz continuous, and all restrictions of* Π_3^{-1} *to bounded sets are Lipschitz continuous. In particular,* Π_3^{-1} *is locally Lipschitz continuous.*

(iii) In case $\mu > 0$, Π_3^{-1} *is Lipschitz continuous.*

Proof 1. Suppose Π_2^{-1} is not Lipschitz continuous. Then there are sequences of solutions $x^n : \mathbb{R} \to \mathbb{R}$ and $y^n : \mathbb{R} \to \mathbb{R}$, $n \in \mathbb{N}$, of equation (1.1) such that $x_0^n \neq y_0^n$ for all $n \in \mathbb{N}$, x_t^n and y_t^n in $\overline{W} \cap S$ and $V(x_t^n - y_t^n) = 2$ for all n and all t, and

$$\frac{|\Pi_2(x_0^n - y_0^n)|_{\mathbb{R}^2}}{||x_0^n - y_0^n||} \to 0 \qquad \text{as } n \to \infty.$$

Due to the compactness of $\overline{W} \cap S$, the solutions x^n and y^n are uniformly bounded, and equation (1.1) yields a bound for their derivatives. So, we can apply the Arzèla–Ascoli Theorem successively on the intervals $[-j, j]$, $1 \leq j \in \mathbb{N}$, to find strictly increasing maps $\chi_j : \mathbb{N} \to \mathbb{N}$, $1 \leq j \in \mathbb{N}$, so that for every integer $j \geq 1$ the subsequences $\left(x^{\chi_1 \circ \cdots \circ \chi_j(n)}\right)_0^\infty$ and $\left(y^{\chi_1 \circ \cdots \circ \chi_j(n)}\right)_0^\infty$ converge uniformly to continuous functions on the interval $[-j, j]$. The subsequences $(\bar{x}^n)_0^\infty = (x^{\chi(n)})_0^\infty$ and $(\bar{y}^n)_0^\infty = (y^{\chi(n)})_0^\infty$ given by $\chi(n) = \chi_1 \circ \cdots \circ \chi_n(n)$ converge to continuous bounded functions $x : \mathbb{R} \to \mathbb{R}$ and $y : \mathbb{R} \to \mathbb{R}$ uniformly on every bounded set. We define functions $z^n : \mathbb{R} \to \mathbb{R}$, $n \in \mathbb{N}$, by

$$z^n(t) = \frac{\bar{x}^n(t) - \bar{y}^n(t)}{||\bar{x}_0^n - \bar{y}_0^n||}.$$

Then

$$||z_0^n|| = 1 \qquad \text{for all } n \in \mathbb{N},$$

$$\Pi_2 z_0^n = \frac{\Pi_2(\bar{x}_0^n - \bar{y}_0^n)}{||\bar{x}_0^n - \bar{y}_0^n||} \to 0 \in \mathbb{R}^2 \qquad \text{as } n \to \infty,$$

$$V(z_t^n) = 2 \qquad \text{for all } n \in \mathbb{N},\ t \in \mathbb{R},$$

$$\dot{z}^n(t) = -\mu z^n(t) + \int_0^1 f'((1-s)\bar{y}^n(t-1) + s\bar{x}^n(t-1))ds\, z^n(t-1) \text{ for } n \in \mathbb{N}, t \in \mathbb{R}.$$

The coefficient functions

$$a^n : \mathbb{R} \ni t \mapsto \int_0^1 f'((1-s)\bar{y}^n(t-1) + s\bar{x}^n(t-1))\, ds \in \mathbb{R}, \qquad n \in \mathbb{N},$$

converge to

$$a : \mathbb{R} \ni t \mapsto \int_0^1 f'((1-s)y(t-1) + sx(t-1))\, ds \in \mathbb{R}$$

uniformly on every bounded set, and clearly there are constants $a_0 > 0$ and $a_1 \geq a_0$ with

$$a_0 \leq a^n(t) \leq a_1 \qquad \text{for all } n \in \mathbb{N} \text{ and all } t \in \mathbb{R}.$$

Using Lemmas VI.5 and VI.6, we obtain a constant $c > 0$ so that for all $n \in \mathbb{N}$ and all $t \in \mathbb{R}$,

$$|z^n(t)| \leq ce^{c|t|}.$$

Therefore, we can apply Lemma VI.4 to find a sequence $(z^{n_k})_{k=0}^{\infty}$ and a C^1-function $z : \mathbb{R} \to \mathbb{R}$ so that the functions z^{n_k} and $\dot{z}^{n_k}$ converge uniformly on every bounded set of $\mathbb{R}$ to z and $\dot{z}$, respectively, and

$$\dot{z}(t) = -\mu z(t) + a(t)z(t-1) \qquad \text{for all } t \in \mathbb{R}.$$

It follows that $||z_0|| = 1$, $\Pi_2 z_0 = \lim_{k\to\infty} \Pi_2 z_0^{n_k} = 0 \in \mathbb{R}^2$.

We claim that $V(z_t) = 2$ for all $t \in \mathbb{R}$. In fact, Lemma VI.1(i) yields $V(z_t) \leq 2$ on $\mathbb{R}$. Suppose $V(z_t) = 0$ for some t. By Lemma VI.2(i) and (iii), we have $V(z_{t+3}) = 0$ and $z_{t+3} \in R$. Lemma VI.1(ii) then gives $V(z_{t+3}^{n_k}) = V(z_{t+3}) = 0$ for k sufficiently large, which contradicts $V(z_s^n) = 2$ for all $n \in \mathbb{N}$ and $s \in \mathbb{R}$. Thus, $V(z_t) = 2$ for all $t \in \mathbb{R}$.

So, Lemma VI.2(iii) implies $z_0 \in R$, which contradicts $\Pi_2 z_0 = 0$.

2. Let $\mu = 0$ and $b > B_0$. Recall from Proposition 4.2 that $\overline{W} \cap \overline{C_b}$ is compact and negatively invariant. Suppose $\Pi_3^{-1}|_{\Pi_3(\overline{W}\cap\overline{C_b})}$ is not Lipschitz continuous. Then there are sequences of solutions $x^n : \mathbb{R} \to \mathbb{R}$ and $y^n : \mathbb{R} \to \mathbb{R}$, $n \in \mathbb{N}$, of equation (1.1) such that $x_0^n \neq y_0^n$ for all $n \in \mathbb{N}$, x_t^n and y_t^n in $\overline{W} \cap \overline{C_b}$ and $V(x_t^n - y_t^n) \leq 2$ for all $n \in \mathbb{N}$ and $t \leq 0$, and

$$\frac{|\Pi_3(x_0^n - y_0^n)|_{\mathbb{R}^3}}{||x_0^n - y_0^n||} \to 0 \qquad \text{as } n \to \infty.$$

As in part 1 there is a strictly increasing map $\chi : \mathbb{N} \to \mathbb{N}$ so that the subsequences given by $\bar{x}^n = x^{\chi(n)}$, $\bar{y}^n = y^{\chi(n)}$ converge uniformly on every bounded subset of $(-\infty, 0]$ to continuous bounded functions $x : (-\infty, 0] \to \mathbb{R}$ and $y : (-\infty, 0] \to \mathbb{R}$, respectively. Using Lemmas VI.4, VI.5 and VI.6, we find a subsequence of the sequence given by

$$z^n(t) = \frac{\bar{x}^n(t) - \bar{y}^n(t)}{||\bar{x}_0^n - \bar{y}_0^n||}, \qquad n \in \mathbb{N},\ t \in \mathbb{R},$$

which converges uniformly on every bounded subset of $(-\infty, 0]$ to a C^1-function $z : (-\infty, 0] \to \mathbb{R}$. Moreover, the derivatives of this sequence converge uniformly on every bounded subset of $(-\infty, 0]$ to $\dot{z}$, and z satisfies

$$\dot{z}(t) = -\mu z(t) + a(t)z(t-1) \qquad \text{for all } t \leq 0,$$

for some continuous bounded function $a : (-\infty, 0] \to \mathbb{R}$ with $\inf_{t\leq 0} a(t) > 0$. It follows that $||z_0|| = 1$, $\Pi_3 z_0 = 0 \in \mathbb{R}^3$, and $V(z_t) \leq 2$ for all $t \leq 0$. In particular, $z(-1) = 0 = z(0)$ and therefore, $z_0 \notin R$. By Lemma VI.2(iii), $V(z_{-3}) > V(z_0)$. Hence, $V(z_0) = 0$ which implies that either $z_0 \geq 0$ or $z_0 \leq 0$. Using these inequalities together with $c_P(z_0) = 0$ and $z_0 \neq 0$, we arrive at a contradiction.

3. Let $\mu = 0$ and let $M \subset \Pi_3 \overline{W}$ be a bounded set. In order to show that $\Pi_3^{-1}|_M$ is Lipschitz continuous it is sufficient to prove that there exists $b > B_0$ such that $M \subset \Pi_3(\overline{W} \cap \overline{C_b})$. The last statement follows if $\Pi_3^{-1}(M) \subset \overline{W}$ is bounded.

Proof of that $\Pi_3^{-1}(M)$ is bounded. Suppose $\Pi_3^{-1}(M) \subset \overline{W}$ is unbounded. Then there is a sequence $(\phi_n)_0^{\infty}$ in $\overline{W}$ so that $(\phi_n(0))_0^{\infty}$ is bounded and $||\phi_n|| \to \infty$ as $n \to \infty$. It follows that there exists $n \in \mathbb{N}$ with $|\phi_k(0)| < ||\phi_k||$ for all integers $k \geq n$. For $n \leq k \in \mathbb{N}$, $F_W(-1, \phi_k)$ has a zero since otherwise equation (1.1) with $\mu = 0$ shows that $|x^{\phi_k}|$ is increasing on $[-1, 0]$ and thus, $|\phi_k(0)| = ||\phi_k||$. Proposition 2.3 yields a bound $B > 0$ for all $|x^{\phi_k}(t)|$ with $n \leq k \in \mathbb{N}$ and $t \leq -2$. Proposition

2.1 then shows that the segments $x_0^{\phi_k} = \phi_k$, $n \le k \in \mathbb{N}$, are also bounded, in contradiction to $||\phi_k|| \to \infty$ as $k \to \infty$.

4. To prove assertion (iii), we simply replace $\overline{W} \cap \overline{C_b}$ in part 2 by the set $\overline{W}$ which is compact for $\mu > 0$. □

Next we construct the desired graph representations. Observe that Π_3 is surjective since $\Pi_3|_{L\oplus P}$ is injective and $\dim L \oplus P = 3$. Choose linearly independent elements ϕ_j, $j \in \{1,2,3\}$, in C with

$$\Pi_3\phi_j = e_j, \quad \text{where } e_{jk} = \delta_{jk} \text{ for } j,k \text{ in } \{1,2,3\}.$$

Let J_3 denote the injective linear map from $\mathbb{R}^3$ into C given by $J_3 e_j = \phi_j$. The continuous linear map

$$P_3 = J_3 \circ \Pi_3$$

satisfies $P_3\phi_j = \phi_j$ for all $j \in \{1,2,3\}$. Consequently,

$$P_3 \circ P_3 = P_3,$$

which means that P_3 is a projection. The space $G_3 = P_3C = \mathbb{R}\phi_1 \oplus \mathbb{R}\phi_2 \oplus \mathbb{R}\phi_3$ is 3-dimensional, and with $E = P_3^{-1}(0)$ we have

$$C = G_3 \oplus E.$$

The restriction of P_3 to $\overline{W}$ is injective. Let $P_3^{-1} : P_3\overline{W} \to C$ be given by the inverse of $\overline{W} \ni \phi \mapsto P_3\phi \in P_3\overline{W}$. Then the map

$$w : P_3\overline{W} \ni \chi \mapsto (\mathrm{id} - P_3) \circ P_3^{-1}(\chi) \in E$$

satisfies

$$\overline{W} = \{\chi + w(\chi) : \chi \in P_3\overline{W}\}.$$

See Figure 6.1.

Consider also the injective linear map J_2 from $\mathbb{R}^2$ into C given by $J_2(1,0)^{tr} = \phi_1$ and $J_2(0,1)^{tr} = \phi_2$. As $\Pi_2\phi_1 = (1,0)^{tr}$ and $\Pi_2\phi_2 = (0,1)^{tr}$, the continuous linear map

$$P_2 = J_2 \circ \Pi_2 : C \to C$$

satisfies $P_2\phi_j = \phi_j$ for $j \in \{1,2\}$, and thus

$$P_2 \circ P_2 = P_2.$$

So, P_2 is a projection. The space $G_2 = P_2C = \mathbb{R}\phi_1 \oplus \mathbb{R}\phi_2$ is a 2-dimensional subspace of G_3, and

$$C = G_2 \oplus P_2^{-1}(0).$$

Set $G_1 = \mathbb{R}\phi_3$, so that

$$G_3 = G_2 \oplus G_1.$$

Proposition 6.3 $P_2^{-1}(0) = G_1 \oplus E$.

Proof Let $\phi \in C$. Then $\phi = \sum_{j=1}^{3} c_j\phi_j + \psi$ with uniquely determined reals c_j and $\psi \in E$. We have $\phi \in P_2^{-1}(0)$ if and only if $\Pi_2\phi = 0$, as J_2 is injective. The last equation is equivalent to $(c_1, c_2) = (0,0)$ which holds if and only if $\phi = c_3\phi_3 + \psi \in G_1 \oplus E$. □

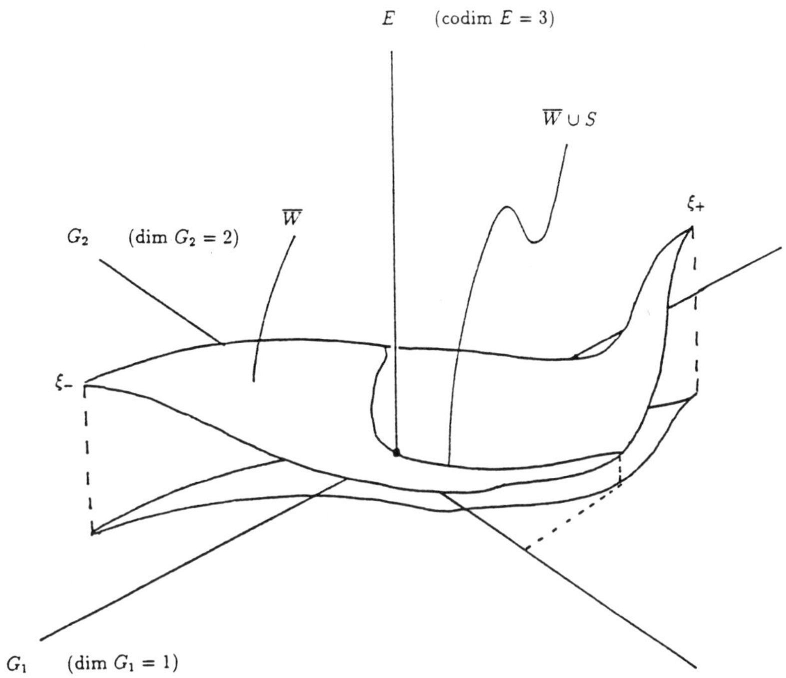

Figure 6.1: The graph representations of $\overline{W} \cap S$ and $\overline{W}$

The injectivity of the restriction of Π_2 to the set $\overline{W} \cap S$ and the injectivity of J_2 combined show that $P_2|_{\overline{W}\cap S}$ is injective. Let $P_2^{-1} : P_2(\overline{W} \cap S) \to C$ be given by the inverse of $\overline{W} \cap S \ni \phi \mapsto P_2\phi \in P_2(\overline{W} \cap S)$. It follows that the map

$$w_S : P_2(\overline{W} \cap S) \ni \chi \mapsto (\mathrm{id} - P_2) \circ P_2^{-1}(\chi) \in G_1 \oplus E$$

satisfies

$$\overline{W} \cap S = \{\chi + w_S(\chi) : \chi \in P_2(\overline{W} \cap S)\}.$$

See Figure 6.1.

Corollary 6.4 *w_S is Lipschitz continuous. In case $\mu = 0$, all restrictions of w to bounded subsets of $P_3\overline{W}$ are Lipschitz continuous, and w is locally Lipschitz continuous. In case $\mu > 0$, w is Lipschitz continuous.*

Proof J_3 defines an isomorphism onto G_3. Let J_3^{-1} denote its inverse. Then $J_3^{-1}P_3\overline{W} = \Pi_3\overline{W}$, and

$$P_3^{-1}(\chi) = \Pi_3^{-1}(J_3^{-1}\chi) \qquad \text{for all } \chi \in P_3\overline{W}.$$

All claimed results regarding w follow from the corresponding results for Π_3^{-1} in Proposition 6.2. Similarly, we get the Lipschitz continuity of w_S. □

In order to obtain more smoothness results, we need to know how Π_3 acts on tangent vectors of $\overline{W}$.

Proposition 6.5 *Let $\phi \in \overline{W}$, and a C^1-curve $c : (-1,1) \to C$ with $c(0) = \phi$, and a sequence $(s_n)_0^\infty$ in $(-1,1) \setminus \{0\}$ be given so that $s_n \to 0$ as $n \to \infty$ and $c(s_n) \in \overline{W}$ for all $n \in \mathbb{N}$.*

(i) If $c'(0) \neq 0$ then $\Pi_3 c'(0) \neq 0$.

(ii) Let a second C^1-curve $d : (-1,1) \to C$ with $d(0) = \phi$ and $|d| \subset \overline{W}$ be given so that $c'(0)$ and $d'(0)$ are linearly independent. Then $\Pi_3 c'(0)$ and $\Pi_3 d'(0)$ are linearly independent.

(iii) Assume that $(s_n)_0^\infty$ has subsequences $(s_{n_j})_0^\infty$ in $(-1,0)$ and $(s_{m_j})_0^\infty$ in $(0,1)$, and $d : (-1,1) \to C$ is a C^1-curve such that $d(0) = \phi$, $d((-1,0]) \subset \overline{W}$, and $c'(0)$ and $d'(0)$ are linearly independent. Then $\Pi_3 c'(0)$ and $\Pi_3 d'(0)$ are linearly independent.

Proof 1. Choose a bounded neighbourhood N of $\Pi_3\phi$ in $\mathbb{R}^3$. According to Proposition 6.2 there is a Lipschitz constant k for $\Pi_3^{-1}|_{N\cap\Pi_3\overline{W}}$. Since $s_n \to 0$ as $n \to \infty$, there exists $n \in \mathbb{N}$ with $\Pi_3 c(s_j) \in N \cap \Pi_3\overline{W}$ for $n \leq j \in \mathbb{N}$. Hence,

$$||c(s_j) - c(0)|| \leq k|\Pi_3 c(s_j) - \Pi_3 c(0)|_{\mathbb{R}^3}$$

for $n \leq j \in \mathbb{N}$. Dividing the last inequality by $|s_j|$ and taking the limit as $j \to \infty$, we obtain $||c'(0)|| \leq k|\Pi_3 c'(0)|$. This proves (i).

2. In order to prove (ii), consider N, k and n as in part 1. By linear independence, $c'(0) \neq 0 \neq d'(0)$, and assertion (i) yields $\Pi_3 c'(0) \neq 0 \neq \Pi_3 d'(0)$. Suppose $\Pi_3 c'(0)$ and $\Pi_3 d'(0)$ are linearly dependent. Then there exists $a \in \mathbb{R} \setminus \{0\}$ with $\Pi_3 c'(0) = a\Pi_3 d'(0)$. Choose $n_d \in \mathbb{N}$ with $n \leq n_d$ so that for $n_d \leq j \in \mathbb{N}$, $|as_j| < 1$ and $\Pi_3 d(as_j) \in N \cap \Pi_3\overline{W}$. For every integer $j \geq n_d$, we get

$$\begin{aligned}&||c(s_j) - c(0) - (d(as_j) - d(0))|| = ||c(s_j) - d(as_j)||\\ &\leq k|\Pi_3 c(s_j) - \Pi_3 d(as_j)|_{\mathbb{R}^3} = k|\Pi_3 c(s_j) - \Pi_3 c(0) - (\Pi_3 d(as_j) - \Pi_3 d(0))|_{\mathbb{R}^3}.\end{aligned}$$

Dividing both sides by $|s_j|$ and then taking the limit as $j \to \infty$, we get

$$||c'(0) - ad'(0)|| \leq k|\Pi_3 c'(0) - a\Pi_3 d'(0)|_{\mathbb{R}^3} = 0,$$

in contradiction to the linear independence of $c'(0)$ and $d'(0)$.

3. Proof of (iii). Suppose $\Pi_3 c'(0)$ and $\Pi_3 d'(0)$ are linearly dependent. As in part 2 we find $a \in \mathbb{R} \setminus \{0\}$ with $\Pi_3 c'(0) = a\Pi_3 d'(0)$. If $a > 0$, then a contradiction can be obtained as in part 2 replacing (s_j) with (s_{n_j}); and if $a < 0$ then a contradiction can be obtained as in part 2 replacing (s_j) with (s_{m_j}). □

Proposition 6.6 *$P_3 W$ is open in G_3, and $w|_{P_3 W}$ is C^1-smooth. $\Pi_3 W$ is open in $\mathbb{R}^3$, and Π_3 defines a C^1-diffeomorphism from the C^1-submanifold W of C onto $\Pi_3 W$.*

Proof 1. For every $\epsilon > 0$, $W = \cup_{n\in\mathbb{N}} F(\{n\} \times (W_{loc} \cap C_\epsilon))$. It follows that it is sufficient to find $\epsilon > 0$ so that for every $n \in \mathbb{N}$ the set $D_n = P_3 F(\{n\} \times (W_{loc} \cap C_\epsilon))$ is open in G_3, and $w|_{D_n}$ is C^1-smooth.

2. The set $W_{loc} \subset W$ is a 3-dimensional C^1-submanifold of C, with $T_0 W_{loc} = L \oplus P$, and P_3 is C^1-smooth. Proposition 6.1 says that $DP_3(0) = P_3$ defines an isomorphism from $L \oplus P$ onto G_3. It follows that there exists $\epsilon > 0$ so that P_3 maps $W_{loc} \cap C_\epsilon$ one-to-one onto an open neighbourhood U of 0 in G_3 and the inverse $\tilde{P}_3^{-1}$ of the map

$$W_{loc} \cap C_\epsilon \ni \phi \mapsto P_3 \phi \in U$$

is C^1-smooth. For every $\chi \in U$,

$$w(\chi) = (\mathrm{id} - P_3) \circ P_3^{-1}(\chi) = (\mathrm{id} - P_3) \circ \tilde{P}_3^{-1}(\chi),$$

and we see that the restriction of w to U is C^1-smooth.

Observe that we have

$$W_{loc} \cap C_\epsilon = \{\chi + w(\chi) : \chi \in U\}.$$

3. Let $n \in \mathbb{N}$ and $\chi \in D_n$ be given. The point $\phi = \chi + w(\chi)$ satisfies $\phi = F(n, \psi)$ with $\psi \in W_{loc} \cap C_\epsilon$. Set $\rho = P_3 \psi$. Then $\chi = P_3 F(n, \rho + w(\rho))$ and $\rho \in U$. Recall $P_3 = J_3 \circ \Pi_3$. Using Proposition 6.5(i), we infer that the derivatives of the C^1-map

$$A : U \ni \tilde{\rho} \mapsto P_3 F(n, \tilde{\rho} + w(\tilde{\rho})) \in G_3$$

are injective. It follows that A maps an open neighbourhood V of ρ in U one-to-one onto an open neighbourhood N of χ in G_3, which belongs to D_n, and the inverse $A^{-1} : N \to G_3$ of the map

$$V \ni \tilde{\rho} \mapsto A(\tilde{\rho}) \in N$$

is C^1-smooth. For every $\tilde{\chi} \in N$,

$$w(\tilde{\chi}) = (\mathrm{id} - P_3) F(n, A^{-1}(\tilde{\chi}) + w(A^{-1}(\tilde{\chi}))),$$

and it becomes obvious that $w|_N$ is C^1-smooth. It follows that D_n is open, and that $w|_{D_n}$ is C^1-smooth.

4. $\Pi_3 W$ is open in $\mathbb{R}^3$ since J_3 is an isomorphism from $\mathbb{R}^3$ onto G_3 and $P_3 W$ is open in G_3. The restriction of the injective linear map Π_3 to the C^1-submanifold W is C^1-smooth. Proposition 6.5(i) guarantees that each derivative $D(\Pi_3|_W)(\phi)$, $\phi \in W$, is an isomorphism from $T_\phi W$ onto $\mathbb{R}^3$. This implies that Π_3 defines a C^1-diffeomorphism onto the open set $\Pi_3 W$ of $\mathbb{R}^3$. □

Corollary 6.7 *W_{loc}, $W_{loc} \cap S$ and $(W_{loc} \cap S) \cap (Q + L_\eta + P_\epsilon) = \{\chi + \mathrm{Sep}_\eta(\chi) : \chi \in L_\eta\}$ with η and ϵ as in Proposition 4.6 are open subsets of W and $W \cap S$, respectively.*

Proof The map $a : W \ni \phi \mapsto P_3 \phi \in G_3$ is a global chart for the 3-dimensional C^1-manifold W, and the map $N_L + N_P \ni \chi \mapsto P_3(\chi + w_{loc}(\chi)) \in G_3$ is C^1-smooth. This means that the map $b : N_L + N_P \ni \chi \mapsto \chi + w_{loc}(\chi) \in W$ is C^1-smooth. The inclusion map $i : W \ni \phi \mapsto \phi \in C$ is C^1-smooth, and for every $\chi \in N_L + N_P$ and all $\tilde{\chi} \in L \oplus P = T_\chi(N_L + N_P)$, we have

$$\begin{aligned} Db(\chi)\tilde{\chi} &= D(i \circ b)(\chi)(\tilde{\chi}) \\ &= D(N_L + N_P \ni \rho \mapsto \rho + w_{loc}(\rho) \in C)(\chi)\tilde{\chi} = \tilde{\chi} + Dw_{loc}(\chi)\tilde{\chi}, \end{aligned}$$

with $Dw_{loc}(\chi)\tilde{\chi} \in Q$. We conclude that each $Db(\chi)$ is injective and maps $T_\chi(N_L + N_P)$ isomorphically onto $T_{b(\chi)}W$. It follows that $b(N_L + N_P) = W_{loc}$ is an open subset of W, i.e., $W_{loc} = W \cap U$ for some open set $U \subset C$. Now the remaining assertions become obvious. □

Corollary 6.8 bd W *is closed.*

Proof Let a sequence $(\phi_n)_0^\infty$ in $\operatorname{bd} W = \overline{W} \setminus W$ converge to some $\phi \in C$. Then $\phi \in \overline{W}$. Suppose $\phi \in W$. Then $P_3\phi$ belongs to the open subset P_3W of G_3. For n sufficiently large, $P_3\phi_n \in P_3W$. The injectivity of $P_3|_{\overline{W}}$ yields $\phi_n \in W$, in contradiction to $\phi_n \in \operatorname{bd} W$. □

Corollary 6.9 *The flow* $\mathbb{R} \times W \ni (t, \phi) \mapsto F_W(t, \phi) \in W$ *is* C^1*-smooth.*

Proof W is a 3-dimensional C^1-submanifold of C. It is sufficient to prove that each map $(t-1, \infty) \times W \ni (s, \psi) \mapsto F_W(s, \psi) \in C$, $t \in \mathbb{R}$, is C^1-smooth.

Let $t \in \mathbb{R}$ be given. Consider the map $A : W \ni \phi \mapsto F_W(t-2, \phi) \in W$. In case $t \geq 2$, A is C^1-smooth since $F(t-2, \cdot)$ is C^1-smooth and maps W into W. In case $t < 2$, the smoothness of the map $F(2-t, \cdot)$, the inclusion $F(2-t, W) \subset W$ and the injectivity of its derivatives combined yield that the inverse of the map $W \ni \phi \mapsto F(2-t, \phi) \in W$, namely A, is C^1-smooth. Observe that for all $s > t-1$ and $\psi \in C$,

$$F_W(s, \psi) = F_W(s+2-t, F_W(t-2, \psi)) = F(s+2-t, A(\psi)).$$

So the C^1-smoothness of the maps

$$(t-1, \infty) \times W \ni (s, \psi) \mapsto (s+2-t, A(\psi)) \in (1, \infty) \times W$$

and $F|_{(1,\infty)\times C}$ yields the assertion. □

We should point out that the information about $W \cap S$ which we have obtained up to this point is not sufficient to prove the openness of $P_2(W \cap S)$ in G_2 and the C^1-smoothness of w_S on $P_2(W \cap S)$ in the same way we derived the results on P_3W and w in Proposition 6.6. However, Proposition 6.5(i), which was used in the proof of Proposition 6.6, has the following analogue for $\overline{W} \cap S$ in place of $\overline{W}$.

Proposition 6.10 *Let* $\phi \in \overline{W} \cap S$. *Let a* C^1*-curve* $c : (-1, 1) \to C$ *with* $c(0) = \phi$ *and a sequence* $(s_n)_0^\infty$ *in* $(-1, 1) \setminus \{0\}$ *be given so that* $s_n \to 0$ *as* $n \to \infty$, *and* $c(s_n) \in \overline{W} \cap S$ *for all* $n \in \mathbb{N}$, *and* $c'(0) \neq 0$. *Then* $\Pi_2 c'(0) \neq 0$.

Proof Compare the proof of Proposition 6.5(i). □

CHAPTER 7

Dynamics on $\overline{W}$ and disk representation of $\overline{W} \cap S$

From Propositions 3.3 and 4.5 we know the limits of the solutions of equation (1.1) with initial values in $\overline{W} \setminus S$ for $t \to \infty$. Proposition 5.3 shows that $\alpha(\phi) = \{0\}$ for all $\phi \in W$. In this chapter we complete the description of the dynamics on $\overline{W}$. Namely, we prove that $\operatorname{bd} W \cap S$ is a periodic orbit, and that this periodic orbit is the ω-limit set of all points in $\overline{W} \cap S \setminus \{0\}$; in case $\mu = 0$ it is the α-limit set of all points in $\operatorname{bd} W$, and in case $\mu > 0$ it is the α-limit set of all points in $\operatorname{bd} W \setminus \{\xi_-, \xi_+\}$. In addition, we show that $\overline{W} \cap S$ is homeomorphic to D^2.

From Propositions 6.1 and 6.2 it follows that Π_2 maps $\overline{W} \cap S$ onto its image $\Pi_2(\overline{W} \cap S)$ homeomorphically. Recall that $\Pi_2^{-1} : \Pi_2(\overline{W} \cap S) \to C$ denotes the map given by the inverse of

$$\overline{W} \cap S \ni \phi \mapsto \Pi_2 \phi \in \Pi_2(\overline{W} \cap S).$$

Let $\chi^0 \in \Pi_2(\overline{W} \cap S)$ and $\psi = \Pi_2^{-1}(\chi^0)$. By the invariance of $\overline{W} \cap S$, $F_W(t, \psi) \in \overline{W} \cap S$ for all $t \in \mathbb{R}$. The curve

$$\mathbb{R} \ni t \mapsto \Pi_2 F_W(t, \psi) \in \mathbb{R}^2$$

is C^1-smooth and has its range in $\Pi_2(\overline{W} \cap S)$. We call this curve the canonical curve through χ^0. Clearly, the images of two canonical curves either coincide or do not intersect.

The images of the closed hyperplane

$$H = \{\phi \in C : \phi(0) = 0\}$$

and of the subsets

$$H^+ = \{\phi \in H : \phi(-1) > 0\}, \quad H^- = \{\phi \in H : \phi(-1) < 0\}$$

under Π_2 are

$$\{(u, v)^{tr} \in \mathbb{R}^2 : u = 0\},$$
$$v_+ = \{(u, v)^{tr} \in \mathbb{R}^2 : u = 0, \ v > 0\} \text{ and } v_- = \{(u, v)^{tr} \in \mathbb{R}^2 : u = 0, \ v < 0\},$$

respectively.

The canonical curves intersect $v_- \cup v_+$ transversally since for every $\chi^0 \in \Pi_2(\overline{W} \cap S)$ and $\psi = \Pi_2^{-1}(\chi_0)$, and for every $t \in \mathbb{R}$ with $\Pi_2 F_W(t, \psi) \in v_+$ $(\in v_-)$, the first component of the tangent vector $D_1 \Pi_2 F_W(t, \psi) 1$ satisfies

$$[D_1 \Pi_2 F_W(t, \psi) 1]_1 = \dot{x}^{\psi}(t) = 0 + f(x^{\psi}(t-1)) > 0 \ (< 0). \tag{7.1}$$

We introduce an intersection map

$$c : (v_- \cup v_+) \cap \Pi_2(\overline{W} \cap S) \to (v_- \cup v_+) \cap \Pi_2(\overline{W} \cap S)$$

as follows. For $\chi^0 \in (v_- \cup v_+) \cap \Pi_2(\overline{W} \cap S)$ and $\psi = \Pi_2^{-1}(\chi^0) \in (H^- \cup H^+) \cap (\overline{W} \cap S)$, Corollary 5.8 shows that there is a smallest zero $z_1 = z_1(\psi)$ of $x^\psi : \mathbb{R} \to \mathbb{R}$ in $(0, \infty)$; we set

$$c(\chi^0) = \Pi_2 F_W(z_1, \psi).$$

Analogously we can use the largest zero $z_{-1} = z_{-1}(\psi)$ of $x^\psi : \mathbb{R} \to \mathbb{R}$ in $(-\infty, 0)$ with $\psi = \Pi_2^{-1}(\chi^0)$, $\chi^0 \in (v_- \cup v_+) \cap \Pi_2(\overline{W} \cap S)$ to define a map

$$c_- : (v_- \cup v_+) \cap \Pi_2(\overline{W} \cap S) \to (v_- \cup v_+) \cap \Pi_2(\overline{W} \cap S)$$

by $c_-(\chi^0) = \Pi_2 F_W(z_{-1}, \psi)$. It follows that c is bijective, with $c^{-1} = c_-$. Moreover, $c(v_+ \cap \Pi_2(\overline{W} \cap S)) = v_- \cap \Pi_2(\overline{W} \cap S)$ and $c(v_- \cap \Pi_2(\overline{W} \cap S)) = v_+ \cap \Pi_2(\overline{W} \cap S)$, and the map

$$\rho : v_+ \cap \Pi_2(\overline{W} \cap S) \ni \chi^0 \mapsto c(c(\chi^0)) \in v_+ \cap \Pi_2(\overline{W} \cap S)$$

is bijective, with

$$\rho^{-1}(\chi^0) = c^{-1}(c^{-1}(\chi^0)).$$

It is convenient to write $\chi^0 \prec \hat{\chi}^0$ for elements in $\{(0, u)^{tr} \in \mathbb{R}^2 : u \in \mathbb{R}\}$ if and only if the second components satisfy $[\chi^0]_2 < [\hat{\chi}^0]_2$.

Proposition 7.1

(i) The map ρ is a homeomorphism.

(ii) For all $\chi^0, \hat{\chi}^0$ in $v_+ \cap \Pi_2(\overline{W} \cap S)$ with $\chi^0 \prec \hat{\chi}^0$,

$$\rho(\chi^0) \prec \rho(\hat{\chi}^0).$$

(iii) For every trajectory $(\chi^n)_{-\infty}^{\infty}$ of ρ in $v_+ \cap \Pi_2(\overline{W} \cap S)$ there exist χ_- and χ_+ in $\overline{v_+} \cap \Pi_2(\overline{W} \cap S)$ so that

$$\chi^n \to \chi_- \text{ as } n \to -\infty, \qquad \chi^n \to \chi_+ \text{ as } n \to +\infty.$$

In case $\chi_- \neq 0$ ($\chi_+ \neq 0$) the solution $x : \mathbb{R} \to \mathbb{R}$ of equation (1.1) with $x_0 = \Pi_2^{-1}(\chi_-)$ ($= \Pi_2^{-1}(\chi_+)$) is periodic. For $\mathcal{O}_{\Pi_2^{-1}(\chi_-)} = \{F_W(t, \Pi_2^{-1}(\chi_-)) : t \in \mathbb{R}\}$ and $\mathcal{O}_{\Pi_2^{-1}(\chi_+)} = \{F_W(t, \Pi_2^{-1}(\chi_+)) : t \in \mathbb{R}\}$,

$$\text{dist}\left(F_W(t, \Pi_2^{-1}(\chi^0)), \mathcal{O}_{\Pi_2^{-1}(\chi_-)}\right) \to 0 \qquad \text{as } t \to -\infty$$

and

$$\text{dist}\left(F_W(t, \Pi_2^{-1}(\chi^0)), \mathcal{O}_{\Pi_2^{-1}(\chi_-)}\right) \to 0 \qquad \text{as } t \to +\infty.$$

Proof 1. In order to obtain (i) it is sufficient to show that c and c^{-1} are continuous. Since Π_2, Π_2^{-1}, and F_W are continuous the latter holds provided the maps z_1 and z_{-1} from $(H^- \cup H^+) \cap (\overline{W} \cap S)$ into $\mathbb{R}$ given by the smallest positive and the largest negative zero of $x^\psi : \mathbb{R} \to \mathbb{R}$, $\psi \in (H^- \cup H^+) \cap (\overline{W} \cap S)$, are continuous.

Let $(\phi_n)_0^\infty$ be a sequence in $(H^- \cup H^+) \cap (\overline{W} \cap S)$ which converges to some $\phi \in (H^- \cup H^+) \cap (\overline{W} \cap S)$. Proposition 4.3 implies that $x^{\phi_n} \to x^\phi$ and $\dot{x}^{\phi_n} \to \dot{x}^\phi$ uniformly on compact sets as $n \to \infty$. Using Corollary 5.8 we obtain $z_1(\phi_n) \to z_1(\phi)$ and $z_{-1}(\phi_n) \to z_{-1}(\phi)$. So, z_1 and z_{-1} are continuous.

2. Assertion (ii) follows from

$$c(\hat{\chi}^0) \prec c(\chi^0) \qquad \text{for all } \chi^0, \hat{\chi}^0 \text{ in } (v_- \cup v_+) \cap \Pi_2(\overline{W} \cap S) \text{ with } \chi^0 \prec \hat{\chi}^0.$$

In order to derive the last statement, consider χ^0 and $\hat{\chi}^0$ in $(v_- \cup v_+) \cap \Pi_2(\overline{W} \cap S)$ with $\chi^0 \prec \hat{\chi}^0$. In case $\chi^0 \in v_-$ and $\hat{\chi}^0 \in v_+$ we have $c(\hat{\chi}^0) \in v_-$ and $c(\chi^0) \in v_+$, or $c(\hat{\chi}^0) \prec c(\chi^0)$.

The case $\chi^0 \in v_+$ and $\hat{\chi}^0 \in v_+$. Let $x : \mathbb{R} \to \mathbb{R}$ and $\hat{x} : \mathbb{R} \to \mathbb{R}$ denote the solutions of equation (1.1) with $x_0 = \Pi_2^{-1}(\chi^0)$ and $\hat{x}_0 = \Pi_2^{-1}(\hat{\chi}^0)$. The canonical curves χ and $\hat{\chi}$ with $\chi(0) = \chi^0$ and $\hat{\chi}(0) = \hat{\chi}^0$ satisfy $\chi(t) = (x(t), x(t-1))^{tr}$ and $\hat{\chi}(t) = (\hat{x}(t), \hat{x}(t-1))^{tr}$ for all $t \in \mathbb{R}$, and $0 = x(0) = \hat{x}(0)$, $0 < x(-1) < \hat{x}(-1)$. Moreover, $\dot{x}(0) > 0$, $\dot{\hat{x}}(0) > 0$, and for the smallest positive zeros z_1 and $\hat{z}_1$ of x and $\hat{x}$, respectively, we have $0 < x$ in $(0, z_1)$, $0 < \hat{x}$ in $(0, \hat{z}_1)$, $\dot{x}(z_1) < 0$, $x(z_1 - 1) < 0$, $\dot{\hat{x}}(\hat{z}_1) < 0$, $\hat{x}(\hat{z}_1 - 1) < 0$. The restriction $\chi|_{[0,z_1]}$ and the line segment $\lambda : [0,1] \ni s \mapsto \chi^0 + s(c(\chi_0) - \chi_0) \in \mathbb{R}^2$ from $\chi(0) = \chi^0$ to $\chi(z_1) = c(\chi^0)$ form a simple closed curve ζ. The set $\{(u,v)^{tr} \in \mathbb{R}^2 : u < 0$, or $u = 0$ and $v < x(z_1 - 1)$, or $u = 0$ and $x(-1) < v\}$ belongs to $\text{ext}(\zeta)$ since each of its points can be connected by parallels to the abscissa in $\mathbb{R}^2 \setminus |\zeta|$ to points with arbitrarily large negative first component. In particular,

$$\hat{\chi}(0) \in \text{ext}(\zeta).$$

Recall that for every C^1-curve $c : (-1,1) \to \mathbb{R}^2$ with $c(0) \in \lambda((0,1))$ and negative first component of $c'(0)$ there exists $\epsilon > 0$ so that $c((-\epsilon, 0)) \subset \text{int}(\zeta)$.

We show

$$\hat{\chi}([0, \hat{z}_1]) \cap \chi([0, z_1]) = \emptyset.$$

Otherwise there exist $\hat{t} \in (0, \hat{z}_1]$ and $t \in (0, z_1]$ with $\hat{x}_{\hat{t}} = F_W(\hat{t}, \Pi_2^{-1}(\hat{\chi}^0)) = F_W(t, \Pi_2^{-1}(\chi^0)) = x_t$, which implies $\hat{x}(s) = x(s + t - \hat{t})$ for all $s \in \mathbb{R}$. In case $\hat{t} < t$, $0 = \hat{x}(0) = x(t - \hat{t})$, a contradiction to $0 < x$ in $(0, z_1)$. In case $t < \hat{t}$, $0 = x(\hat{t} - t + t - \hat{t}) = \hat{x}(\hat{t} - t)$, a contradiction to $0 < \hat{x}$ in $(0, \hat{z}_1)$. Therefore $\hat{x} = x$; in particular, $\hat{x}_0 = x_0$, and $\hat{\chi}^0 = \chi^0$, in contradiction to $\chi^0 \prec \hat{\chi}^0$.

Using $0 < \hat{x}$ on $(0, \hat{z}_1)$ and $\hat{\chi}(0) \in \text{ext}(\zeta)$ we infer $\hat{\chi}([0, \hat{z}_1)) \subset \mathbb{R}^2 \setminus |\zeta|$, and thereby

$$\hat{\chi}([0, \hat{z}_1)) \subset \text{ext}(\zeta).$$

Now we derive $c(\hat{\chi}^0) \prec c(\chi^0)$ in the case considered. The injectivity of c yields $c(\hat{\chi}^0) \neq c(\chi^0)$. Suppose $c(\chi^0) \prec c(\hat{\chi}^0)$. Then $\hat{\chi}(\hat{z}_1) = (0, \hat{x}(\hat{z}_1 - 1))^{tr} \in \lambda((0,1))$, and the first component $\dot{\hat{x}}(z_1)$ of $D\hat{\chi}(\hat{z}_1)1$ is negative. Therefore $\hat{\chi}((\hat{z}_1 - \epsilon, \hat{z}_1)) \subset \text{int}(\zeta)$ for some $\epsilon > 0$, in contradiction to $\hat{\chi}([0, \hat{z}_1)) \subset \text{ext}(\zeta)$.

In case $\chi^0 \in v_-$ and $\hat{\chi}^0 \in v_-$ one can proceed analogously, with the roles of χ^0 and $\hat{\chi}^0$ interchanged.

3. Let a trajectory $(\chi^n)_{-\infty}^{\infty}$ of ρ in $v_+ \cap \Pi_2(\overline{W} \cap S)$ be given. The statement in (ii) implies that the second components form a monotone sequence in $\mathbb{R}$. As $\overline{W} \cap S$ is compact and Π_2 is continuous, the set $\Pi_2(\overline{W} \cap S)$ is compact in $\mathbb{R}^2$. It follows that there exist χ_- and χ_+ in $\overline{v_+} \cap \Pi_2(\overline{W} \cap S)$ such that

$$\chi^n \to \chi_- \text{ as } n \to -\infty, \quad \chi^n \to \chi_+ \text{ as } n \to +\infty.$$

Observe $\overline{v_+} = v_+ \cup \{0\}$.

If $\chi_- \neq 0$ then $\rho(\chi_-) = \chi_-$ by the continuity of ρ. From the definition of ρ, it follows that the solution through $\Pi_2^{-1}(\chi_-)$ is a nontrivial periodic solution of equation (1.1). The argument in case $\chi_+ \neq 0$ is the same.

Finally, consider the canonical curve $\chi : \mathbb{R} \ni t \mapsto \Pi_2 F_W(t, \Pi_2^{-1}(\chi^0)) \in \mathbb{R}^2$. The definition of ρ implies that there exists a sequence $(s^n)_{-\infty}^{\infty}$ such that $\chi(s^n) = \chi^n$

and $1 < s^{n+1} - s^n < 2$ for every $n \in \mathbb{Z}$. By the continuity of Π_2^{-1}, we have

$$F_W(s^n, \Pi_2^{-1}(\chi^0)) = \Pi_2^{-1}(\chi^n) \to \Pi_2^{-1}(\chi_+) \qquad \text{as } n \to \infty.$$

Let $\epsilon > 0$. By the continuous dependence of the solutions of equation (1.1) on initial data, there is $\delta > 0$ such that $\phi \in C$ and $||\phi - \Pi_2^{-1}(\chi_+)|| < \delta$ imply

$$\text{dist}\left(\mathcal{O}_{\Pi_2^{-1}(\chi_+)}, \{F_W(s,\phi) : 0 \le s \le 2\}\right) < \epsilon.$$

Choose $n_0 \in \mathbb{N}$ so that for $n_0 \le n \in \mathbb{N}$,

$$||\Pi_2^{-1}(\chi^n) - \Pi_2^{-1}(\chi_+)|| < \delta.$$

Thus

$$\text{dist}\left(\mathcal{O}_{\Pi_2^{-1}(\chi_+)}, \{F_W(s, \Pi_2^{-1}(\chi^n)) : 0 \le s \le 2\}\right) < \epsilon \qquad \text{for all integers } n \ge n_0.$$

Since $1 < s^{n+1} - s^n < 2$ for all $n \in \mathbb{Z}$, it follows that

$$\text{dist}\left(\mathcal{O}_{\Pi_2^{-1}(\chi_+)}, F_W(t, \Pi_2^{-1}(\chi^0))\right) < \epsilon \qquad \text{for all } t \ge s^{n_0}.$$

The assertion for $\mathcal{O}_{\Pi_2^{-1}(\chi_-)}$ can be shown analogously. □

Proposition 7.2

(i) There is a periodic solution $p : \mathbb{R} \to \mathbb{R}$ of equation (1.1) such that $p_0 \in H^+$ and

$$p_t \in \overline{W \cap S},\ V(p_t) = 2,\ V(\dot{p}_t) = 2,\ p_t \in R,\ \dot{p}_t \in R \qquad \textit{for all } t \in \mathbb{R}.$$

The minimal period ω of p coincides with the second zero of p in $(0, \infty)$; $1 < \omega < 2$.

(ii) For every $\chi \in v^+ \cap \Pi_2(W \cap S)$, $\chi \prec \rho(\chi)$, and for each $\phi \in W \cap S \setminus \{0\}$ the phase curve $\mathbb{R}^+ \ni t \mapsto F(t, \phi) \in C$ tends to the periodic orbit $\mathcal{O} = \{p_t : t \in [0, \omega]\}$ as $t \to \infty$.

Proof 1. Let $\psi \in W \cap S \setminus \{0\}$. By the invariance of $W \cap S$ and Corollary 5.8, there is $s \ge 0$ with $F_W(s, \psi) \in H^+ \cap W \cap S$. Let $\chi^0 = \Pi_2 F_W(s, \psi)$ and consider the trajectory $(\chi^n)_{-\infty}^{\infty}$ of ρ in $v_+ \cap \Pi_2(W \cap S)$. By Proposition 7.1(ii), the sequence $(\chi^n)_{-\infty}^{\infty}$ is monotone. By Corollary 5.6, $F_W(t, \psi) \to 0$ as $t \to -\infty$. This fact implies $\chi^n \to 0$ as $n \to -\infty$. Therefore, the sequence $(\chi^n)_{-\infty}^{\infty}$ is strictly increasing with respect to the ordering of v_+. Then $\chi_+ = \lim_{n \to \infty} \chi^n$ satisfies $\chi_+ \in v_+ \cap \Pi_2(\overline{W \cap S})$. From Proposition 7.1(iii) it follows that the solution $p : \mathbb{R} \to \mathbb{R}$ of equation (1.1) with $p_0 = \Pi_2^{-1}(\chi_+)$ is periodic. Let $\omega > 0$ be its minimal period. As $\chi_+ \in v_+ \cap \Pi_2 \overline{W \cap S}$ and $\overline{W \cap S}$ is invariant, we obtain that $p_0 \in H^+$ and $p_t \in \overline{W \cap S}$ for all $t \in \mathbb{R}$. $V(p_t) = 2$ and $p_t \in R$ for all $t \in \mathbb{R}$ follow from Proposition 5.7 and $\overline{W \cap S} \subseteq \overline{W} \cap S$. As p_t, p_s are in $\overline{W \cap S}$ and $p_t \ne p_s$ for $0 \le t < s < \omega$, Proposition 5.7 yields

$$V\left(\frac{p_t - p_s}{t - s}\right) = 2 \qquad \text{for } 0 \le t < s < \omega.$$

Hence, using Lemma VI.1 and the periodicity of p, we obtain $V(\dot{p}_t) \le 2$ for all $t \in \mathbb{R}$. On the other hand, from Corollary 5.8 it follows that $V(\dot{p}_t) \ge 2$ for all $t \in \mathbb{R}$. Thus $V(\dot{p}_t) = 2$ for all $t \in \mathbb{R}$. Lemma VI.2(iii) implies $\dot{p}_t \in R$ for all $t \in \mathbb{R}$.

The statement about the minimal period ω is a consequence of Corollary 5.8 and $\rho(\chi_+) = \chi_+$.

2. Let $\phi \in W \cap S \setminus \{0\}$. There exist $\hat{s} \geq 0$ and a trajectory $(\hat{\chi}^n)_{-\infty}^{\infty}$ of ρ in $v_+ \cap \Pi_2(W \cap S)$ such that $F_W(\hat{s}, \phi) \in H^+$ and $\hat{\chi}^0 = \Pi_2 F_W(\hat{s}, \phi)$. In the same way as for $(\chi^n)_{-\infty}^{\infty}$, it follows that $\hat{\chi}^n \to 0$ as $n \to -\infty$, and $(\hat{\chi}^n)_{-\infty}^{\infty}$ is an increasing sequence. The monotonicity of ρ and $\lim_{n\to-\infty} \chi^n = \lim_{n\to-\infty} \hat{\chi}^n = 0$ imply that there is an integer k such that either

$$\chi^n = \hat{\chi}^{n+k} \qquad \text{for all } n \in \mathbb{Z},$$

or

$$\chi^n \prec \hat{\chi}^{n+k} \prec \chi^{n+1} \prec \hat{\chi}^{n+k+1} \qquad \text{for all } n \in \mathbb{Z}.$$

In both cases $\lim_{n\to\infty} \hat{\chi}^n = \lim_{n\to\infty} \chi^n = \chi_+$. Then Proposition 7.1(iii) implies

$$F(t, \phi) \to \mathcal{O} \qquad \text{as } t \to \infty.$$

□

The C^1-maps $\eta : [0, \omega] \ni t \mapsto p_t \in C$ and $\Pi_2 \circ \eta$ are simple closed curves in C and $\mathbb{R}^2$, respectively. Obviously, $\mathcal{O} = \eta([0, \omega])$.

Proposition 7.3

(i) $\overline{W \cap S} \setminus (W \cap S) = \mathcal{O}$.

(ii) $\Pi_2(\overline{W \cap S} \setminus (W \cap S)) = |\Pi_2 \circ \eta|$.

(iii) $\Pi_2(W \cap S) = \mathrm{int}(\Pi_2 \circ \eta)$.

(iv) *Let z_1 denote the first positive zero of p. Then* $\Pi_2(H \cap \overline{W \cap S}) = \{(0, v)^{tr} \in \mathbb{R}^2 : p(z_1 - 1) \leq v \leq p(-1)\}$, $\Pi_2(H \cap W \cap S) = \{(0, v)^{tr} \in \mathbb{R}^2 : p(z_1 - 1) < v < p(-1)\}$, $\Pi_2(H \cap \mathcal{O}) = \{(0, p(z_1 - 1))^{tr}, (0, p(-1))^{tr}\}$.

Proof 1. Let z_1 be the first positive zero of p. As $p_0 \in H^+$, Corollary 5.8 implies that $p(z_1 - 1) < 0$ and $p(-1) > 0$. We show that

$$\{(u, v)^{tr} \in \mathbb{R}^2 : u = 0,\ p(z_1 - 1) < v < p(-1)\} \subset \mathrm{int}(\Pi_2 \circ \eta)$$

and

$$\{(u, v)^{tr} \in \mathbb{R}^2 : u = 0,\ v > p(-1)\} \subset \mathrm{ext}(\Pi_2 \circ \eta).$$

Corollary 5.8 and the fact that the minimal period of p is given by three consecutive zeros show that the arcs $\{(u, v)^{tr} \in \mathbb{R}^2 : u = 0,\ p(z_1 - 1) < v < p(-1)\}$ and $\{(u, v)^{tr} \in \mathbb{R}^2 : u = 0,\ v > p(-1)\}$ in $\mathbb{R}^2$ do not intersect $|\Pi_2 \circ \eta|$.

Using that the curve $\Pi_2 \circ \eta$ is C^1-smooth and intersects v_+ transversally, we find $a > 0$ and $b > 0$ such that the intersection of the rectangle

$$N_{a,b} = \{(u, v)^{tr} \in \mathbb{R}^2 : |u| < a,\ |v - p(-1)| < b\}.$$

with $|\Pi_2 \circ \eta|$ is a C^1-function defined on $(-a, a)$ with values in $(p(-1) - b, p(-1) + b)$. Let N^- and N^+ be the subsets of $N_{a,b}$ containing the points of $N_{a,b}$ below and above the graph, respectively. N^- and N^+ are disjoint connected subsets of $\mathbb{R}^2 \setminus |\Pi_2 \circ \eta|$. By the Jordan Curve Theorem, the points of $N_{a,b} \cap |\Pi_2 \circ \eta|$ are boundary points of both connected components of $\mathbb{R}^2 \setminus |\Pi_2 \circ \eta|$. Hence, the sets N^- and N^+ are subsets of different connected components of $\mathbb{R}^2 \setminus |\Pi_2 \circ \eta|$. It follows that, for small $\epsilon > 0$, the two points $(0, p(-1) - \epsilon)^{tr}$ and $(0, p(-1) + \epsilon)^{tr}$ belong to different connected components of $\mathbb{R}^2 \setminus |\Pi_2 \circ \eta|$. Consequently, the arcs $\{(u, v)^{tr} \in \mathbb{R}^2 : u = 0,\ p(z_1 - 1) < v < p(-1)\}$ and $\{(u, v)^{tr} \in \mathbb{R}^2 : u = 0,\ v > p(-1)\}$ belong to different connected components of $\mathbb{R}^2 \setminus |\Pi_2 \circ \eta|$. As $\{(u,v)^{tr} \in \mathbb{R}^2 : u = 0,\ v > p(-1)\}$ is unbounded, the first and second inclusions follow. In particular, $0 \in \text{int}(\Pi_2 \circ \eta)$.

It is now obvious how to derive the additional inclusion

$$\{(u,v)^{tr} \in \mathbb{R}^2 : u = 0,\ v < p(z_1 - 1)\} \subset \text{ext}(\Pi_2 \circ \eta).$$

2. Proposition 7.2 and the fact that the α-limit set of all points in W is $\{0\}$, imply that $\mathcal{O} \subseteq \overline{W \cap S} \setminus (W \cap S)$. We show that $\Pi_2(W \cap S) \subseteq \text{int}(\Pi_2 \circ \eta)$. Let $\phi \in W \cap S$. Then $F_W(t,\phi) \to 0$ as $t \to -\infty$. This implies that

$$\Pi_2 F_W(t,\phi) \to 0 \in \mathbb{R}^2 \qquad \text{as } t \to -\infty.$$

From this fact and from $0 \in \text{int}(\Pi_2 \circ \eta)$ it follows that $\Pi_2 F_W(t,\phi) \in \text{int}(\Pi_2 \circ \eta)$ for all sufficiently large negative t. As $\{\Pi_2 F_W(t,\phi) : t \in \mathbb{R}\} \subset \Pi_2(W \cap S)$ and $|\Pi_2 \circ \eta| = \Pi_2(\mathcal{O})$ are disjoint sets,

$$\{\Pi_2 F_W(t,\phi) : t \in \mathbb{R}\} \subseteq \text{int}(\Pi_2 \circ \eta)$$

follows. In particular, $\Pi_2 \phi \in \text{int}(\Pi_2 \circ \eta)$.

3. Proof of (i). According to $\mathcal{O} \subseteq \overline{W \cap S} \setminus (W \cap S)$, it remains to show that $\overline{W \cap S} \setminus (W \cap S) \subseteq \mathcal{O}$. Let $\phi \in \overline{W \cap S} \setminus (W \cap S)$ be given. There exist $\tilde{s} \geq 0$ and a trajectory $(\tilde{\chi}^n)_{-\infty}^{\infty}$ of ρ in $v_+ \cap \Pi_2 \overline{W \cap S}$ such that $F_W(\tilde{s},\phi) \in H^+$ and $\tilde{\chi}^0 = \Pi_2 F_W(\tilde{s},\phi)$. The sequence $(\tilde{\chi}^n)_{-\infty}^{\infty}$ is monotone on v_+ and $\lim_{n \to -\infty} \tilde{\chi}^n \neq 0$, since, by Proposition 7.1(iii), $\lim_{n \to -\infty} \tilde{\chi}^n = 0$ would imply $F_W(t,\phi) \to 0$ as $t \to -\infty$, and so Corollary 5.6 would imply $\phi \in W \cap S$. The results of parts 1,2 and $\Pi_2 \overline{W \cap S} \subseteq \overline{\Pi_2(W \cap S)}$ imply

$$\begin{aligned} v_+ \cap \Pi_2 \overline{W \cap S} &\subseteq v_+ \cap \overline{\Pi_2(W \cap S)} \\ &\subseteq v_+ \cap (\text{int}(\Pi_2 \circ \eta) \cup |\Pi_2 \circ \eta|) \\ &\subseteq \{(u,v)^{tr} \in \mathbb{R}^2 : u = 0,\ 0 < v \leq p(-1)\}. \end{aligned}$$

It follows that either $0 \prec \tilde{\chi}^n \prec (0, p(-1))^{tr}$ for all $n \in \mathbb{Z}$ or $0 \prec \tilde{\chi}^n = (0, p(-1))^{tr}$ for all $n \in \mathbb{Z}$. Suppose that $0 \prec \tilde{\chi}^n \prec (0, p(-1))^{tr}$ for all $n \in \mathbb{Z}$. Choose a trajectory $(\chi^n)_{-\infty}^{\infty}$ of ρ in $v_+ \cap \Pi_2(\overline{W \cap S})$ with $\lim_{n \to -\infty} \chi^n = 0$ and $\lim_{n \to \infty} \chi^n = (0, p(-1))^{tr}$, as in the proof of Proposition 7.2. The monotonicity of ρ implies that there is an integer k such that either

$$\chi^n = \tilde{\chi}^{n+k} \qquad \text{for all } n \in \mathbb{Z},$$

or

$$\chi^n \prec \tilde{\chi}^{n+k} \prec \chi^{n+1} \prec \tilde{\chi}^{n+k+1} \qquad \text{for all } n \in \mathbb{Z}.$$

Hence, $\lim_{n \to -\infty} \tilde{\chi}^n = \lim_{n \to -\infty} \chi^n = 0$, a contradiction. Therefore, we have $\tilde{\chi}^n = (0, p(-1))^{tr}$ for all $n \in \mathbb{Z}$, and thus $\phi \in \mathcal{O}$ follows.

4. Assertion (ii) is a consequence of (i) and of $\Pi_2 \mathcal{O} = |\Pi_2 \circ \eta|$.

5. Proof of (iii). By part 2, $\text{int}(\Pi_2 \circ \eta) \supseteq \Pi_2(W \cap S)$. Define

$$U = \text{int}(\Pi_2 \circ \eta) \setminus \Pi_2(W \cap S).$$

Let $(\psi_n)_0^{\infty}$ be a sequence in $\Pi_2(W \cap S)$ and $\psi \in \text{int}(\Pi_2 \circ \eta)$ such that $\psi_n \to \psi$ as $n \to \infty$. Then $\psi_n \in \Pi_2 \overline{W \cap S}$, and the compactness of $\Pi_2 \overline{W \cap S}$ implies $\psi \in \Pi_2 \overline{W \cap S}$. From $\psi \in \text{int}(\Pi_2 \circ \eta)$ and (ii) it follows that $\psi \in \Pi_2(W \cap S)$. Therefore, $\Pi_2(W \cap S)$ is closed in $\text{int}(\Pi_2 \circ \eta)$ in the relative topology. Consequently, U is open in $\text{int}(\Pi_2 \circ \eta)$ in the relative topology. If the nonempty set $\Pi_2(W \cap S)$ is open in $\mathbb{R}^2$, then it is open in $\text{int}(\Pi_2 \circ \eta)$ in the relative topology, and the equalities

$$\text{int}(\Pi_2 \circ \eta) = U \cup \Pi_2(W \cap S),$$
$$U \cap \Pi_2(W \cap S) = \emptyset,$$

combined with the connectedness of $\text{int}(\Pi_2 \circ \eta)$ yield $U = \emptyset$ which in turn gives the desired equality $\text{int}(\Pi_2 \circ \eta) = \Pi_2(W \cap S)$. That is, it is sufficient to show that $\Pi_2(W \cap S)$ is open in $\mathbb{R}^2$.

Recall that, by Proposition 4.6, there exist $\eta > 0$ and a Lipschitz continuous map $\text{Sep}_\eta : L_\eta \to Q \oplus P$ with $\text{Sep}_\eta(0) = 0$ and

$$\{\xi + \text{Sep}_\eta(\xi) : \xi \in L_\eta\} \subset W \cap S.$$

The continuous mapping $L_\eta \ni \xi \mapsto \Pi_2(\xi + \text{Sep}_\eta(\xi)) \in \mathbb{R}^2$ is injective. Therefore its image

$$\Pi_2\left(\{\xi + \text{Sep}_\eta(\xi) : \xi \in L_\eta\}\right)$$

is an open subset of $\mathbb{R}^2$. The equality $\Pi_2(0 + \text{Sep}_\eta(0)) = 0$ implies that the sets $\Pi_2(\{\xi + \text{Sep}_\eta(\xi) : \ \xi \in L_\eta\}) \subset \Pi_2(W \cap S)$ contain an open disk $\mathbb{R}^2_r$, $r > 0$.

Let $\chi^0 \in \Pi_2(W \cap S)$. We want to show that there is an open disk in $\Pi_2(W \cap S)$ centered at χ^0. There is $s \le 0$ such that the canonical curve $\chi = \Pi_2 F_W(\cdot, \psi)$ with $\chi(0) = \Pi_2(\psi) = \chi^0$ satisfies $\chi(s) \in \mathbb{R}^2_r$, since $\psi \in W$ implies $\chi(t) \to 0$ as $t \to -\infty$. There is a $\delta > 0$ such that $\chi(s) + \mathbb{R}^2_\delta \subset \Pi_2(W \cap S)$. Then $\Pi_2^{-1}(\chi(s) + \mathbb{R}^2_\delta) \subset W \cap S$. We have $F_W(t, \psi) = \Pi_2^{-1}(\chi(t))$ for all $t \in \mathbb{R}$. The continuous mapping

$$h : \Pi_2(W \cap S) \ni \xi \mapsto \Pi_2 F_W(-s, \Pi_2^{-1}(\xi)) \in \mathbb{R}^2$$

is injective, with values in $\Pi_2(W \cap S)$. It follows that the image of the open set $\chi(s) + \mathbb{R}^2_\delta$ under h is open, and contained in $\Pi_2(W \cap S)$. Note $\chi^0 = h(\chi(s)) \in h(\chi(s) + \mathbb{R}^2_\delta)$. Thus, $\Pi_2(W \cap S)$ is an open subset of $\mathbb{R}^2$.

6. Assertion (iv) easily follows from the results of part 1 and assertions (ii), (iii). □

Corollary 7.4 *$P_2(W \cap S)$ is open in G_2.*

Proof $P_2(W \cap S) = J_2(\Pi_2(W \cap S))$ is open in G_2, since $\Pi_2(W \cap S)$ is open in $\mathbb{R}^2$ by Proposition 7.3(iii) and J_2 defines an isomorphism from $\mathbb{R}^2$ onto G^2. □

Proposition 7.5 $\overline{W} \cap S = \overline{W \cap S}$.

Proof The closedness of S implies $\overline{W \cap S} \subseteq \overline{W} \cap S$. Suppose that there exists $\phi \in \overline{W} \cap S \setminus \overline{W \cap S}$. As Π_2 maps $\overline{W} \cap S$ injectively into $\mathbb{R}^2$ and $\Pi_2 \overline{W \cap S} = \text{int}(\Pi_2 \circ \eta) \cup |\Pi_2 \circ \eta|$ by Proposition 7.3, it follows that

$$\Pi_2 \phi \in \text{ext}(\Pi_2 \circ \eta).$$

The map $c : [0, \omega] \ni s \mapsto \phi - p_s \in C^1$ is continuous and $c([0, \omega])$ is a compact subset of C^1. By Proposition 5.7, $c([0, \omega]) \subset R \cap V^{-1}(2)$. Using Lemma VI.1(ii) and compactness, we find $\epsilon > 0$ so that for every χ in

$$N_\epsilon = \{\hat{\chi} \in C^1 : \inf_{0 \le s \le \omega} ||\hat{\chi} - c(s)||_{C^1} < \epsilon\},$$

$V(\chi) = 2$. Since $\phi \in \overline{W}$ and $\Pi_2\phi \in \text{ext}(\Pi_2 \circ \eta)$, Proposition 4.3 permits to find $\xi \in W$ such that $||\xi - \phi||_{C^1} < \epsilon$ and $\Pi_2\xi \in \text{ext}(\Pi_2 \circ \eta)$. Then $||\xi - \phi||_{C^1} < \epsilon$ implies that $\xi - p_s \in N_\epsilon$ for all $s \in [0, \omega]$, and thus

$$V(\xi - p_s) = 2, \qquad 0 \le s \le \omega.$$

Using the nonincreasing property of V, the relations $F_W(t, \xi) \in \overline{W}$ for all $t \le 0$, $p_s \in \overline{W}$ for all $s \in \mathbb{R}$, and Proposition 5.2, we obtain

$$V(F_W(t, \xi) - p_s) = 2 \qquad \text{for all } t \le 0 \text{ and } s \in [0, \omega].$$

Then Lemma VI.2(iii) implies

$$F_W(t, \xi) - p_s \in R \qquad \text{for all } t \le 0 \text{ and } s \in [0, \omega],$$

and hence

$$\Pi_2 F_W(t, \xi) \ne \Pi_2 p_s \qquad \text{for all } t \le 0 \text{ and } s \in [0, \omega].$$

Therefore, the curves $(-\infty, 0] \ni t \mapsto \Pi_2 F_W(t, \xi) \in \mathbb{R}^2$ and $[0, \omega] \ni s \mapsto \Pi_2 p_s \in \mathbb{R}^2$ do not intersect.

On the other hand, $\xi \in W$ implies that $F_W(t, \xi) \to 0$ as $t \to -\infty$, and thus

$$\Pi_2 F_W(t, \xi) \to 0 \in \mathbb{R}^2 \qquad \text{as } t \to -\infty.$$

By Proposition 7.3(iii) and (iv), $0 \in \text{int}(\Pi_2 \circ \eta)$. It follows that $\Pi_2 F_W(t, \xi) \in \text{int}(\Pi_2 \circ \eta)$ for large negative t. Hence, as the curves $(-\infty, 0] \ni t \mapsto \Pi_2 F_W(t, \xi)$ and $\Pi_2 \circ \eta$ do not intersect, it follows that $\Pi_2\xi = \Pi_2 F_W(0, \xi) \in \text{int}(\Pi_2 \circ \eta)$. This is a contradiction. □

Proposition 7.6 *There is a homeomorphism h from $\overline{W} \cap S$ onto D^2 with $h((\overline{W} \cap S) \setminus (W \cap S)) = S^1$.*

Proof Proposition 7.5, assertions (i), (ii) and (iii) in Proposition 7.3, the fact that $\Pi_2 \circ \eta$ is a simple closed curve in $\mathbb{R}^2$, and Theorem VIII.1(i) combined guarantee the existence of a homeomorphism $\tilde{h}$ from $\Pi_2(\overline{W} \cap S)$ onto D^2 with $\tilde{h}(\Pi_2((\overline{W} \cap S) \setminus (W \cap S))) = S^1$. As Π_2 restricted to $\overline{W} \cap S$ defines a homeomorphism from $\overline{W} \cap S$ onto $\Pi_2(\overline{W} \cap S)$ (see Proposition 6.2) the map

$$\overline{W} \cap S \ni \phi \mapsto \tilde{h}(\Pi_2\phi) \in D^2$$

is clearly a homeomorphism with the desired properties. □

Corollary 7.7 $\text{bd}\, W \cap S = \mathcal{O}$.

Proof From Propositions 7.3 and 7.5 we have $\text{bd}\, W \cap S = (\overline{W} \setminus W) \cap S = \overline{W} \cap S \setminus (W \cap S) = \overline{W \cap S} \setminus (W \cap S) = \mathcal{O}$. □

Corollary 7.8 *$\mathcal{O}$ is the only nontrivial periodic orbit in $\overline{W}$.*

Proof Let $\hat{\mathcal{O}}$ be a nontrivial periodic orbit in $\overline{W}$. Propositions 3.3 and 4.5 imply $\hat{\mathcal{O}} \subset \overline{W} \cap S$. From Proposition 5.3 we obtain $\hat{\mathcal{O}} \subset \text{bd}\, W$. Consequently, $\hat{\mathcal{O}} \subseteq \text{bd}\, W \cap S$. From Corollary 7.7 it follows that $\hat{\mathcal{O}} = \mathcal{O}$. □

Proposition 7.9 *If $\mu = 0$ and $\phi \in \text{bd}\, W$, or $\mu > 0$ and $\phi \in \text{bd}\, W \setminus \{\xi_-, \xi_+\}$, then $\alpha(\phi) = \mathcal{O}$.*

Proof Consider the cases $\phi \in \operatorname{bd} W$ and $\mu = 0$, and $\phi \in \operatorname{bd} W \setminus \{\xi_-, \xi_+\}$ and $\mu > 0$. Propositions 4.1 and 4.2 imply that $\{F_W(t, \phi) : t \leq 0\}$ is bounded. From Chapter 4 and Corollary 6.8 it follows that $\operatorname{bd} W$ is invariant and closed. Consequently, $\alpha(\phi) \neq \emptyset$ and $\alpha(\phi) \subset \operatorname{bd} W$.

We claim that $\alpha(\phi) \subset S$. Suppose there exists $\psi \in \alpha(\phi)$ such that $\psi \notin S$. By the monotonicity of the semiflow and the invariance of $\alpha(\phi)$, there exists $\psi^* \in \alpha(\phi)$ such that either $0 \ll \psi^*$ or $\psi^* \ll 0$. Assume $0 \ll \psi^*$. In case $\mu > 0$, Lemma 4.1 permits to find $\xi \in W$ such that $\xi \ll \psi^*$ and $\omega(\xi) = \{\xi_+\}$. In case $\mu = 0$, Lemma 4.1 permits to find $\xi \in W$ so that $\xi \ll \psi^*$ and $x^\xi(t) \to \infty$ as $t \to \infty$. Then $\psi^* \in \alpha(\phi)$ implies the existence of a sequence $(t_n)_0^\infty$ in $(-\infty, 0)$ so that $t_n \to -\infty$ as $n \to \infty$ and $\xi \ll F_W(t_n, \phi)$ for all $n \in \mathbb{N}$. By the monotonicity of the semiflow

$$\phi = F(-t_n, F_W(t_n, \phi)) \geq F(-t_n, \xi) \qquad \text{for all } n \in \mathbb{N}.$$

In case $\mu > 0$, it follows that $\phi = \xi_+$, in contradiction to the assumption. If $\mu = 0$ then we also get a contradiction. The case $\psi^* \ll 0$ is similar. Therefore, $\alpha(\phi) \subset S$.

Consequently, $\alpha(\phi) \subset \operatorname{bd} W \cap S$. The invariance of $\alpha(\phi)$ and Corollary 7.7 imply $\alpha(\phi) = \mathcal{O}$. □

CHAPTER 8

Minimal linear instability of the periodic orbit $\mathcal{O}$

Let p denote the periodic solution of equation (1.1) given in Proposition 7.2(i). Recall that $\{p_t : t \in \mathbb{R}\} = \mathcal{O} = \text{bd}\, W \cap S$, and that the minimal period ω of p is larger than 1.

Information about stability properties of p and $\mathcal{O}$ can be obtained from the Floquet multipliers which are the eigenvalues of finite multiplicity forming the spectrum away from 0 of the compact linearized period map

$$M = D_2F(\omega, p_0).$$

Recall that for $\phi \in C$ we have $M\phi = v_\omega$, with the solution $v : [-1, \infty) \to \mathbb{R}$ of the variational equation

$$\dot{v}(t) = -\mu v(t) + f'(p(t-1))v(t-1) \tag{8.1}$$

satisfying

$$v_0 = \phi.$$

So we are in a special case of the situation studied in Appendix VII. As $\dot{p}$ is a solution of equation (8.1) and is different from the zero solution, we infer that 1 is a Floquet multiplier, and

$$M\dot{p}_0 = \dot{p}_0.$$

The periodic solution p is called hyperbolic if the realified generalized eigenspace of the eigenvalue 1 of M is one-dimensional, and if there is no other Floquet multiplier on the unit circle. Recall also that

$$\dot{p}_0 \in C \setminus K.$$

Our first result is about the instability.

Proposition 8.1 *There exist a Floquet multiplier $\lambda > 1$ and $\phi \in \overset{\circ}{K}$ with $M\phi = \lambda\phi$.*

Proof Proposition VII.1(i) says that there are $\lambda > 0$ and $\phi \in \overset{\circ}{K}$ with $M\phi = \lambda\phi$. Suppose that $\lambda < 1$. Then $\phi + \epsilon\dot{p}_0 \in \overset{\circ}{K}$ for some $\epsilon > 0$, and $M^j(\phi + \epsilon\dot{p}_0) = \lambda^j\phi + \epsilon\dot{p}_0 \to \epsilon\dot{p}_0 \in C \setminus K$ as $j \to \infty$, in contradiction to $M(K) \subset K$ (see Proposition 2.2). Suppose $\lambda = 1$. As $\dot{p}(t) < 0$ for some $t \in [-1, 0]$, there exists $r > 0$ with $\phi + r\dot{p}_0 \in C \setminus K$. It follows that there exists $s \in (0, r)$ with $\phi + s\dot{p}_0 \in K$ and $\phi + t\dot{p}_0 \in C \setminus K$ for $s < t \leq r$. We have $\phi + s\dot{p}_0 \neq 0$ since $\phi \in \overset{\circ}{K}$ while $-s\dot{p}_0 \in C \setminus K$. Hence, $M^2(\phi + s\dot{p}_0) = \phi + s\dot{p}_0 \in K \setminus \overset{\circ}{K}$, in contradiction to $M^2(K \setminus \{0\}) \subset \overset{\circ}{K}$. $\square$

Let σ denote the spectrum of M. For $0 \neq \lambda \in \sigma$ with $\operatorname{Im}\lambda \geq 0$, $G_{\mathbb{R}}(\lambda)$ stands for the realified generalized eigenspace of the eigenvalue λ of M. If $r > 0$ is given and if there exists $\lambda \in \sigma$ with $r < |\lambda|$, then $C_{\leq r}$ and $C_{r<}$ denote the realified generalized eigenspaces of M associated with the nonempty disjoint compact spectral sets $\{\lambda \in \sigma : |\lambda| \leq r\}$ and $\{\lambda \in \sigma : r < |\lambda|\}$, respectively, so that $C = C_{\leq r} \oplus C_{r<}$ and

$$C_{r<} = \bigoplus_{\lambda\in\sigma, r<|\lambda|, 0\leq \operatorname{Im}\lambda} G_{\mathbb{R}}(\lambda).$$

Analogously, we shall consider realified generalized eigenspaces $C_{<r}$ and $C_{r\leq}$.

Let

$$T = V^{-1}(\{0,2\}).$$

Using Lemma VI.1(i), one finds $\overline{T} = T \cup \{0\}$. Theorem VII.9 shows that there exists $r_M > 0$ so that

$$\sigma \cap \{\lambda \in \sigma : r_M < |\lambda|\} \neq \emptyset,\ C_{\leq r_M} \cap T = \emptyset,\ C_{r_M<} \subset \overline{T},\ \dim C_{r_M<} \leq 3.$$

By $V(\dot{p}_0) = 2$, $\dot{p}_0 \in T$. We have $\dot{p}_0 \in G_{\mathbb{R}}(1)$. It follows that $r_M < 1$. Using $\dim C_{r_M<} \leq 3$ and Proposition 8.1, we obtain

$$1 \leq \dim C_{1<} \leq 2$$

and

$$\mathbb{R}\phi \subset C_{1<} \qquad \text{for some } \phi \in \overset{\circ}{K}.$$

Theorem 8.2 $\dim C_{1<} = 1$.

Proof 1. Suppose $\dim C_{1<} = 2$. Using $r_M < 1$, $1 \leq \dim G_{\mathbb{R}}(1)$ and $\dim C_{r_M<} \leq 3$, we find that p is hyperbolic, and that $C_{\leq r_M}$ coincides with the linear stable space $C_{<1}$ of M, which thereby is disjoint from the set T.

2. There exists a solution $x : \mathbb{R} \to \mathbb{R}$ of equation (1.1) such that $x_t \in W \cap S$ for all $t \in \mathbb{R}$, $x_t \to 0$ as $t \to -\infty$ and $x_t \to \mathcal{O}$ as $t \to \infty$. Proposition 5.7 implies $p_t - x_s \in T$ for all t, s in $\mathbb{R}$. The function $z : \mathbb{R} \ni u \mapsto p(t+u) - x(s+u) \in \mathbb{R}$ satisfies

$$\dot{z}(u) = -\mu z(u) + \int_0^1 f'((1-\theta)x(s+u-1) + \theta p(t+u-1))\, d\theta\, z(u-1)$$

for all $u \in \mathbb{R}$. Applying Lemma VI.3 to this equation with

$$\begin{aligned} b(u) &= \int_0^1 f'((1-\theta)x(s+u-1) + \theta p(t+u-1))\, d\theta, \\ b_0 &= \inf_{\theta\in[0,1], \chi\in\mathbb{R}, \eta\in\mathbb{R}} f'((1-\theta)x(\chi) + \theta p(\eta)), \\ b_1 &= \sup_{\theta\in[0,1], \chi\in\mathbb{R}, \eta\in\mathbb{R}} f'((1-\theta)x(\chi) + \theta p(\eta)), \end{aligned}$$

we get a constant $c > 0$ such that the a-priori estimate $||p_t - x_s|| \leq c||p_{t+1} - x_{s+1}||$ holds for all t, s in $\mathbb{R}$.

3. Choose a closed complementary subspace Y of $\mathbb{R}\dot{p}_0$ in C. Then there are an open neighbourhood U of p_0 in C and a C^1-map $\tau : U \to \mathbb{R}$ with $\tau(p_0) = \omega$ and $F(\tau(\phi), \phi) \in p_0 + Y$ for all $\phi \in U$. The Poincaré map

$$P_Y : U \cap (p_0 + Y) \ni \phi \mapsto F(\tau(\phi), \phi) \in p_0 + Y$$

is C^1-smooth. Proposition I.5 implies that the fixed point p_0 of P_Y is hyperbolic, i.e., no point from the spectrum of $DP_Y(p_0) : Y \to Y$ is on the unit circle. Consider a local stable manifold W^s of P_Y at p_0 in the sense of Theorem I.2. According to the remark following Theorem I.2 there is a neighbourhood U^s of p_0 in U so that for every trajectory $(\phi_n)_0^\infty$ of P_Y in $U^s \cap (p_0 + Y)$ we have $\phi_0 \in W^s$. As x_t converges to $\mathcal{O}$ for $t \to \infty$, we can apply Proposition I.7 to x_0 and U^s. It follows that there exists $t_P \in \mathbb{R}$ such that $x_{t_P} \in W^s$, and $W^s \ni P_Y^j(x_{t_P}) \to p_0$ as $j \to \infty$.

4. The phase curve $\mathbb{R} \ni t \mapsto p_t \in C$ is continuously differentiable and periodic, hence Lipschitz continuous. Recall Proposition 2.1(i). Theorem I.6 on the asymptotic phase is applicable and yields $\alpha > 0$ so that $\phi_0 = F(\alpha, x_{t_P})$ belongs to a local stable manifold W_ω^s of the period map $F(\omega, \cdot)$ at the fixed point p_0. Therefore, the points

$$\phi_n = F(n\omega, \phi_0), \qquad n \in \mathbb{N},$$

converge to p_0 as $n \to \infty$. Since $C_{\leq r_M} = T_{p_0} W_\omega^s$, we have $\mathrm{Pr}_{C_{\leq r_M}}(\phi_n - p_0) \neq 0$ for sufficiently large integers n, and

$$\frac{||\mathrm{Pr}_{C_{r_M <}}(\phi_n - p_0)||}{||\mathrm{Pr}_{C_{\leq r_M}}(\phi_n - p_0)||} \to 0 \qquad \text{as } n \to \infty.$$

5. From part 2 we know that

$$\frac{1}{||\phi_n - p_0||}(\phi_n - p_0) \in T \qquad \text{for all } n \in \mathbb{N}.$$

Let $c^* = \sup\{f'(\xi) : |\xi| \leq \sup|x| + \sup|p|\}$. Applying the a-priori estimate in part 2 and the differential equation (1.1), we obtain that for $s \in [-1, 0]$,

$$\begin{aligned} |\dot{\phi}_n(s) - \dot{p}(s)| &\leq \mu||\phi_n - p_0|| + c^*|\phi_n(s-1) - p(s-1)| \\ &\leq \mu||\phi_n - p_0|| + c^*||x_{-1}^{\phi_n} - p_{-1}|| \\ &\leq (\mu + cc^*)||\phi_n - p_0||. \end{aligned}$$

Therefore, the sequence of normed differences $\frac{\phi_n - p_0}{||\phi_n - p_0||}$ satisfies the hypotheses of the Arzèla–Ascoli Theorem and thus has a subsequence which converges to some unit vector $\chi \in \overline{T}$. The result in part 4 shows that $\chi \in C_{\leq r_M}$. Therefore, we arrive at $0 \neq \chi \in C_{\leq r_M} \cap \overline{T}$, a contradiction to $C_{\leq r_M} \cap T = \emptyset$. □

Proposition 8.1 and Theorem 8.2 show that there exists exactly one Floquet multiplier λ_u outside the unit circle; λ_u is real and is in $(1, \infty)$, and there is an element $\psi_u \in \overset{\circ}{K}$ with $||\psi_u|| = 1$ and

$$G_{\mathbb{R}}(\lambda_u) = \mathbb{R}\psi_u.$$

Proposition 8.3 $-1 \notin \sigma$.

Proof 1. Suppose -1 is a Floquet multiplier. According to Appendix VII there exist solutions $v : \mathbb{R} \to \mathbb{R}$ and $y : \mathbb{R} \to \mathbb{R}$ of equation (8.1) with $Mv_0 = -v_0$ and $y_0 = \psi_u$. Fix three consecutive zeros $\alpha_1 < \alpha_2 < \alpha_3$ of $q = \dot{p}$. Then $\omega = \alpha_3 - \alpha_1$. As $v(\omega + \cdot) : \mathbb{R} \ni t \mapsto v(\omega + t) \in \mathbb{R}$ is also a solution of equation (8.1), and $v(\omega + \cdot)_0 = v_\omega = -v_0$, we obtain

$$v_{\omega+t} = -v_t \quad \text{for all } t \in \mathbb{R}.$$

In particular, $v_{\alpha_3} = -v_{\alpha_1}$. Also, v is bounded.

2. We show that $V(c_1 v_t + c_2 q_t) = 2$ for all $(c_1, c_2) \in \mathbb{R} \times \mathbb{R} \setminus \{(0,0)\}$ and all $t \in \mathbb{R}$. By Proposition VII.5 and $0 \neq c_1 v_0 + c_2 q_0 \in C_{r_M <} \subset \overline{T}$, $V(c_1 v_t + c_2 q_t) \leq 2$ for all $t \in \mathbb{R}$. Suppose $V(c_1 v_{t_0} + c_2 q_{t_0}) \neq 2$ for some $t_0 \in \mathbb{R}$. Then $V(c_1 v_{t_0} + c_2 q_{t_0}) = 0$. Therefore, $0 \neq c_1 v_{t_0} + c_2 q_{t_0} \geq 0$ or $0 \neq c_1 v_{t_0} + c_2 q_{t_0} \leq 0$. Proposition 2.2(ii) shows that the solution $z = c_1 v + c_2 q$ of equation (8.1) satisfies $z_{t_0+2} \in \overset{\circ}{K} \cup (-\overset{\circ}{K})$. Without loss of generality, we may assume $z_{t_0+2} \in \overset{\circ}{K}$. Then there exists $\delta > 0$ such that $z_{t_0+2} - \delta y_{t_0+2} \in \overset{\circ}{K}$. The function $\mathbb{R} \ni t \mapsto z(t) - \delta y(t) \in \mathbb{R}$ is a solution of equation (8.1). Proposition 2.2(ii) yields $z(t) > \delta y(t)$ for all $t \geq t_0 + 2$. This implies $z(t) \to \infty$ as $t \to \infty$, a contradiction to the boundedness of v and q.

3. The last result, combined with (iii) of Lemma VI.2, ensures that for every $(c_1, c_2) \in \mathbb{R} \times \mathbb{R} \setminus \{(0,0)\}$, $c_1 v + c_2 q$ can not have a double zero.

We establish a certain synchronization property of periodic solutions. Let β_1 be the largest zero of v which is less than or equal to α_1. Then, $\beta_1 < \alpha_1$, for otherwise $q - \frac{\dot{q}(\alpha_1)}{\dot{v}(\alpha_1)} v$ has a double zero at α_1. We claim that the next zero β_2 of v is in (α_1, α_2). Otherwise, either $\beta_2 = \alpha_2$ or $\beta_2 > \alpha_2$. But if $\beta_2 = \alpha_2$ then $q - \frac{\dot{q}(\alpha_2)}{\dot{v}(\alpha_2)} v$ has a double zero at α_2, a contradiction. If $\beta_2 > \alpha_2$ then there exist $c \in \mathbb{R} \setminus \{0\}$, $t^* \in (\alpha_1, \alpha_2)$ such that $|cq(t)| \leq |v(t)|$ for $t \in (\alpha_1, \alpha_2)$ and $|cq(t^*)| = |v(t^*)|$ with $\operatorname{sign}(cq(t^*)) = \operatorname{sign} v(t^*)$. It follows that $v - cq$ has a double zero at t^*, again a contradiction.

4. In a similar way, we can show that the next zeros β_3, β_4 of v satisfy

$$\beta_3 \in (\alpha_2, \alpha_3), \quad \alpha_3 \in (\beta_3, \beta_4).$$

So we must have

$$\operatorname{sign} v(\alpha_3) = \operatorname{sign} \dot{v}(\beta_3) = \operatorname{sign} \dot{v}(\beta_1) = \operatorname{sign} v(\alpha_1),$$

a contradiction to $v(\alpha_3) = -v(\alpha_1)$. □

We should mention that a similar result was earlier established by Walther **[68]** for a delay differential equation with negative feedback.

Corollary 8.4
(i) For every $\lambda \in \sigma \setminus \{0, 1, \lambda_u\}$, $|\lambda| < 1$.
(ii) $1 \leq \dim G_{\mathbb{R}}(1) \leq 2$.
(iii) $C_{\leq 1} \cap (K \cup (-K)) = \{0\}$.
(iv) In case $\dim G_{\mathbb{R}}(1) = 2$, $C_{1\leq} \subset \overline{T}$ *and* $C_{<1} \cap T = \emptyset$.

Proof 1. Suppose $1 \le |\lambda|$ for some $\lambda \in \sigma \setminus \{0, 1, \lambda_u\}$. Theorem 8.2 gives $|\lambda| = 1$. Proposition 8.3 excludes $\lambda = -1$. It follows that $\lambda \in S^1_{\mathbb{C}} \setminus \mathbb{R}$. Hence, $\overline{\lambda} \in \sigma \cap S^1_{\mathbb{C}} \setminus \mathbb{R}$, and therefore, $4 \le \dim C_{1\le} \le \dim C_{r_M<}$, in contradiction to $\dim C_{r_M<} \le 3$.

2. Recall $1 \le \dim G_{\mathbb{R}}(1)$. By $r_M < 1 < \lambda_u$, we have $\dim G_{\mathbb{R}}(1) + 1 = \dim G_{\mathbb{R}}(1) + \dim G_{\mathbb{R}}(\lambda_u) \le \dim C_{r_M<} \le 3$, from which it follows that $\dim G_{\mathbb{R}}(1) \le 2$.

3. Applying Proposition VII.1(ii) to $C_{\le 1} = C_{<\lambda_u}$, we get assertion (iii).

4. Recall that $r_M < 1$, $C_{\le r_M} \cap T = \emptyset$, $C_{r_M<} \subset \overline{T}$, $G_{\mathbb{R}}(1) \cup C_{1<} \subset C_{r_M<}$ and $\dim C_{1<} = 1$. In case $\dim G_{\mathbb{R}}(1) = 2$, we obtain $3 \le \dim C_{1\le} \le \dim C_{r_M<} \le 3$. It follows that $\sigma \cap \{\lambda \in \mathbb{C} : r_M < |\lambda| < 1\} = \emptyset$. Hence, $C_{\le r_M} = C_{<1}$. □

It is convenient here to discuss further relations between solutions of equation (8.1) with low oscillation frequency and the realified generalized eigenspaces $C_{r_M<}$, $C_{r_M<} \cap C_{\le 1}$ and $C_{1<} = G_{\mathbb{R}}(\lambda_u)$.

Proposition 8.5

(i) Let $v : \mathbb{R} \to \mathbb{R}$ be a solution of equation (8.1) with $v_0 \neq 0$. If $V(v_t) = 2$ (≤ 2) [$= 0$] for all $t \in \mathbb{R}$ then $v_0 \in C_{r_M<} \cap C_{\le 1}$ $(\in C_{r_M<})$ [$\in C_{1<}$].

(ii) For every $\phi \in C_{r_M<} \cap C_{\le 1} \setminus \{0\}$ there is a solution $v : \mathbb{R} \to \mathbb{R}$ of equation (8.1) with $V(v_t) = 2$ for all $t \in \mathbb{R}$.

Proof 1. Let $v : \mathbb{R} \to \mathbb{R}$ be a solution of equation (8.1) with $v_0 \neq 0$. For $j \in \mathbb{Z}$ set $\phi_j = v_{j\omega}$. Then $\phi_{j+1} = M\phi_j$ for every $j \in \mathbb{Z}$. If $V(v_t) \le 2$ for all $t \in \mathbb{R}$, then Lemma VI.3 yields a constant $c > 0$ with $||\phi_j|| \le c||\phi_{j+1}||$ for all $j \in \mathbb{Z}$. Fix $s \in (0, \frac{1}{c})$ and $\delta > 0$ with $s + \delta < \frac{1}{c}$. We claim that $\mathrm{Pr}_{C_{\le s}} \phi_0 = 0$. To verify the claim, we note that the spectrum of the map $C_{\le s} \ni \chi \mapsto M\chi \in C_{\le s}$ is bounded by s. It follows that there is a constant $c_s > 0$ with

$$||M^j \chi|| \le c_s (s+\delta)^j ||\chi|| \qquad \text{for all } j \in \mathbb{N},\ \chi \in C_{\le s}.$$

Consequently,

$$\begin{aligned} ||\mathrm{Pr}_{C_{\le s}} v_0|| &\le c_s (s+\delta)^j || \mathrm{Pr}_{C_{\le s}} \phi_{-j}|| \\ &\le c_s || \mathrm{Pr}_{C_{\le s}} ||(s+\delta)^j c^j ||v_0|| \end{aligned}$$

for every $j \in \mathbb{N}$, and hence $\mathrm{Pr}_{C_{\le s}} v_0 = 0$.

2. The last result implies that v_0 belongs to the finite-dimensional space $C_{s<}$. There exists a smallest number $t > s$ with $\sigma \cap t S^1_{\mathbb{C}} \neq \emptyset$ so that the spectral projection ψ_0 of $v_0 = \phi_0$ onto the realified generalized eigenspace $G \subset C$ of the spectral set $\sigma \cap t S^1_{\mathbb{C}}$ does not vanish. All ϕ_j, $j \in \mathbb{Z}$, belong to $C_{t\le} \subset C_{s<}$. Suppose $\phi_0 \notin G$. There exists a trajectory $(\psi_j)_{-\infty}^0$ of M in $G \subset C_{s\le}$, and there exists $u > t$ so that $\phi_j - \psi_j \in C_{u\le}$ for all integers $j \le 0$. As we are in finite-dimensional spaces there exists a strictly decreasing sequence $(j_n)_{-\infty}^0$ of negative integers so that the vectors $\frac{1}{||\phi_{j_n}||}\phi_{j_n}$ converge as $n \to -\infty$ to some unit vector ϕ, while the vectors $\frac{1}{||\psi_{j_n}||}\psi_{j_n}$ converge to some $\psi \in G$. We now claim that $\phi = \psi$. To show this, we note that for every integer $j \le 0$,

$$\begin{aligned} &\left\| \frac{1}{||\phi_j||}\phi_j - \frac{1}{||\psi_j||}\psi_j \right\| \le \frac{1}{||\psi_j||}||\phi_j - \psi_j|| + \left| \frac{1}{||\phi_j||} - \frac{1}{||\psi_j||} \right| ||\phi_j|| \\ &= \frac{1}{||\psi_j||}||\phi_j - \psi_j|| + \frac{|||\psi_j|| - ||\phi_j|||}{||\phi_j||||\psi_j||}||\phi_j|| \le 2\frac{||\phi_j - \psi_j||}{||\psi_j||}. \end{aligned}$$

Since the spectra of the isomorphisms induced by M and M^{-1} on G and $C_{u\leq}$ are bounded by t and $\frac{1}{u}$, respectively, we obtain $\delta > 0$ with $\frac{t+\delta}{u-\delta} < 1$ and $c_\delta \geq 0$ with

$$\frac{||\phi_j - \psi_j||}{||\psi_j||} \leq c_\delta \left(\frac{t+\delta}{u-\delta}\right)^{-j} \frac{||\phi_0 - \psi_0||}{||\psi_0||}$$

for all integers $j \leq 0$. This shows that $\phi = \psi$.

3. Suppose that $V(v_t) = 2$ for all $t \in \mathbb{R}$. In case $\phi_0 \in G$, $\phi_0 \in G \cap T$. In case $\phi_0 \notin G$, $V(\phi) \leq 2$ (see Lemma VI.1(i)), and thus $\phi = \psi \in G \cap T$. It follows that $r_M < t$, and thereby $v_0 \in C_{r_M<}$. It remains to show that $\mathrm{Pr}_{C_{1<}} v_0 = 0$. Assume the contrary. Arguments similar to those above show that there is a trajectory $(\xi_n)_0^\infty$ of M in $C_{1<}$ with $\xi_0 = \mathrm{Pr}_{C_{1<}}$, and there exist a subsequence $(\phi_{j_n})_0^\infty$ of $(\phi_j)_0^\infty$ in $C_{t\leq}$ and a subsequence $(\xi_{j_n})_0^\infty$ of $(\xi_n)_0^\infty$ in $C_{1<}$ so that the unit vectors $\frac{1}{||\phi_{j_n}||}\phi_{j_n}$ and $\frac{1}{||\xi_{j_n}||}\xi_{j_n}$ both converge as $n \to \infty$ to a vector $\phi = \xi \in C_{1<} \setminus \{0\} \subset \overset{\circ}{K}$. It follows that for some integer j, ϕ_j has no zero, a contradiction to $V(\phi_j) = V(v_{j\omega}) = 2$.

4. The proofs of the statements in brackets in assertion (i) are similar and simpler.

5. We now prove assertion (ii). As $M(C_{r_M<} \cap C_{\leq 1}) = C_{r_M<} \cap C_{\leq 1}$, there is a trajectory $(\phi_j)_{-\infty}^\infty$ of M in $C_{r_M<} \cap C_{\leq 1} \setminus \{0\}$ with $\phi_0 = \phi$. Let $v^{(j)} : [-1, \infty) \to \mathbb{R}$ be the solution of (8.1) with $v_0^{(j)} = \phi_j$. Define $v : \mathbb{R} \to \mathbb{R}$ by

$$v(t) = v^{(j)}(t + j\omega)$$

for $j\omega - 1 \leq t < (j+1)\omega - 1$ and $j \in \mathbb{Z}$. Then v is a solution of equation (8.1) with $v_0 = \phi$ and $v_{j\omega} = \phi_j$ for all $j \in \mathbb{Z}$. By $C_{r_M<} \cap C_{\leq 1} \subset C_{r_M<} \subset \overline{T}$, $V(v_{j\omega}) = V(\phi_j) \leq 2$ for all $j \in \mathbb{Z}$, and the monotonicity of V implies $V(v_t) \leq 2$ for all $t \in \mathbb{R}$. Suppose $V(v_t) = 0$ for some $t \in \mathbb{R}$. Then $\phi_j \in K \cup (-K)$ for all integers $j \geq \frac{t}{\omega}$, and Corollary 8.4(iii) yields a contradiction to $\phi_j \neq 0$. Hence, $V(v_t) = 2$ for all $t \in \mathbb{R}$. $\square$

CHAPTER 9

Smoothness of $W \cap S$ in case $\mathcal{O}$ is hyperbolic

In this chapter, we assume that p is hyperbolic. We show that in this case the map $w_S|_{P_2(W\cap S)}$, which satisfies

$$W \cap S = \{\chi + w_S(\chi) : \chi \in P_2(W \cap S)\},$$

is C^1-smooth.

We begin with a Poincaré return map. Recall from the last chapter that $C_{1<} = G_{\mathbb{R}}(\lambda_u)$ is one-dimensional. Due to the hyperbolicity of p, the closed subspace

$$Y = C_{<1} \oplus C_{1<}$$

has codimension 1. Moreover, $D_1F(\omega, p_0)1 = \dot{p}_0 \in C \setminus Y$ since $\dot{p}_0 \in G_{\mathbb{R}}(1)$. There exist $\epsilon_p > 0$, a convex open neighbourhood U of p_0 in C, and a C^1-map $\tau : U \to \mathbb{R}$ with the following properties:

$$\begin{aligned}
&1 < \omega - \epsilon_p,\ \tau(U) \subset (\omega - \epsilon_p,\ \omega + \epsilon_p),\ \tau(p_0) = \omega;\\
&\text{for every } (t, \phi) \in (\omega - \epsilon_p, \omega + \epsilon_p) \times U,\ F(t, \phi) \in p_0 + Y \text{ if and only if } t = \tau(\phi);\\
&D_1F(\tau(\phi), \phi)1 \in C \setminus Y \text{ for all } \phi \in U.
\end{aligned}$$

See Appendix I for the construction of ϵ_p and τ. Using $p_0 \neq 0$ we can also achieve that

$$0 \notin U.$$

The continuity of the map $\overline{W} \ni \phi \mapsto \dot{\phi} \in C$ (Proposition 4.3) and the relation $\dot{p}_0 \notin Y$ permit to find U so that in addition

$$\dot{\phi} \in C \setminus Y \qquad \text{for all } \phi \in U \cap \overline{W}.$$

Our Poincaré map $P_Y : U \cap (p_0 + Y) \to p_0 + Y$ is given by

$$P_Y(\phi) = F(\tau(\phi), \phi).$$

Fix a continuous linear functional $\phi^* : C \to \mathbb{R}$ so that $Y = (\phi^*)^{-1}(0)$. For every ϕ in the domain of P_Y^j, $1 \leq j \in \mathbb{N}$, define

$$\tau_j(\phi) = \sum_{k=0}^{j-1} \tau(P_Y^k(\phi)).$$

Then

$$DP_Y^j(\phi)\chi = D_2F(\tau_j(\phi),\phi)\chi - \frac{\phi^*(D_2F(\tau_j(\phi),\phi)\chi)}{\phi^*(D_1F(\tau_j(\phi),\phi)1)}D_1F(\tau_j(\phi),\phi)1$$

for every $\chi \in T_\phi(U \cap (p_0 + Y)) = Y$ (see Appendix I). In particular,

$$DP_Y(p_0)\chi = M\chi - \frac{\phi^*(M\chi)}{\phi^*(\dot{p}_0)}\dot{p}_0 \qquad \text{for all } \chi \in Y,$$

and it follows that

$$DP_Y(p_0)\chi = M\chi \qquad \text{for all } \chi \in Y$$

since $MY \subset Y = (\phi^*)^{-1}(0)$. We conclude that on $C_{1<} = G_{\mathbb{R}}(\lambda_u) = \mathbb{R}\psi_u$,

$$DP_Y(p_0)\psi_u = \lambda_u\psi_u,$$

while the spectrum of the map $C_{<1} \ni \phi \mapsto DP_Y(p_0)\phi \in C_{<1}$ is contained in a compact subset of the open unit disk. So we have a local stable manifold W^s of P_Y at p_0 in the sense of Theorem I.2, with codimension 1 in $U \cap (p_0 + Y)$ and $T_{p_0}W^s = C_{<1}$. It follows that W^s is a C^1-submanifold of codimension 2 in the ambient space C.

Proposition 9.1
(i) $W^s \cap W \subset W \cap S$.
(ii) For every $\phi \in W \cap S \setminus \{0\}$ *there exists* $t > 0$ *with* $F_W(t,\phi) \in W^s$.

Proof 1. For $\phi \in W^s \cap W$, $P_Y^j\phi \to p_0$ as $j \to \infty$. Propositions 3.3 and 4.5 yield $\phi \in S$. This proves (i).

2. Let $0 \neq \phi \in W \cap S$. Then $F_W(t,\phi) \to \mathcal{O}$ as $t \to \infty$. According to the version of Theorem I.2 in case of hyperbolicity there is an open neighbourhood V^s of p_0 in $U \cap (p_0 + Y)$ so that for every trajectory $(\phi_j)_0^\infty$ of P_Y in V^s, $\phi_0 \in W^s$. Apply Proposition I.7. □

Proposition 9.2 $W^s \cap W$ *is a one-dimensional* C^1*-submanifold of* W.

Proof 1. Proposition 9.1 guarantees $W^s \cap W \neq \emptyset$. For $W^s \cap W$ to be a one-dimensional C^1-submanifold of W, we need to show that the inclusion map $i : W \ni \phi \mapsto \phi \in C$ and W^s are transversal, i.e., for every $\phi \in W$ with $\phi = i(\phi) \in W^s$ the space $Di(\phi)T_\phi W = T_\phi W$ contains a closed complementary subspace of $T_{i(\phi)}W^s = T_\phi W^s$ in C, and $Di(\phi)^{-1}T_{i(\phi)}W^s = T_\phi W \cap T_\phi W^s$ has a complementary subspace in $T_\phi W$. The last statement is true, of course, as $\dim T_\phi W = 3 < \infty$. As a first step towards the first statement, we observe that due to the choice of U, $\dot{\phi} \in T_\phi W \setminus Y$ and hence, $\dot{\phi} \in T_\phi W \setminus T_\phi W^s$.

2. Next we find a vector $\chi \in T_\phi W \cap \overset{\circ}{K}$. Recall from Corollary 6.7 that W_{loc} is an open neighbourhood of 0 in W. As $F_W(t,\phi) \to 0$ for $t \to -\infty$ there exists $t_0 \in \mathbb{R}$ with $F_W(t,\phi) \in W_{loc}$ for $t \leq t_0$. Recall that $\chi_0(\theta) = e^{\lambda_0\theta}, -1 \leq \theta \leq 0$. Then $\chi_0 \in P \cap \overset{\circ}{K}$. For $t_0 \geq t \to -\infty$, the vectors $\xi_t = \chi_0 +$

$Dw_{loc}(\mathrm{Pr}_{L\oplus P} F_W(t,\phi))\chi_0 \in T_{F_W(t,\phi)}W_{loc} \subset T_{F_W(t,\phi)}W$ converge to χ_0, and consequently there exists $t \le t_0$ so that $\xi_t \in T_{F_W(t,\phi)}W \cap \overset{\circ}{K}$. We have $F_W(-t,W) \subset W$ and $D_2F_W(-t,F_W(t,\phi))\overset{\circ}{K} \subset \overset{\circ}{K}$, see Proposition 2.2. It follows that

$$\chi = D_2F_W(-t,F_W(t,\phi))\xi_t \in T_\phi W \cap \overset{\circ}{K}\,.$$

3. $\dot\phi$ and χ are linearly independent since $\phi \in W \cap S$ implies that $\dot\phi$ has a sign change (Corollary 5.8) while χ is strictly positive.

4. It remains to show that

$$T_\phi W^s \cap (\mathbb{R}\dot\phi \oplus \mathbb{R}\chi) = \{0\}.$$

Suppose $0 \ne a\dot\phi + b\chi \in T_\phi W^s$ for some reals a and b. As $\dot\phi \notin T_\phi W^s$, $b \ne 0$. Set $c = \frac{a}{b}$. Then $0 \ne c\dot\phi + \chi \in T_\phi W^s$. Let $v : [-1,\infty) \to \mathbb{R}$ be the solution of equation (2.2) with $v_0 = \chi$. Set $x = x^\phi$, so that $\dot x$ is the solution of equation (2.2) with $\dot x_0 = \dot\phi$. For integers $j > 0$, set $\tau_j = \tau_j(\phi)$. Then

$$\begin{aligned} T_{P_Y^j(\phi)}W^s \ni DP_Y^j(\phi)(c\dot\phi + \chi) &= c\dot x_{\tau_j} + v_{\tau_j} - \frac{\phi^*(c\dot x_{\tau_j} + v_{\tau_j})}{\phi^*(\dot x_{\tau_j})}\dot x_{\tau_j} \\ &= v_{\tau_j} - \frac{\phi^*(v_{\tau_j})}{\phi^*(\dot x_{\tau_j})}\dot x_{\tau_j}. \end{aligned}$$

The compactness of $\overline{W \cap S}$ implies that x^ϕ is bounded. By Proposition 5.7, $V(x_t^\phi) = 2$ for all $t \in \mathbb{R}$. An application of Lemma VI.3 to equation (2.2) and its strictly positive solution v yields $r \ge 0$ and $t \ge 0$ with

$$||v_{s-1}|| \le r||v_s|| \qquad \text{for all } s \ge t.$$

This estimate and equation (2.2) combined show that the derivatives $\frac{1}{||v_{\tau_j}||}\dot v_{\tau_j}$, $1 \le j \in \mathbb{N}$, are uniformly bounded. The Arzèla–Ascoli Theorem yields a subsequence $(\frac{1}{||v_{\tau_{j_k}}||}v_{\tau_{j_k}})_0^\infty$ which converges to a unit vector $\rho \in K$. From $x_{\tau_j} = P_Y^j(\phi) \to p_0$, we get $\dot x_{\tau_j} \to \dot p_0$ (see Proposition 4.3), and it follows that the vectors $\frac{1}{||v_{\tau_{j_k}}||}DP_Y^j(\phi)(c\dot\phi + \chi) \in T_{P_Y^j(\phi)}W^s$ converge as $k \to \infty$ to the vector

$$\rho - \frac{\phi^*(\rho)}{\phi^*(\dot p_0)}\dot p_0 \in T_{p_0}W^s = C_{<1} \subset C_{\le 1}.$$

Since we have $\dot p_0 \in G_{\mathbb{R}}(1) \subset C_{\le 1}$, we conclude that

$$0 \ne \rho \in K \cap C_{\le 1},$$

in contradiction to Corollary 8.4(iii). □

Proposition 9.3 *The restriction of w_S to the open set $P_2(W \cap S) \setminus \{0\}$ is C^1-smooth.*

Proof 1. Let $\chi \in P_2(W \cap S) \setminus \{0\}$. Set $\phi = \chi + w_S(\chi) \in (W \cap S) \setminus \{0\}$. There exists $t \in \mathbb{R}$ so that $\psi = F_W(t, \phi) \in W^s \cap W$.

2. We show that close to ψ the flow extends $W^s \cap W$ to a smooth 2-dimensional submanifold. Consider the 2-dimensional C^1-submanifold $\mathbb{R} \times (W^s \cap W)$ of $\mathbb{R} \times C$. Choose a basis vector χ of $T_\psi(W^s \cap W)$ and a C^1-curve $\gamma : (-1, 1) \to C$ with range in $W^s \cap W$ such that $\gamma(0) = \psi$ and $D\gamma(0)1 = \chi$. Then the space $T_{(0,\psi)}(\mathbb{R} \times (W^s \cap W))$ is spanned by the tangent vectors at 0 of the curves $(-1, 1) \ni s \mapsto (s, \psi) \in \mathbb{R} \times C$ and $(-1, 1) \ni s \mapsto (0, \gamma(s)) \in \mathbb{R} \times C$. The flow F_W defines a C^1-map A of $\mathbb{R} \times (W^s \cap W)$ into W, and $DA(0, \psi)$ has range $\mathbb{R}\dot{\psi} + \mathbb{R}\chi \subset T_\psi W$. The vectors $\dot{\psi} \in C \setminus Y$ and $\chi \in T_\psi(W^s \cap W) \subset Y$ are linearly independent. Lemma I.8 shows that there exists $\delta > 0$ so that the set

$$W_\psi = \{F_W(t, \rho) : |t| < \delta,\ \rho \in W^s \cap W \cap (\psi + C_\delta)\}$$

is a 2-dimensional C^1-submanifold of W. Due to $W^s \cap W \subset W \cap S$ and the invariance, we have $W_\psi \subset W \cap S$.

3. The map $F_W(-t, \cdot)$ defines a C^1-diffeomorphism of W onto itself. Using Proposition 6.10, we infer that the C^1-map B from W_ψ into G_2 given by $B(\rho) = P_2 F_W(-t, \rho)$ has an injective derivative at ψ. By the Inverse Function Theorem, there exist an open neighbourhood V of $B(\psi) = \chi$ in $P_2(W \cap S) \setminus \{0\}$ and a C^1-inverse $B_V^{-1} : V \to W_\psi$ of the restriction of B to $B^{-1}(V)$. Observe that the restriction of w_S to V is given by

$$w_S(\tilde{\chi}) = (\mathrm{id} - P_2) F_W(-t, B_V^{-1}(\tilde{\chi})).$$

It follows that $w_S|_V$ is C^1-smooth. The rest is obvious. □

Choose ϵ and η according to Proposition 4.6. In order to show that w_S is differentiable at 0 and that Dw_S is continuous at 0, we employ the Lipschitz map

$$\mathrm{Sep}_\eta : L_\eta \to Q \oplus P.$$

Recall that

$$\begin{aligned} M &= (W_{loc} \cap S) \cap (Q + L_\eta + P_\epsilon) \\ &= \{\chi + \mathrm{Sep}_\eta(\chi) : \chi \in L_\eta\} \end{aligned}$$

is an open subset of $W \cap S$ (Corollary 6.7). Using Proposition 9.3, we infer that $M \setminus \{0\}$ is a 2-dimensional C^1-submanifold of C.

Corollary 9.4 $\mathrm{Sep}_\eta|_{L_\eta \setminus \{0\}}$ *is C^1-smooth.*

Proof Using Proposition 4.7, we deduce that for every $\phi \in M \setminus \{0\}$ the restriction of the derivative $D\,\mathrm{Pr}_L(\phi) = \mathrm{Pr}_L$ to the 2-dimensional space $T_\phi M$ is injective. It follows that the inverse $A : L_\eta \setminus \{0\} \to M \setminus \{0\}$ of the map $M \setminus \{0\} \ni \phi \mapsto \mathrm{Pr}_L\, \phi \in L$ is C^1-smooth. The conclusion then follows from

$$\mathrm{Sep}_\eta = (\mathrm{id} - \mathrm{Pr}_L) \circ A.$$

□

Proposition 9.5 Sep_η *is differentiable at* 0, $D\mathrm{Sep}_\eta(0) = 0$, *and* $D\mathrm{Sep}_\eta$ *is continuous at* 0.

Proof 1. We show that for every $\tilde{\epsilon} > 0$ there exists $\delta > 0$ so that for all ϕ and ψ in $(W \cap S) \cap C_\delta$ with $\mathrm{Pr}_L(\phi - \psi) \neq 0$,

$$\frac{||\mathrm{Pr}_{Q\oplus P}(\phi - \psi)||}{||\mathrm{Pr}_L(\phi - \psi)||} < \tilde{\epsilon}.$$

Suppose the contrary. Then there exist $\tilde{\epsilon} > 0$ and sequences $(\phi_n)_0^\infty$, $(\psi_n)_0^\infty$ in $W \cap S$ which converge to 0 and satisfy $\mathrm{Pr}_L(\phi_n - \psi_n) \neq 0$ and

$$\tilde{\epsilon} \leq \frac{||\mathrm{Pr}_{Q\oplus P}(\phi_n - \psi_n)||}{||\mathrm{Pr}_L(\phi_n - \psi_n)||} \qquad \text{for all } n \in \mathbb{N}.$$

The functions $x^{\phi_n} : \mathbb{R} \to \mathbb{R}$ and $x^{\psi_n} : \mathbb{R} \to \mathbb{R}$ are uniformly bounded since $\overline{W} \cap S$ is compact. We have $V(x_t^{\phi_n} - x_t^{\psi_n}) = 2$ for all $t \in \mathbb{R}$, see Proposition 5.7. Using the flow F_W, we obtain that on every compact interval the functions x^{ϕ_n} and x^{ψ_n} converge uniformly to 0.

For every $n \in \mathbb{N}$, the function

$$y^n = \frac{1}{||\phi_n - \psi_n||}(x^{\phi_n} - x^{\psi_n})$$

is a solution of the equation

$$\dot{y}(t) = -\mu y(t) + \int_0^1 f'(x^{\psi_n}(t-1) + s(x^{\phi_n}(t-1) - x^{\psi_n}(t-1)))\, ds\, y(t-1),$$

with $||y_0^n|| = 1$ and $V(y_t^n) = 2$ for all $t \in \mathbb{R}$. We apply Lemmas VI.5, VI.6 and VI.4 to obtain a subsequence $(y^{n_k})_0^\infty$ which converges uniformly on every compact interval of $\mathbb{R}$ to a solution $y : \mathbb{R} \to \mathbb{R}$ of equation (1.2), with $||y_0|| = 1$ and $V(y_t) \leq 2$ for all $t \in \mathbb{R}$. By Proposition 5.1, $y_0 \in L \oplus P$. Suppose $\mathrm{Pr}_P\, y_0 \neq 0$. Then there are reals $a \neq 0$, b, θ so that

$$y(t) = ae^{\lambda_0 t} + be^{\mathrm{Re}\,\lambda_1 t}\cos(\mathrm{Im}\,\lambda_1 t + \theta) \qquad \text{for all } t \in \mathbb{R}.$$

It follows that $y_t \in \overset{\circ}{K}$ for some $t \in \mathbb{R}$. Consequently, $x_t^{\phi_n} - x_t^{\psi_n} \in \overset{\circ}{K}$ for some $n \in \mathbb{N}$, in contradiction to $V(x_t^{\phi_n} - x_t^{\psi_n}) = 2$. Therefore, $\mathrm{Pr}_P\, y_0 = 0$ and $0 \neq y_0 \in L$, in contradiction to $\tilde{\epsilon} \leq ||\mathrm{Pr}_{Q\oplus P}\, y_0|| / ||\mathrm{Pr}_L\, y_0||$.

2. Part 1 yields

$$\lim_{0 \neq \chi \to 0} \frac{||\mathrm{Sep}_\eta(\chi)||}{||\chi||} = 0$$

so that Sep_η is differentiable at 0 with $D\mathrm{Sep}_\eta(0) = 0$. Also, given $\tilde{\epsilon} > 0$ and δ according to part 1, choose $\tilde{\eta} \in (0, \eta)$ so small that for all $\chi \in L_{\tilde{\eta}}$, $\chi + \mathrm{Sep}_\eta(\chi) \in C_\delta$. For such χ and for all $\hat{\chi} \in L_{\tilde{\eta}} \setminus \{\chi\}$, we get

$$\frac{||\mathrm{Sep}_\eta(\hat{\chi}) - \mathrm{Sep}_\eta(\chi)||}{||\hat{\chi} - \chi||} < \tilde{\epsilon}$$

which implies $||D\mathrm{Sep}_\eta(\chi)|| \leq \tilde{\epsilon}$. □

Corollary 9.6 *$w_S|_{W\cap S}$ is C^1-smooth.*

Proof Using Corollary 9.4, Propositions 9.5 and 6.10 we infer that the map

$$a : L_\eta \ni \chi \mapsto P_2(\chi + \mathrm{Sep}_\eta \chi) \in G_2$$

defines a C^1-diffeomorphism from L_η onto an open neighbourhood V of 0 in G_2. Let a^{-1} denote its inverse. For all $\chi \in V$,

$$w_S(\chi) = (\mathrm{id} - P_2)(a^{-1}(\chi) + \mathrm{Sep}_\eta(a^{-1}(\chi))),$$

so that $w_S|_V$ is C^1-smooth. Using Proposition 9.3, we then complete the proof.

□

CHAPTER 10

Smoothness of $W \cap S$ in case $\mathcal{O}$ is not hyperbolic

In this chapter, it is assumed that the periodic solution p is not hyperbolic. We show that also in this case the map $w_S|_{P_2(W\cap S)}$ is C^1-smooth.

Corollary 8.4 permits us to choose a unit vector $\xi \in C$ with

$$G_{\mathbb{R}}(1) = \mathbb{R}\dot{p}_0 \oplus \mathbb{R}\xi.$$

This corollary also yields

$$C_{<1} \cap T = \emptyset. \tag{10.1}$$

Set

$$Y = C_{<1} \oplus \mathbb{R}\xi \oplus C_{1<}$$

so that $\dot{p}_0 \in C \setminus Y$. Then there exist $\epsilon_p > 0$, a convex open neighbourhood U of p_0 in C and a C^1-map $\tau : U \to \mathbb{R}$ with the same properties as in the previous chapter and, in addition, with

$$V(\phi) \geq 2 \qquad \text{for all } \phi \in U;$$

the last inequality can be achieved since p_0 has a sign change.

The Poincaré map

$$P_Y : U \cap (p_0 + Y) \ni \phi \mapsto F(\tau(\phi), \phi) \in Y$$

is C^1-smooth and has p_0 as a fixed point. Proposition I.5 implies that the points $\lambda \neq 0$ in the spectrum of the derivative $DP_Y(p_0)$ coincide with the Floquet multipliers of the periodic solution p. Moreover, we infer

$$DP_Y(p_0)\xi = \xi.$$

Fix a continuous linear functional $\phi^* : C \to \mathbb{R}$ with

$$Y = (\phi^*)^{-1}(0).$$

Using $MC_{<1} \subset C_{<1} \subset Y$, $MC_{1<} \subset C_{1<} \subset Y$, the last equation and the formula

$$DP_Y(p_0)\phi = M\phi - \frac{\phi^*(M\phi)}{\phi^*(\dot{p}_0)}\dot{p}_0 \qquad \text{for all } \phi \in Y,$$

we find

$$DP_Y(p_0)\phi = M\phi \qquad \text{on } C_{<1} \cup C_{1<}.$$

In particular,

$$DP_Y(p_0)\phi = \lambda_u \phi \qquad \text{for all } \phi \in C_{1<} = \mathbb{R}\psi_u,$$

and the spectrum of the map $C_{<1} \ni \phi \mapsto DP_Y(p_0)\phi \in C_{<1}$ is contained in a compact subset of the open unit disk. Observe that Proposition I.5(iii) yields $DP_Y(p_0)\xi = \xi$.

So we have a C^1-smooth local center-stable manifold W^{cs} of P_Y at p_0 in the sense of Theorem II.1, with codimension 1 in $U \cap (p_0 + Y)$, $T_{p_0}W^{cs} = C_{<1} \oplus \mathbb{R}\xi$, and there is an open neighbourhood N of p_0 in C with $P_Y(W^{cs} \cap N) \subset W^{cs}$. The C^1-map

$$P_Y^{cs} : W^{cs} \cap N \ni \phi \mapsto P_Y(\phi) \in W^{cs}$$

satisfies

$$DP_Y^{cs}(p_0)\phi = DP_Y(p_0)\phi \qquad \text{for all } \phi \in T_{p_0}(W^{cs} \cap N) = C_{<1} \oplus \mathbb{R}\xi.$$

The projection $Y \to Y$ along $C_{1<}$ onto $C_{<1} \oplus \mathbb{R}\xi$ defines a C^1-diffeomorphism A of the C^1-smooth graph $W^{cs} \cap N - p_0$ (see Theorem II.1) onto an open neighbourhood of 0 in $C_{<1} \oplus \mathbb{R}\xi$, with

$$DA(0)\phi = \phi \qquad \text{for all } \phi \in T_0(W^{cs} \cap N - p_0) = C_{<1} \oplus \mathbb{R}\xi.$$

We apply Theorem I.2 to the C^1-map given by

$$A(P_Y^{cs}(A^{-1}(\phi) + p_0) - p_0)$$

for ϕ in an open neighbourhood of 0 in $C_{<1} \oplus \mathbb{R}\xi$ and obtain, after transformation (see Appendix IV), a C^1-submanifold $W^{cs,s}$ of $W^{cs} \cap N$ of codimension 1, with $p_0 \in W^{cs,s}$, $T_{p_0}W^{cs,s} = C_{<1}$ and $P_Y^{cs}(W^{cs,s}) \subset W^{cs,s}$.

Proposition 10.1 *Let a trajectory $(\phi_j)_0^\infty$ of P_Y in $W^{cs} \cap W$ be given with $\phi_j \to p_0$ as $j \to \infty$. Then there exists $k \in \mathbb{N}$ with $\phi_j \notin W^{cs,s}$ for all integers $j \geq k$.*

Proof 1. Propositions 3.3 and 4.5 yield $\phi_0 \in W \cap S$. It follows that $x^{\phi_0} : \mathbb{R} \to \mathbb{R}$ is bounded.

The solutions $x^{\phi_j} - p : \mathbb{R} \to \mathbb{R}$, $j \in \mathbb{N}$, of the equations

$$\dot{z}(t) = -\mu z(t) + \int_0^1 f'(p(t-1) + s(x^{\phi_j}(t-1) - p(t-1)))\, ds\, z(t-1)$$

satisfy $x_t^{\phi_j} - p_t \neq 0$ (since $\phi_j \in W$ and $W \cap \mathcal{O} = \emptyset$) and $V(x_t^{\phi_j} - p_t) \leq 2$ for all $j \in \mathbb{N}$ and all $t \in \mathbb{R}$. Lemma VI.3 yields $k_0 > 0$ so that

$$||F_W(-1, \phi_j) - F_W(-1, p_0)|| \leq k_0 ||\phi_j - p_0|| \qquad \text{for all } j \in \mathbb{N}.$$

Set

$$c = \max\left\{ f'(\tilde{\xi}) : |\tilde{\xi}| \leq \sup_{t \in \mathbb{R}} |x^{\phi_0}(t)| + \max_{t \in \mathbb{R}} |p(t)| \right\}.$$

Then

$$||\dot{\phi}_j - \dot{p}_0|| \leq (\mu + ck_0) ||\phi_j - p_0|| \qquad \text{for all } j \in \mathbb{N}.$$

2. Suppose there is a subsequence $(\phi_{j_k})_0^\infty$ in $W^{cs,s}$. The last estimate permits to apply the Arzèla–Ascoli Theorem to the normed differences $\psi_k = \frac{1}{||\phi_{j_k} - p_0||}(\phi_{j_k} - p_0)$, $k \in \mathbb{N}$. We obtain a subsequence $(\psi_{k_m})_0^\infty$ which converges to some unit vector ρ, with $V(\rho) \leq 2$. $W^{cs,s}$ is also a C^1-submanifold of Y. The relations $T_{p_0}W^{cs,s} = C_{<1}$ and $\phi_{j_k} \in W^{cs,s}$ for all $k \in \mathbb{N}$ combined imply

$$\Pr_{\mathbb{R}\xi \oplus C_{1<}} \rho = \lim_{m \to \infty} \Pr_{\mathbb{R}\xi \oplus C_{1<}} \psi_{k_m} = 0.$$

Therefore, $0 \neq \rho \in C_{<1}$ and $V(\rho) \leq 2$, in contradiction to (10.1). □

Proposition 10.2 *For every $\phi \in W \cap S \setminus \{0\}$ there exist $t \geq 0$, a trajectory $(\phi_n)_0^\infty$ of P_Y in W^{cs}, and a neighbourhood N_t of ϕ_0 in C so that $\phi_0 = F(t, \phi)$, $\phi_n \to p_0$ as $n \to \infty$, and $W^{cs} \cap N_t \subset S$.*

Proof According to Theorem II.1 there is a neighbourhood V^{cs} of p_0 in $U \cap (p_0 + Y)$ so that for every trajectory $(\phi_n)_0^\infty$ of P_Y in V^{cs} we have $\psi_0 \in W^{cs}$. We apply Proposition I.7 to ϕ and V^{cs} to obtain $s \geq 0$ and a trajectory $(\phi_n^*)_0^\infty$ of P_Y in V^{cs} with $F(s, \phi) = \phi_0^*$, and $\phi_n^* \to p_0$ as $n \to \infty$. We infer $\phi_n^* \in W^{cs}$ for all $n \in \mathbb{N}$. Proposition 10.1 yields $k \in \mathbb{N}$ so that for all integers $j \geq k$, $\phi_j^* \in W^{cs} \setminus W^{cs,s}$. From Proposition IV.4 we obtain an integer $j \geq k$ and an open set N_t of C with $\phi_j^* \in N_t$ and $P_Y(W^{cs} \cap N_t) \subset W^{cs} \cap N_t \subset U$. For $n \in \mathbb{N}$, set $\phi_n = \phi_{n+j}^*$. Recall $2 \leq V(\psi)$ for all $\psi \in U$. It follows that for every $\psi \in W^{cs} \cap N_t$ and for all integers $n \geq 1$, $2 \leq V(F(\tau_n(\psi), \psi))$, where

$$\tau_n(\psi) = \sum_{m=0}^{n-1} \tau(P_Y^m(\psi)) \geq n(\omega - \epsilon_p).$$

Consequently, $W^{cs} \cap N_t \subset S$.

□

Proposition 10.3 *Suppose $\phi \in W^{cs} \cap W$, and there is a trajectory $(\phi_n)_0^\infty$ of P_Y in W^{cs} so that $\phi_0 = \phi$ and $\phi_n \to p_0$ as $n \to \infty$. Then there exists an open neighbourhood N_ϕ of ϕ in C so that $W^{cs} \cap W \cap N_\phi$ is a one-dimensional C^1-submanifold of $W \cap N_\phi$ and W.*

Proof 1. It is sufficient to show that the inclusion map $i : W \ni \phi \mapsto \phi \in C$ and the C^1-submanifold W^{cs} of C are transversal at ϕ, i.e., $Di(\phi)T_\phi W = T_\phi W$ contains a closed complementary subspace of $T_{i(\phi)}W^{cs} = T_\phi W^{cs}$ in C, and $Di(\phi)^{-1}T_{i(\phi)}W^{cs} = T_\phi W \cap T_\phi W^{cs}$ has a closed complementary subspace in $T_\phi W$. The last statement is true as $\dim T_\phi W = 3$.

2. Propositions 3.3 and 4.5 combined with $\lim_{n\to\infty} \phi_n = p_0$ imply $\phi \in S$. Now derive the remaining part of transversality as in the proof of Proposition 9.2, with W^{cs} in place of W^s and $T_{p_0}W^{cs} = C_{<1} \oplus \mathbb{R}\xi \subset C_{\leq 1}$ in place of $T_{p_0}W^s = C_{<1} \subset C_{\leq 1}$.

□

Proposition 10.4 *The restriction of w_S to the open set $P_2(W \cap S)$ is C^1-smooth.*

Proof 1. Let $\chi \in P_2(W \cap S) \setminus \{0\}$. Set $\phi = \chi + w_S(\chi) \in (W \cap S) \setminus \{0\}$. Proposition 10.2 guarantees the existence of $t \geq 0$, a trajectory $(\phi_n)_0^\infty$ of P_Y in W^{cs}, and a neighbourhood N_t of ϕ_0 in C so that $\phi_0 = F(t, \phi)$, $\phi_n \to p_0$ as $n \to \infty$, and $W^{cs} \cap N_t \subset S$. Of course, $\phi_0 \in W$. By Proposition 10.3, there is an open neighbourhood N_{ϕ_0} of ϕ_0 in C so that $W^{cs} \cap W \cap N_{\phi_0}$ is a one-dimensional C^1-submanifold of W which is contained in S. Arguing as in part 2 of the proof of Proposition 9.3, we obtain $\delta > 0$ so that the set

$$W_{\phi_0} = \{F_W(t, \rho) : |t| < \delta, \ \rho \in W^{cs} \cap W \cap (\phi_0 + C_\delta)\}$$

is a 2-dimensional C^1-submanifold of W, and $\phi_0 + C_\delta \subset N_{\phi_0} \cap N_t$. By invariance, $W_{\phi_0} \subset W \cap S$. As in part 3 of the proof of Proposition 9.3 we deduce that the restriction of w_S to a neighbourhood of χ is C^1-smooth, and it follows that $w_S|_{P_2(W \cap S) \setminus \{0\}}$ is C^1-smooth.

2. We continue as in Chapter 9 to show that the map Sep_η given by Proposition 4.6 is C^1-smooth, and it follows that the restriction of w_S to a neighbourhood of 0 is C^1-smooth.

$\square$

CHAPTER 11

The unstable set of $\mathcal{O}$ contains the nonstationary points of bd W

The aim of this chapter is to show that the unstable set of the periodic orbit $\mathcal{O}$ contains the nonstationary points of bd W, that is, all points of bd W if $\mu = 0$, and the points of bd $W \setminus \{\xi_-, \xi_+\}$ if $\mu > 0$.

Recall from Chapter 8 that the monodromy operator $M = D_2F(\omega, p_0)$ has exactly one Floquet multiplier λ_u outside the unit circle, and

$$\lambda_u \in (1, \infty),\ C_{1<} = \mathbb{R}\psi_u,\ \psi_u \in \overset{\circ}{K},\ ||\psi_u|| = 1.$$

Choose $\lambda \in (0,1)$ so that

$$\lambda > \max\left\{\frac{1}{\lambda_u}, \max_{\zeta\in\sigma, |\zeta|<1} |\zeta|\right\}.$$

Theorem I.3 guarantees the existence of a local strong unstable manifold of the period-ω map $F(\omega, \cdot)$ at its fixed point p_0; namely, there are convex open neighbourhoods $N_{1<}$ of 0 in $C_{1<}$ and $N_{\leq 1}$ of 0 in the realified generalized eigenspace $C_{\leq 1}$ of M associated with the spectral set $\{\zeta \in \sigma : |\zeta| \leq 1\}$ and a C^1-map

$$w^u : N_{1<} \to C_{\leq 1}$$

so that

$$w^u(0) = 0,\ Dw^u(0) = 0,\ w^u(N_{1<}) \subset N_{\leq 1},$$

and with $N^u = N_{\leq 1} + N_{1<}$ the shifted graph

$$W^u(p_0, F(\omega, \cdot), N^u) = \{p_0 + \chi + w^u(\chi) : \chi \in N^u\}$$

is equal to the set

$$\{\chi \in p_0 + N^u : \text{There is a trajectory } (\chi^n)^0_{-\infty} \text{ of } F(\omega,\cdot) \text{ with } \chi^0 = \chi,$$
$$\lambda^n(\chi^n - p_0) \in N^u \text{ for all } n \in -\mathbb{N}, \text{ and } \lambda^n(\chi^n - p_0) \to 0 \text{ as } n \to -\infty\}.$$

Define the unstable set $W^u(\mathcal{O})$ of the periodic orbit $\mathcal{O}$ by

$$W^u(\mathcal{O}) = \bigcup_{t\geq 0} F(t, W^u(p_0, F(\omega, \cdot), N^u)).$$

Proposition 11.1 *If* $\mu = 0$ *then* $\operatorname{bd} W \subseteq W^u(\mathcal{O})$, *and if* $\mu > 0$ *then* $\operatorname{bd} W \setminus \{\xi_-, \xi_+\} \subseteq W^u(\mathcal{O})$.

Proof 1. Let $\phi \in \operatorname{bd} W$ if $\mu = 0$ and $\phi \in \operatorname{bd} W \setminus \{\xi_-, \xi_+\}$ if $\mu > 0$. Consider the solution $x^\phi : \mathbb{R} \to \mathbb{R}$ of equation (1.1). It suffices to show that there exists $t \leq 0$ such that $x_t^\phi \in W^u(p_0, F(\omega, \cdot), N^u)$. Suppose $x_t^\phi \notin W^u(p_0, F(\omega, \cdot), N^u)$ for all $t \leq 0$. Then, in particular, $\phi \notin \mathcal{O}$ follows.

Assume that $s \leq 0$ has the property that the sequence $(\phi_n)_{-\infty}^0$, defined by $\phi_n = x_{s+n\omega}^\phi$ for all $n \in -\mathbb{N}$, has a subsequence converging to p_0 as $n \to -\infty$. We claim that $V(\phi_n - p_0) = 2$ for all sufficiently large negative integers n.

We first show that $(\lambda^n(\phi_n - p_0))_{-\infty}^0$ does not converge to 0 as $n \to -\infty$. Assume $\lambda^n(\phi_n - p_0) \to 0$ as $n \to -\infty$. For every negative integer k, $\eta_n = \phi_{k+n}$ satisfies

$$\lambda^n(\eta_n - p_0) = \lambda^n(\phi_{k+n} - p_0) = \lambda^{-k}\lambda^{k+n}(\phi_{k+n} - p_0).$$

Therefore, if k is a sufficiently large negative integer, then $\lambda^n(\eta_n - p_0) \in N^u$ for all $n \in -\mathbb{N}$, and $\lim_{n \to -\infty} \lambda^n(\eta_n - p_0) = 0$. Thus $\eta_0 = \phi_k = x_{s+k\omega}^\phi \in W^u(p_0, F(\omega, \cdot), N^u)$, which is a contradiction to our assumption.

Choose $\delta > 0$ so that $C_\delta \subset N^u$. Choose $\lambda^* \in (0, \lambda)$ such that $\lambda^* \lambda_u > 1$. We construct a subsequence $(\phi_{n_k})_{-\infty}^0$ of $(\phi_n)_{-\infty}^0$ such that $\phi_{n_k} \to p_0$ as $k \to -\infty$ and

$$\frac{||\lambda^{*n_k - 1}(\phi_{n_k - 1} - p_0)||}{||\lambda^{*n_k}(\phi_{n_k} - p_0)||} \geq 1 \tag{11.1}$$

holds for all $k \in -\mathbb{N}$. In order to define n_0, choose an integer $m \in -\mathbb{N}$ such that $||\phi_m - p_0|| < \delta = \frac{\delta}{2^0}$. If (11.1) holds with $n_k = m$, then let $n_0 = m$. Otherwise, we note that if (11.1) does not hold with $n_k = m, m-1, \ldots, m-j$ for some $j \in \mathbb{N}$, then

$$||\phi_{m-j-1} - p_0|| < \lambda^* ||\phi_{m-j} - p_0|| < \lambda^{*2} ||\phi_{m-j+1} - p_0|| < \ldots < \lambda^{*j+1} ||\phi_m - p_0||. \tag{11.2}$$

If this is the case for all $j \in \mathbb{N}$, then with $\eta_n = \phi_{m+n}$, $n \in -\mathbb{N}$, it follows that

$$||\lambda^n(\eta_n - p_0)|| = ||\lambda^n(\phi_{m+n} - p_0)|| \leq \left(\frac{\lambda}{\lambda^*}\right)^n ||\phi_m - p_0|| < \delta$$

for all $n \in -\mathbb{N}$, and $\lim_{n \to -\infty} \lambda^n(\eta_n - p_0) = 0$. Hence

$$\phi_m = x_{s+m\omega}^\phi \in W^u(p_0, F(\omega, \cdot), N^u)$$

follows, a contradiction. Therefore, there is a maximal $j \in \mathbb{N}$ such that (11.2) holds. Now define $n_0 = m - j - 1$. Then (11.1) holds with $k = 0$. Assume that $n_0, n_{-1}, \ldots, n_l$ are defined for some $l \in -\mathbb{N}$. In order to define n_{l-1}, we choose $m \in -\mathbb{N}$ such that $m < n_l$ and

$$||\phi_m - p_0|| < \frac{\delta}{2^{-l+1}}.$$

If (11.1) holds with $n_k = m$, then let $n_{l-1} = m$. If (11.1) does not hold with $n_k = m$ then the same argument as above shows that there exists a maximal $j \in \mathbb{N}$ such that (11.2) holds. In this case define $n_{l-1} = m - j - 1$. Then $(\phi_{n_k})_{-\infty}^{0}$ is defined by induction and has the desired properties.

For every $k \in -\mathbb{N}$ the function

$$z^k = \frac{1}{||\phi_{n_k} - p_0||}\left(x^{\phi_{n_k}} - p\right)$$

is a solution of the equation

$$\dot{z}(t) = -\mu z(t) + \int_0^1 f'\left(ux^{\phi_{n_k}}(t-1) + (1-u)p(t-1)\right) du\, z(t-1)$$

with $||z_0^k|| = 1$ and $V(z_t^k) \leq 2$ for all $t \in \mathbb{R}$. As $\phi_{n_k} \to p_0$ and F_W is a continuous flow on $\overline{W}$,

$$b^k(t) = \int_0^1 f'\left(ux^{\phi_{n_k}}(t-1) + (1-u)p(t-1)\right) du \to f'(p(t-1)) \qquad \text{as } k \to -\infty$$

uniformly on compact subsets of $\mathbb{R}$. Using $\alpha(\phi) = \mathcal{O}$, it follows that x^ϕ is bounded on $(-\infty, 0]$. So, we can find positive constants b_0 and b_1 such that $b_0 \leq b^k(t) \leq b_1$ for all $t \leq 0$ and $k \in -\mathbb{N}$. Then, by Lemma VI.5, there is $c > 0$ with $||z_t^k|| \leq ce^{c|t|}$ for all $t \leq 0$ and $k \in -\mathbb{N}$. Hence, Lemma VI.4 with $I_0 = (-\infty, 0]$ implies that $(z^k)_{-\infty}^0$ has a subsequence $(z^{k_i})_{-\infty}^0$ and there is a C^1-function $z : (-\infty, 0] \to \mathbb{R}$ such that $z^{k_i} \to z$ as $i \to -\infty$ uniformly on compact subsets of $(-\infty, 0]$, and z satisfies

$$\dot{z}(t) = -\mu z(t) + f'(p(t-1))z(t-1) \tag{8.1}$$

for all $t \leq 0$. It also follows that $||z_0|| = 1$. From

$$||z_{-\omega}^k|| = \frac{||x_{-\omega}^{\phi_{n_k}} - p_{-\omega}||}{||\phi_{n_k} - p_0||} = \frac{||\lambda^{*n_k - 1}(\phi_{n_k - 1} - p_0)||}{||\lambda^{*n_k}(\phi_{n_k} - p_0)||}\lambda^*$$

and property (11.1) of $(\phi_{n_k})_{-\infty}^0$, we infer

$$||z_{-\omega}|| \geq \lambda^*.$$

Suppose the statement

$$V(\phi_n - p_0) = 2 \qquad \text{for all sufficiently large negative integers } n$$

is false. Then, from $V(\phi_n - p_0) \leq 2$ for all $n \in -\mathbb{N}$ and from the monotonicity of V, we get $V(\phi_n - p_0) = 0$ for all $n \in -\mathbb{N}$. Extending z to a solution $v : \mathbb{R} \to \mathbb{R}$ of (8.1) and using the monotonicity of V, we find

$$V(v_t) = 0 \qquad \text{for all } t \in \mathbb{R}.$$

Then, Proposition 8.5(i) implies $z_{-\omega} \in C_{1<} = \mathbb{R}\psi_u$. Let $z_{-\omega} = c\psi_u$ for some $c \in \mathbb{R}$. Then

$$z_0 = M(z_{-\omega}) = M(c\psi_u) = cM(\psi_u) = c\lambda_u\psi_u,$$

and hence

$$1 = ||z_0|| = ||c\lambda_u\psi_u|| = \lambda_u||z_{-\omega}|| \geq \lambda_u\lambda^* > 1,$$

a contradiction. So, we have $V(\phi_n - p_0) = 2$ for all sufficiently large negative integers n.

2. We prove that there exists $T_0 < 0$ with $V(x_t^\phi - p_0) = 2$ for all $t \leq T_0$.

Assume that there exists a sequence $(t_n)_{-\infty}^0$ in $(-\infty, 0)$ with $t_n \to -\infty$ as $n \to -\infty$ and

$$V(x_{t_n}^\phi - p_0) = 0 \qquad \text{for all } n \in -\mathbb{N}.$$

We claim that $x_{t_n}^\phi \to p_0$ as $n \to -\infty$. If not, then there exist a subsequence $(t_{n_k})_{-\infty}^0$ of $(t_n)_{-\infty}^0$ and $\tau \in (0, \omega)$ such that $x_{t_{n_k}}^\phi \to p_\tau$ as $k \to -\infty$ since $\alpha(\phi) = \mathcal{O}$ by Proposition 7.9. As $\tau \in (0, \omega)$, we have $V(p_\tau - p_0) = 2$, and hence $V(x_{t_{n_k}}^\phi - p_0) = 2$ for all sufficiently large negative integers k, a contradiction.

Each t_n can be uniquely written as

$$t_n = m_n\omega + \tau_n$$

for some $m_n \in -\mathbb{N}$ and $\tau_n \in [0, \omega)$. We may assume (replacing $(t_n)_{-\infty}^0$ with a subsequence if necessary)

$$\tau_n \to \tau^* \in [0, \omega] \qquad \text{as } n \to -\infty$$

and

$$\text{either } x_{t_n}^\phi > p_0 \text{ for all } n \in -\mathbb{N} \quad \text{or} \quad x_{t_n}^\phi < p_0 \text{ for all } n \in -\mathbb{N}.$$

We consider only the case when $x_{t_n}^\phi > p_0$ for all $n \in -\mathbb{N}$. The other case is analogous.

We claim that

$$x^\phi(\tau^* + u) > p(u) \qquad \text{for all } u \in \mathbb{R}. \tag{11.3}$$

If $x^\phi(\tau^* + u) \geq p(u)$ for all $u \in \mathbb{R}$, then the relation $\phi \notin \mathcal{O}$ and Proposition 2.2 imply that (11.3) holds. Thus, if (11.3) is not satisfied, then there exists $\hat{u} \in \mathbb{R}$ such that

$$x^\phi(\tau^* + \hat{u}) < p(\hat{u}).$$

By continuity, there is $\epsilon \in (0, 1)$ so that

$$x^\phi(\tau + \hat{u}) < p(\hat{u}) \qquad \text{for all } \tau \in \mathbb{R} \quad \text{with } |\tau - \tau^*| < \epsilon.$$

Choose $n \in -\mathbb{N}$ such that $|\tau_n - \tau^*| < \epsilon$ and $t_n = m_n\omega + \tau_n < \hat{u}$. From $x_{t_n}^\phi > p_0$ and the monotonicity of the semiflow, we obtain

$$x^\phi(t_n + \theta) \geq p(\theta) \qquad \text{for all } \theta \geq -1.$$

Taking $\theta = \hat{u} - m_n\omega$, we get

$$x^\phi(\tau_n + \hat{u}) = x^\phi(t_n + \hat{u} - m_n\omega) \geq p(\hat{u} - m_n\omega) = p(\hat{u}),$$

which is a contradiction since $|\tau_n - \tau^*| < \epsilon$. Therefore, (11.3) holds.

We also notice that

$$x^{\phi}_{\tau^*+m_n\omega} \to p_0 \quad \text{as} \quad n \to -\infty,$$

since $x^{\phi}_{\tau^*+m_n\omega} = F_W(\tau^* - \tau_n, F_W(m_n\omega + \tau_n, \phi))$ and $\tau_n \to \tau^*$, $F_W(m_n\omega + \tau_n, \phi) = x^{\phi}_{t_n} \to p_0$ as $n \to -\infty$. Then the result of part 1 with $s = \tau^* - \omega$ implies that $V(x^{\phi}_{\tau^*+n\omega} - p_0) = 2$ for all sufficiently large negative integers n. This contradicts (11.3) and proves the existence of T_0.

3. We show that $V(x^{\phi}_t - \psi) = 2$ for all $\psi \in \mathcal{O}$ and for all $t \leq T_0 - \omega$.

Let $u \in [-\omega, 0]$. Then, from part 2 and the monotonicity of V it follows that

$$2 = V(x^{\phi}_t - p_0) \leq V(x^{\phi}_{t+u} - p_u) \qquad \text{for all } t \leq T_0.$$

On the other hand, $V(x^{\phi}_{t+u} - p_u) \leq 2$ since x^{ϕ}_{t+u} and p_u are in $\overline{W}$. This proves the assertion.

4. We claim that there exists $T_1 \leq T_0 - \omega$ such that $V(x^{\phi}_t - \psi) = 2$ for all $\psi \in W \cap S$ and for all $t \leq T_1$.

Let $\hat{\phi} = x^{\phi}_{T_0-\omega}$. The map $c : [0, \omega] \ni t \mapsto p_t - \hat{\phi} \in C^1$ is continuous and $c([0, \omega])$ is a compact subset of C^1. By part 3 and Lemma VI.2(iii), $c([0, \omega]) \subset R \subset C^1$. Using Lemma VI.1(ii) and compactness, we find $\epsilon > 0$ so that $V(\kappa) = 2$ for every κ in

$$N_\epsilon = \left\{ \hat{\kappa} \in C^1 : \inf_{0 \leq s \leq \omega} ||\hat{\kappa} - c(s)||_{C^1} < \epsilon \right\}.$$

If the claim is not true, then there exist a sequence $(t_n)^0_{-\infty}$ in $(-\infty, 0)$ with $t_n \to -\infty$ as $n \to -\infty$, and a sequence $(\psi_n)^0_{-\infty}$ in $W \cap S$ such that $V(x^{\hat{\phi}}_{t_n} - \psi_n) = 0$ for all $n \in -\mathbb{N}$. There exist a strictly increasing sequence $(n_j)^0_{-\infty}$ and $\xi \in \mathcal{O}$, $\zeta \in \overline{W \cap S}$ so that

$$x^{\hat{\phi}}_{t_{n_j}} \to \xi \text{ and } \psi_{n_j} \to \zeta \qquad \text{as } j \to -\infty,$$

since $\alpha(\phi) = \alpha(\hat{\phi}) = \mathcal{O}$ by Proposition 7.9 and $\overline{W \cap S}$ is compact. If $\xi \neq \zeta$, then Proposition 5.7 implies $V(\xi - \zeta) = 2$. By Lemma VI.1(i), we have

$$2 = V(\xi - \zeta) \leq \liminf_{j \to -\infty} V(x^{\hat{\phi}}_{t_{n_j}} - \psi_{n_j}) = 0,$$

a contradiction. Therefore, $\xi = \zeta$.

The set $N_\epsilon + \hat{\phi}$ is a C^1-neighbourhood of $\mathcal{O}$. Proposition 4.3 implies that $\mathcal{O}$ has a neighbourhood U_1 in C such that $U_1 \cap \overline{W} \subset N_\epsilon + \hat{\phi}$. By Proposition 7.2 and Corollary 5.8, for the first positive zero z_1 of p, we have $p(z_1 - 1) < 0 < p(-1)$. Hence Proposition 7.3(iv) implies $\{(0, v)^{tr} \in \mathbb{R}^2 : 0 \leq v \leq p(-1)\} \subset \Pi_2(\overline{W \cap S})$. By the continuous dependence of solutions on initial data and the continuity of Π_2^{-1}, we can choose $v^* \in (0, p(-1))$ such that for every χ in the set

$$A = \{(0, v)^{tr} \in \mathbb{R}^2 : v^* < v \leq p(-1)\},$$

$F(t, \Pi_2^{-1}(\chi)) \in U_1$ for $0 \leq t \leq 2$. Let $\chi \in A \backslash \{(0, p(-1))^{tr}\}$. If $\rho(\chi) = \chi$ or $\rho(\chi) \prec \chi$ then Proposition 7.1 can be used to show that the ω-limit set $\omega\left(\Pi_2^{-1}(\chi)\right)$ of $\Pi_2^{-1}(\chi)$

is different from $\mathcal{O}$. As $\Pi_2^{-1}(\chi) \in W \cap S$, this is a contradiction. Therefore, $\chi \prec \rho(\chi)$. $\rho(\chi) \prec (0, p(-1))^{tr}$ also holds because of the monotonicity of ρ and $\rho((0,p(-1))^{tr}) = (0,p(-1))^{tr}$. Thus, also applying Corollary 5.8 and the definition of ρ, for every $\chi \in A$ there is $u \in (1,2)$ such that $\rho(\chi) = \Pi_2 F(u, \Pi_2^{-1}(\chi)) \in A$. Consequently,

$$F(t, \Pi_2^{-1}(\chi)) \in U_1 \cap \overline{W} \subset N_\epsilon + \hat{\phi} \qquad \text{for all } \chi \in A \text{ and } t \geq 0. \tag{11.4}$$

Set

$$B = \{(0,v)^{tr} \in \mathbb{R}^2 : 0 \leq v \leq v^*\}.$$

We have $B \cap \Pi_2(\mathcal{O}) = \emptyset$ by Proposition 7.3. Then we can find a neighbourhood U_2 of $\mathcal{O}$ in C such that $U_2 \subset U_1$ and

$$\Pi_2^{-1}(B) \cap U_2 = \emptyset.$$

By the continuous dependence of solutions on initial data, there is a neighbourhood U_3 of $\zeta \in \mathcal{O}$ in C such that

$$x_t^\kappa \in U_2 \cap \overline{W} \subset N_\epsilon + \hat{\phi} \qquad \text{for all } \kappa \in U_3 \cap \overline{W} \text{ and for all } t \in [0,2]. \tag{11.5}$$

Let $\kappa \in U_3 \cap W \cap S$. The choice of U_3 implies $\Pi_2 x_t^\kappa \notin B$ for $0 \leq t \leq 2$. Corollary 5.8 and the invariance of $W \cap S$ imply that $\Pi_2 x_u^\kappa \in A$ for some $u \in [0,2)$. Then from (11.4) and (11.5) we conclude

$$x_t^\kappa \in N_\epsilon + \hat{\phi} \qquad \text{for all } \kappa \in U_3 \cap W \cap S \text{ and for all } t \geq 0.$$

Consequently, there exists $n_0 \in -\mathbb{N}$ such that

$$x_t^{\psi_n} \in N_\epsilon + \hat{\phi} \qquad \text{for all } t \geq 0 \text{ and for all integers } n \leq n_0.$$

Then $x_t^{\psi_n} - \hat{\phi} \in N_\epsilon$ for all $t \geq 0$ and for all integers $n \leq n_0$. Thus, by the monotonicity of V,

$$0 = V(x_{t_n}^{\hat{\phi}} - \psi_n) \geq V(\hat{\phi} - x_{-t_n}^{\psi_n}) = 2 \qquad \text{for all integers } n \leq n_0,$$

a contradiction. This shows the existence of T_1.

5. It follows from parts 3 and 4, from the invariance of $\overline{W \cap S}$ and from Lemma VI.2(iii) that

$$\Pi_2 x_t^\phi \neq \Pi_2 \psi \qquad \text{for all } t \leq T_1 \text{ and for all } \psi \in \overline{W \cap S}.$$

Consequently,

$$\Pi_2 x_t^\phi \in \operatorname{ext}(\Pi_2 \circ \eta) \qquad \text{for all } t \leq T_1,$$

since $\Pi_2 \overline{W \cap S} = |\Pi_2 \circ \eta| \cup \operatorname{int}(\Pi \circ \eta)$ by Proposition 7.3.

We prove that we can choose a sequence $(\chi_n)_0^\infty$ in W such that $\chi_n \to x_{T_1}^\phi$ as $n \to \infty$, and $V(x_t^{\chi_n} - \psi) = 2$ for all $t \leq 0$, for all $\psi \in \mathcal{O}$ and for all $n \in \mathbb{N}$.

The map $d : [0,\omega] \ni t \mapsto p_t - x_{T_1}^\phi \in C^1$ is continuous and $d([0,\omega])$ is a compact subset of C^1. By part 3 and Lemma VI.2(iii), $d([0,\omega]) \subset R \subset C^1$. Lemma VI.1(ii) and the compactness of $d([0,\omega])$ ensure the existence of $\delta > 0$ such that $V(\chi) = 2$ for every χ in

$$M_\delta = \left\{ \hat{\chi} \in C^1 : \inf_{0 \le s \le \omega} ||\hat{\chi} - d(s)||_{C^1} < \delta \right\}.$$

Then it follows that for every $\chi \in C^1$ with $||\chi - x^{\phi}_{T_1}||_{C^1} < \delta$ and for all $\psi \in \mathcal{O}$, $V(\chi - \psi) = 2$. Since the map $\overline{W} \ni \phi \mapsto \dot{\phi} \in C$ is continuous and $x^{\phi}_{T_1} \in \mathrm{bd}\, W$, there is a sequence $(\chi_n)_0^\infty$ in W with

$$||\chi_n - x^{\phi}_{T_1}||_{C^1} < \delta \quad \text{for all } n \in \mathbb{N}, \text{ and } \chi_n \to x^{\phi}_{T_1} \quad \text{as } n \to \infty.$$

Then

$$V(\chi_n - \psi) = 2 \qquad \text{for all } \psi \in \mathcal{O} \text{ and for all } n \in \mathbb{N}.$$

Using the monotonicity of V and Proposition 5.2, it also follows that

$$V(x^{\chi_n}_t - \psi) = 2 \qquad \text{for all } t \le 0, \text{ for all } \psi \in \mathcal{O} \text{ and for all } n \in \mathbb{N}.$$

6. The results in part 5 and Lemma VI.2(iii) combined show that for any $n \in \mathbb{N}$, the curve $\Pi_2 \circ \eta$ and the map $(-\infty, 0] \ni t \mapsto \Pi_2 x^{\chi_n}_t \in \mathbb{R}^2$ have disjoint sets of values.

If $n \in \mathbb{N}$ is large enough then

$$\Pi_2 \chi_n \in \mathrm{ext}(\Pi_2 \circ \eta),$$

since $\Pi_2 \chi_n \to \Pi_2 x^{\phi}_{T_1} \in \mathrm{ext}(\Pi_2 \circ \eta)$ as $n \to \infty$. On the other hand, $x^{\chi_n}_t \to 0$ as $t \to -\infty$, and thus

$$\Pi_2 x^{\chi_n}_t \to 0 \in \mathrm{int}(\Pi_2 \circ \eta) \qquad \text{as } t \to -\infty.$$

This yields a contradiction.

Consequently, $x^{\phi}_t \notin W^u(p_0, F(\omega, \cdot), N^u)$ for all $t \le 0$ is impossible, and hence, $\phi \in W^u(\mathcal{O})$ follows. □

CHAPTER 12

bd W contains the unstable set of the periodic orbit $\mathcal{O}$

In this chapter after a series of technical results we conclude that the unstable set $W^u(\mathcal{O})$ of the periodic orbit $\mathcal{O}$ is contained in bd W. This fact and the result in Chapter 11 imply that $W^u(\mathcal{O}) = \text{bd}\, W$ if $\mu = 0$; and $W^u(\mathcal{O}) = \text{bd}\, W \setminus \{\xi_-, \xi_+\}$ if $\mu > 0$.

Recall that p is a periodic solution of (1.1) with the minimal period $\omega \in (1, 2)$, $\mathcal{O} = \{p_t : 0 \le t \le \omega\}$, $H = \{\phi \in C : \phi(0) = 0\}$, $H^+ = \{\phi \in H : \phi(-1) > 0\}$, and $p_0 \in H^+$. For $l : C \in \phi \mapsto \phi(0) \in \mathbb{R}$, $H = l^{-1}(0)$.

By the continuity of the map $C \ni \phi \mapsto -\mu\phi(0) + f(\phi(-1)) \in \mathbb{R}$ and by $\dot{p}(0) = -\mu p(0) + f(p(-1)) > 0$, there exists $\hat{\delta} > 0$ such that

$$-\mu\phi(0) + f(\phi(-1)) > 0 \qquad \text{for all } \phi \in p_0 + C_{\hat{\delta}}.$$

Select $\tilde{\epsilon} > 0$ so that $p_t \in p_0 + C_{\hat{\delta}}$ for all $t \in [0, \tilde{\epsilon}]$. By the continuous dependence of solutions of equation (1.1) on initial data there exists $\tilde{\delta} > 0$ so that

$$x_t^\phi \in p_0 + C_{\hat{\delta}} \qquad \text{for all } \phi \in p_0 + C_{\tilde{\delta}} \text{ and for all } t \in [0, \tilde{\epsilon}].$$

It follows that

$$\dot{x}^\phi(t) > 0 \qquad \text{for all } \phi \in p_0 + C_{\tilde{\delta}} \text{ and for all } t \in (0, \tilde{\epsilon}].$$

Note that by $l(\dot{p}_0) = f(p(-1)) > 0$ we have $\dot{p}_0 \notin H$, and hence there exist a convex bounded open neighborhood N_{p_0} of p_0 in $p_0 + C_{\tilde{\delta}}$, $\epsilon_H \in (0, \min\{\omega - 1, \tilde{\epsilon}/2\})$, and a C^1-map $\tau_H : N_{p_0} \to (\omega - \epsilon_H, \omega + \epsilon_H)$ such that $\tau_H(p_0) = \omega$; and for every $(t, \phi) \in (\omega - \epsilon_H, \omega + \epsilon_H) \times N_{p_0}$, $F(t, \phi) \in H$ if and only if $t = \tau_H(\phi)$; and $D_1F(\tau_H(\phi), \phi)1 \notin H$ for all $\phi \in N_{p_0}$. Let P_{τ_H} denote the C^1-map $N_{p_0} \to H$ given by $P_{\tau_H}(\phi) = F(\tau_H(\phi), \phi)$. The Poincaré map $P_H : N_{p_0} \cap H \ni \phi \mapsto F(\tau_H(\phi), \phi) \in H$ is C^1-smooth, and p_0 is a fixed point of P_H. See Appendix I for more details.

Let σ^{P_H} denote the spectrum of the derivative $DP_H(p_0) : H \to H$. By Lemma I.5, $\sigma \setminus \{0, 1\} = \sigma^{P_H} \setminus \{0, 1\}$. In particular, $\{\zeta \in \sigma : |\zeta| < 1\} = \{\zeta \in \sigma^{P_H} : |\zeta| < 1\}$ and $\{\lambda_u\} = \{\zeta \in \sigma : 1 < |\zeta|\} = \{\zeta \in \sigma^{P_H} : 1 < |\zeta|\}$.

If $r > 0$ is given and if there exists $\zeta \in \sigma^{P_H}$ with $r < |\zeta|$, then $H_{\le r}$ and $H_{r<}$ denote the realified generalized eigenspaces of $DP_H(p_0)$ associated with the nonempty disjoint compact spectral sets $\{\zeta \in \sigma^{P_H} : |\zeta| \le r\}$ and $\{\zeta \in \sigma^{P_H} : r < |\zeta|\}$, respectively. For such an r, $H = H_{\le r} \oplus H_{r<}$.

Recall the choice of $\lambda \in (0, 1)$ with

$$\lambda > \max\left\{\frac{1}{\lambda_u}, \sup_{\zeta \in \sigma, |\zeta| < 1} |\zeta|\right\}$$

in Chapter 11. We apply Theorem I.3 to P_H and its fixed point p_0 and obtain convex open neighbourhoods $N_{H,\leq 1}$ of 0 in $H_{\leq 1}$, $N_{H,1<}$ of 0 in $H_{1<}$ and a C^1-map $w_H^u : N_{H,1<} \to H_{\leq 1}$ with $w_H^u(0) = 0$, $Dw_H^u(0) = 0$, $w_H^u(N_{H,1<}) \subset N_{H,\leq 1}$, so that for the neighbourhood $N_H^u = N_{H,1<} + N_{H,\leq 1}$ of 0 in H we have that the shifted graph

$$W^u(p_0, P_H, N_H^u) = \{p_0 + \chi + w_H^u(\chi) : \chi \in N_{H,1<}\}$$

coincides with the set

$$\{\phi \in N_H^u : \text{There is a trajectory } (\phi_n)_{-\infty}^0 \text{ of } P_H \text{ with } \phi = \phi_0,$$
$$\lambda^n(\phi_n - p_0) \in N_H^u \text{ for all } n \in -\mathbb{N} \text{ and } \lambda^n(\phi_n - p_0) \to 0 \text{ as } n \to -\infty\}.$$

Furthermore, there exist an open neighbourhood $\widetilde{N_H^u}$ of 0 in N_H^u, a constant $\alpha_{H,u} \in (0, \lambda)$, and a norm $||\cdot||_{H,u}$ on H, equivalent to $\|\cdot\|\|_H$ with

$$||\phi||_{H,u} = \max\{||\operatorname{Pr}_{H_{1<}} \phi||_{H,u}, ||\operatorname{Pr}_{H_{\leq 1}} \phi||_{H,u}\} \qquad \text{for all } \phi \in H$$

so that the restriction of P_H to $W^u(p_0, P_H, N_H^u) \cap (p_0 + \widetilde{N_H^u})$ defines a C^1-smooth diffeomorphism $(P_H)_u$ onto $W^u(p_0, P_H, N_H^u)$ whose inverse satisfies

$$||(P_H)_u^{-1}(\phi) - (P_H)_u^{-1}(\psi)||_{H,u} \leq \alpha_{H,u}||\phi - \psi||_{H,u} \qquad \text{for all } \phi, \psi \text{ in } W^u(p_0, P_H, N_H^u).$$

Recall $C_{1<} = \mathbb{R}\psi_u$, $\psi_u \in \overset{\circ}{K}$, $||\psi_u|| = 1$. The projection $\operatorname{Pr}_{H,\dot{p}_0} : C \to C$ onto H along $\dot{p}_0$ is given by

$$\operatorname{Pr}_{H,\dot{p}_0} \phi = \phi - \frac{l(\phi)}{l(\dot{p}_0)}\dot{p}_0.$$

Proposition I.5 says that the vector

$$\operatorname{Pr}_{H,\dot{p}_0} \psi_u$$

spans the realified generalized eigenspace $H_{1<}$ associated with $\sigma^{P_H} \setminus \overline{\mathbb{C}_1} = \{\lambda_u\}$. Let

$$\psi_H = \frac{1}{||\operatorname{Pr}_{H,\dot{p}_0} \psi_u||_{H,u}} \operatorname{Pr}_{H,\dot{p}_0} \psi_u.$$

Then

$$H_{1<} = \mathbb{R}\psi_H \quad \text{and} \quad ||\psi_H||_{H,u} = 1.$$

The formula for the derivatives of the maps P_{τ_H} and P_H in Appendix I yields

$$DP_{\tau_H}(p_0)\psi_u = \operatorname{Pr}_{H,\dot{p}_0} D_2F(\omega, p_0)\psi_u = \operatorname{Pr}_{H,\dot{p}_0} \lambda_u\psi_u = \lambda_u \operatorname{Pr}_{H,\dot{p}_0} \psi_u.$$

Since $H_{1<} = \mathbb{R}\psi_H$ and $N_{H,1<}$ is a convex open neighbourhood of 0 in $H_{1<}$, it follows that

$$N_{H,1<} = \{s\psi_H : -\beta_1 < s < \beta_2\}$$

for some constants $\beta_1 > 0$ and $\beta_2 > 0$. Choose $\beta_0 \in (0, \min\{\beta_1, \beta_2\})$ so that $||Dw_H^u(s\psi_H)\psi_H||_{H,u} < 1$ for $|s| \leq \beta_0$. For $0 < \beta < \beta_0$, define

$$W_\beta^u(p_0, P_H) = \{p_0 + s\psi_H + w_H^u(s\psi_H) : |s| < \beta\}.$$

Clearly, $W^u_\beta(p_0, P_H) \subset W^u(p_0, P_H, N^u_H)$. If $\phi \in W^u_\beta(p_0, P_H)$, that is, $\phi = p_0 + s\psi_H + w^u_H(s\psi_H)$ with $|s| < \beta$, then $(P_H)^{-1}_u(\phi) \in W^u(p_0, P_H, N^u_H)$ and

$$||(P_H)^{-1}_u(\phi) - p_0||_{H,u} \le \alpha_{H,u}||\phi - p_0||_{H,u}.$$

Let $(P_H)^{-1}_u(\phi) = p_0 + \hat{s}\psi_H + w^u_H(\hat{s}\psi_H)$ for some $\hat{s} \in (-\beta_1, \beta_2)$. Then, by the choice of β_0 and the definition of the equivalent norm $||\cdot||_{H,u}$, we have

$$\begin{aligned}|\hat{s}| &= ||\hat{s}\psi_H||_{H,u} \le ||p_0 + \hat{s}\psi_H + w^u_H(\hat{s}\psi_H) - p_0||_{H,u}\\ &\le \alpha_{H,u}||p_0 + s\psi_H + w^u_H(s\psi_H) - p_0||_{H,u}\\ &= \alpha_{H,u}\max\{|s|, ||w^u_H(s\psi_H)||_{H,u}\} = \alpha_{H,u}|s| < \alpha_{H,u}\beta.\end{aligned}$$

As $\alpha_{H,u} \in (0,\lambda) \subset (0,1)$, it follows that $(P_H)^{-1}_u(\phi) \in W^u_\beta(p_0, P_H)$ and $||(P_H)^{-1}_u(\phi) - p_0||_{H,u} < \alpha_{H,u}\beta$.

Recall that in Chapter 11 we applied Theorem I.3 to the period-ω map $F(\omega,\cdot)$ at its fixed point p_0. In addition to the properties of $N_{1<}$, $N_{\le 1}$, N^u, w^u stated in Chapter 11 there exist an open neighbourhood $\widetilde{N^u}$ of 0 in N^u, a constant $\alpha_u \in (0,\lambda)$ and a norm $||\cdot||_u : C \to \mathbb{R}$, equivalent to $||\cdot||$, with $||\phi||_u = \max\{||\operatorname{Pr}_{C_{1<}}\phi||_u, ||\operatorname{Pr}_{C_{\le 1}}\phi||_u\}$ for all $\phi \in C$, so that the restriction of $F(\omega,\cdot)$ to $W^u(p_0, F(\omega,\cdot), N^u) \cap (p_0 + \widetilde{N^u})$ defines a C^1-diffeomorphism $F(\omega,\cdot)_u$ onto $W^u(p_0, F(\omega,\cdot), N^u)$ whose inverse satisfies

$$||(F(\omega,\cdot)_u)^{-1}(\phi) - (F(\omega,\cdot)_u)^{-1}(\psi)||_u \le \alpha_u||\phi - \psi||_u \text{ for } \phi, \psi \text{ in } W^u(p_0, F(\omega,\cdot), N^u).$$

Choose $\delta_0 > 0$ such that

$$\{s\psi_u : |s| < \delta_0\} \subset N_{1<}$$

and $||Dw^u(s\psi_u)\psi_u||_u < 1$ for all $|s| \le \delta_0$. Then, in the same way as for $W^u_\beta(p_0, P_H)$, $0 < \beta < \beta_0$, it can be shown that for every ϕ in

$$W^u_\delta(p_0, F(\omega,\cdot)) = \{p_0 + s\psi_u + w^u(s\psi_u) : |s| < \delta\}, \qquad 0 < \delta < \delta_0,$$

we have $(F(\omega,\cdot)_u)^{-1}(\phi) \in W^u_\delta(p_0, F(\omega,\cdot))$ and $||(F(\omega,\cdot)_u)^{-1}(\phi) - p_0||_u < \alpha_u\delta$.

Define the unstable set of the fixed point p_0 of the period-ω map $F(\omega,\cdot)$ by

$$W^u(p_0, F(\omega,\cdot)) = \bigcup_{n\in\mathbb{N}} F(n\omega, W^u(p_0, F(\omega,\cdot), N^u)).$$

Observe that, for $0 < \delta < \delta_0$, we have

$$W^u(p_0, F(\omega,\cdot)) = \bigcup_{n\in\mathbb{N}} F(n\omega, W^u_\delta(p_0, F(\omega,\cdot))).$$

Recall that the unstable set of the periodic orbit $\mathcal{O}$ is defined by

$$W^u(\mathcal{O}) = \bigcup_{t\ge 0} F(t, W^u(p_0, F(\omega,\cdot), N^u)).$$

Proposition 12.1

(i) *There exist* $\delta_1 \in (0, \delta_0)$, δ_2 *and* δ_3 *in* $(0, \delta_1)$, *and a continuous strictly increasing function* $g : [-\delta_2, \delta_3] \to [-\delta_1, \delta_1]$ *such that* $g(-\delta_2) = -\delta_1$, $g(0) = 0$, $g(\delta_3) = \delta_1$, $|g(s)| > |s|$ *for all* $s \in [-\delta_2, \delta_3] \setminus \{0\}$, *and*

$$F(\omega, p_0 + s\psi_u + w^u(s\psi_u)) = p_0 + g(s)\psi_u + w^u(g(s)\psi_u) \quad \text{forall } s \in [-\delta_2, \delta_3]. \tag{12.1}$$

(ii) *If* $-\delta_1 \le s_1 < s_2 \le \delta_1$, *then*

$$p_0 + s_1\psi_u + w^u(s_1\psi_u) \ll p_0 + s_2\psi_u + w^u(s_2\psi_u).$$

Proof We first claim that there exists $\epsilon > 0$ with

$$F(\omega, p_0 + s\psi_u + w^u(s\psi_u)) \gg p_0 + s\psi_u + w^u(s\psi_u) \qquad \text{for } 0 < s < \epsilon,$$
$$F(\omega, p_0 + s\psi_u + w^u(s\psi_u)) \ll p_0 + s\psi_u + w^u(s\psi_u) \qquad \text{for } -\epsilon < s < 0.$$

In fact, for every $s \in (-\delta_0, \delta_0)$ we have

$$\begin{aligned} &F(\omega, p_0 + s\psi_u + w^u(s\psi_u)) - (p_0 + s\psi_u + w^u(s\psi_u)) \\ &= F(\omega, p_0 + s\psi_u + w^u(s\psi_u)) - F(\omega, p_0) - (s\psi_u + w^u(s\psi_u)) \\ &= \int_0^1 D_2F(\omega, p_0 + \theta s\psi_u + \theta w^u(s\psi_u))(s\psi_u + w^u(s\psi_u))\, d\theta - (s\psi_u + w^u(s\psi_u)). \end{aligned}$$

Hence

$$\begin{aligned} &\frac{1}{s}[F(\omega, p_0 + s\psi_u + w^u(s\psi_u)) - (p_0 + s\psi_u + w^u(s\psi_u))] \\ &\qquad \to D_2F(\omega, p_0)\psi_u - \psi_u = (\lambda_u - 1)\psi_u \gg 0 \qquad \text{as } s \to 0, \end{aligned}$$

from which the claim follows.

Fix $\delta_1 \in (0, \min\{\epsilon, \delta_0\})$. The choice of δ_0 shows that for each $s \in [-\delta_1, \delta_1]$ there exists $\hat{g}(s) \in [-\alpha_u\delta_1, \alpha_u\delta_1]$ such that

$$(F(\omega, \cdot)_u)^{-1}(p_0 + s\psi_u + w^u(s\psi_u)) = p_0 + \hat{g}(s)\psi_u + w^u(\hat{g}(s)\psi_u).$$

This defines a map $\hat{g} : [-\delta_1, \delta_1] \to [-\alpha_u\delta_1, \alpha_u\delta_1]$. Since $(F(\omega, \cdot)_u)^{-1}$ is a C^1-diffeomorphism of $W^u(p_0, F(\omega, \cdot), N^u)$ onto its image $W^u(p_0, F(\omega, \cdot), N^u) \cap (p_0 + \widetilde{N^u})$, it can be easily seen that $\hat{g}$ is continuous and injective. Clearly, $\hat{g}(0) = 0$. If $-\delta_1 \le s_1 < s_2 \le \delta_1$, then from $||Dw^u(s\psi_u)\psi_u||_u < 1$ for $|s| \le \delta_0$ and from $\psi_u \in \overset{\circ}{K}$ it follows that

$$p_0 + s_1\psi_u + w^u(s_1\psi_u) \ll p_0 + s_2\psi_u + w^u(s_2\psi_u).$$

Since $\hat{g}(s_1), \hat{g}(s_2)$ belong to $[-\delta_1, \delta_1]$, it also follows that in case $\hat{g}(s_1) > \hat{g}(s_2)$ we have

$$p_0 + \hat{g}(s_1)\psi_u + w^u(\hat{g}(s_1)\psi_u) \gg p_0 + \hat{g}(s_2)\psi_u + w^u(\hat{g}(s_2)\psi_u),$$

while in case $\hat{g}(s_1) < \hat{g}(s_2)$ we have

$$p_0 + \hat{g}(s_1)\psi_u + w^u(\hat{g}(s_1)\psi_u) \ll p_0 + \hat{g}(s_2)\psi_u + w^u(\hat{g}(s_2)\psi_u).$$

Therefore, using Proposition 2.2, we obtain that for $-\delta_1 \le s_1 < s_2 \le \delta_1$ we have $\hat{g}(s_1) < \hat{g}(s_2)$. From $\delta_1 < \epsilon$ and the claim above, it can be obtained that $|\hat{g}(s)| < |s|$ for $0 < |s| \le \delta_1$. Defining $\delta_2 = -\hat{g}(-\delta_1)$, $\delta_3 = \hat{g}(\delta_1)$ and $g = \hat{g}^{-1}$, we get the desired function. □

Proposition 12.2 *There exists a homeomorphism $j : \mathbb{R} \to W^u(p_0, F(\omega, \cdot))$ with $j(0) = p_0$ and*

$$j(s) = p_0 + s\psi_u + w^u(s\psi_u) \quad \text{for all } s \in (-\delta_2, \delta_3),$$

which is order-preserving in the sense that $j(s_1) \ll j(s_2)$ for $s_1 < s_2$.

Proof Extend g to a continuous function $G : \mathbb{R} \to \mathbb{R}$ such that G is strictly increasing and $|G(s)| > |s|$ for all $s \neq 0$. Then for each $s \in \mathbb{R}$ there is a minimal nonnegative integer $k = k(s)$ such that $G^{-k}(s) \in (-\delta_2, \delta_3)$. Consider the map

$$j : \mathbb{R} \ni s \mapsto F(k(s)\omega, p_0 + G^{-k(s)}(s)\psi_u + w^u(G^{-k(s)}(s)\psi_u)) \in W^u(p_0, F(\omega, \cdot)).$$

Clearly, $j(0) = p_0$. Let $0 < s_1 < s_2 < \infty$. Then $k(s_2) \geq k(s_1)$ and $0 < G^{-k(s_2)}(s_1) < G^{-k(s_2)}(s_2)$. Therefore, Proposition 12.1(ii) and Proposition 2.2 combined imply

$$0 \ll F(k(s_2)\omega, p_0 + G^{-k(s_2)}(s_1)\psi_u + w^u(G^{-k(s_2)}(s_1)\psi_u)) \ll j(s_2).$$

On the other hand, (12.1) gives

$$j(s_1) = F(k(s_2)\omega, p_0 + G^{-k(s_2)}(s_1)\psi_u + w^u(G^{-k(s_2)}(s_1)\psi_u)).$$

Thus, $0 \ll j(s_1) \ll j(s_2)$. Similarly, $-\infty < s_1 < s_2 < 0$ implies $j(s_1) \ll j(s_2) \ll 0$. Consequently, j is order-preserving and injective. Since every $\phi \in W^u(p_0, F(\omega, \cdot))$ can be written as $\phi = F(k\omega, p_0 + s\psi_u + w^u(s\psi_u))$ for some $k \in \mathbb{N}$ and $s \in (-\delta_2, \delta_3)$, it also follows easily that $j(\mathbb{R}) = W^u(p_0, F(\omega, \cdot))$.

To prove the continuity of j, let $s \in \mathbb{R}$ and a sequence $(s^n)_0^\infty$ in $\mathbb{R}$ with $s^n \to s$ as $n \to \infty$ be given. If $k(s) = 0$ or if $k(s) \geq 1$ and $G^{1-k(s)}(s) \notin \{-\delta_2, \delta_3\}$, then $k(s^n) = k(s)$ for all sufficiently large $n \in \mathbb{N}$. Then the continuity of the functions $G^{-k(s)}$, w^u and F implies $j(s^n) \to j(s)$ as $n \to \infty$. Assume $k(s) \geq 1$ and $G^{1-k(s)}(s) = \delta_3$. If $s^n \geq s$ and $n \in \mathbb{N}$ is sufficiently large, then $k(s^n) = k(s)$ and $j(s^n) = F(k(s)\omega, p_0 + G^{-k(s)}(s^n)\psi_u + w^u(G^{-k(s)}(s^n)\psi_u))$. If $s^n < s$ and $n \in \mathbb{N}$ is sufficiently large, then $k(s^n) = k(s) - 1$ and $G^{-k(s)}(s^n) \in (-\delta_2, \delta_3)$. Then (12.1) can be applied and yields

$$\begin{aligned} &F(\omega, p_0 + G^{-k(s)}(s^n)\psi_u + w^u(G^{-k(s)}(s^n)\psi_u)) \\ &= p_0 + G^{1-k(s)}(s^n)\psi_u + w^u(G^{1-k(s)}(s^n)\psi_u)). \end{aligned}$$

Hence it follows in this case that

$$\begin{aligned} j(s^n) &= F((k(s) - 1)\omega, p_0 + G^{1-k(s)}(s^n)\psi_u + w^u(G^{1-k(s)}(s^n)\psi_u)) \\ &= F((k(s) - 1)\omega, F(\omega, p_0 + G^{-k(s)}(s^n)\psi_u + w^u(G^{-k(s)}(s^n)\psi_u))) \\ &= F(k(s)\omega, p_0 + G^{-k(s)}(s^n)\psi_u + w^u(G^{-k(s)}(s^n)\psi_u)). \end{aligned}$$

Thus, the continuity of the functions $G^{-k(s)}$, w^u and F implies $j(s^n) \to j(s)$ as $n \to \infty$. The case $k(s) \geq 1$ and $G^{1-k(s)}(s) = -\delta_2$ is analogous. Therefore, j is continuous.

Let ψ and a sequence $(\psi^n)_0^\infty$ in $W^u(p_0, F(\omega, \cdot))$ be given with $\psi^n \to \psi$ as $n \to \infty$. Set $s = j^{-1}(\psi)$, and $s^n = j^{-1}(\psi^n)$ for all $n \in \mathbb{N}$. If $(s^n)_0^\infty$ does not converge to s as $n \to \infty$, then there exist $\epsilon > 0$ and a subsequence $(s^{n_i})_{i=0}^\infty$ of $(s^n)_0^\infty$ such that $|s^{n_i} - s| > \epsilon$ for all $i \in \mathbb{N}$. The order-preserving property of j implies that, for every $i \in \mathbb{N}$, either $j(s^{n_i}) \ll j(s - \epsilon) \ll j(s)$ or $j(s) \ll j(s + \epsilon) \ll j(s^{n_i})$ holds. This contradicts $\psi^n \to \psi$ as $n \to \infty$. Therefore, j^{-1} is also continuous, and thus j is a homeomorphism. $\square$

Proposition 12.3 *For every $\beta \in (0, \beta_0)$ there exist $\delta = \delta(\beta) \in (0, \delta_0)$ and $\epsilon = \epsilon(\beta) > 0$ such that*

$$P_{\tau_H}\left((p_0 + C_\delta) \cap W^u(p_0, F(\omega, \cdot), N^u)\right) \subset W^u_\beta(p_0, P_H)$$

and

$$H \cap \operatorname{bd} W \cap (p_0 + C_\epsilon) \subset W^u_\beta(p_0, P_H).$$

Proof Applying Proposition V.2 we find $\delta^* > 0$ such that

$$P_{\tau_H}\left((p_0 + C_{\delta^*}) \cap W^u(p_0, F(\omega, \cdot), N^u)\right) \subset W^u(p_0, P_H, N^u_H).$$

So, for each $\phi \in (p_0 + C_{\delta^*}) \cap W^u(p_0, F(\omega, \cdot), N^u)$, there exists a unique $s = s(\phi) \in \mathbb{R}$ such that $s(\phi)\psi_H \in H_{H,1<}$ and

$$P_{\tau_H}(\phi) = p_0 + s\psi_H + w^u_H(s\psi_H).$$

Observe $s(p_0) = 0$. By the continuity of P_{τ_H} at p_0 and by $P_{\tau_H}(p_0) = p_0$, the function

$$(p_0 + C_{\delta^*}) \cap W^u(p_0, F(\omega, \cdot), N^u) \ni \phi \mapsto s(\phi) \in \mathbb{R}$$

is continuous at $s = 0$. Thus, for every $\beta \in (0, \beta_0)$ we can find $\delta = \delta(\beta) \in (0, \delta^*)$ such that for all $\phi \in (p_0 + C_\delta) \cap W^u(p_0, F(\omega, \cdot), N^u)$,

$$|s(\phi)| = ||s(\phi)\psi_H||_{H,u} < \beta.$$

Therefore, for every $\beta \in (0, \beta_0)$, there exists $\delta = \delta(\beta) \in (0, \delta^*)$ such that

$$P_{\tau_H}\left((p_0 + C_\delta) \cap W^u(p_0, F(\omega, \cdot), N^u)\right) \subset W^u_\beta(p_0, P_H).$$

Thus, using Proposition 11.1 and the fact that $H \cap \{\xi_-, \xi_+\} = \emptyset$ in case $\mu > 0$, we see that it suffices to find, for every $\delta \in (0, \delta^*)$, a real $\epsilon = \epsilon(\delta) > 0$ such that

$$H \cap W^u(\mathcal{O}) \cap (p_0 + C_\epsilon) \subset P_{\tau_H}\left((p_0 + C_\delta) \cap W^u(p_0, F(\omega, \cdot), N^u)\right).$$

Assume that there is no such $\epsilon > 0$ for some $\delta \in (0, \delta^*)$. Then there exists a sequence $(\phi^n)_0^\infty$ in $H \cap W^u(\mathcal{O})$ such that $\phi^n \to p_0$ as $n \to \infty$ and

$$\phi^n \notin P_{\tau_H}\left((p_0 + C_\delta) \cap W^u(p_0, F(\omega, \cdot), N^u)\right) \qquad \text{for all } n \in \mathbb{N}.$$

From the relation $(F(\omega, \cdot)_u)^{-1}(W^u(p_0, F(\omega, \cdot), N^u)) \subset W^u(p_0, F(\omega, \cdot), N^u)$ and from the definitions of $W^u(\mathcal{O})$ and $W^u(p_0, F(\omega, \cdot))$, it can be seen that, for every $n \in \mathbb{N}$, there exist $t^n \in [\frac{\omega}{2}, \frac{3\omega}{2})$ and $\psi^n \in W^u(p_0, F(\omega, \cdot))$ with $\phi^n = F(t^n, \psi^n)$. Recall the map j from Proposition 12.2 and set $s^n = j^{-1}(\psi^n)$ for all $n \in \mathbb{N}$.

We first claim that $s^n \to 0$ as $n \to \infty$. If not, then there exist $c > 0$, a strictly increasing sequence $(n_k)_0^\infty$ in $\mathbb{N}$, $m \in \{-1, 1\}$, and $t^* \in [\frac{\omega}{2}, \frac{3\omega}{2}]$ so that $|s^{n_k}| > c$, and $\operatorname{sign} s^{n_k} = \operatorname{sign} m$ for all $k \in \mathbb{N}$, and $t^{n_k} \to t^*$ as $k \to \infty$. By Proposition 12.2, either $\psi^{n_k} = j(s^{n_k}) \ll j(-c) \ll p_0$ for all $k \in \mathbb{N}$ or $\psi^{n_k} = j(s^{n_k}) \gg j(c) \gg p_0$ for all $k \in \mathbb{N}$. For the sake of definiteness, assume that $j(s^{n_k}) \gg j(c) \gg p_0$ for all $k \in \mathbb{N}$. The other case is similar. It follows that $\phi^{n_k} = F(t^{n_k}, \psi^{n_k}) \gg F(t^{n_k}, j(c))$. Letting $k \to \infty$, we obtain that $p_0 \geq F(t^*, j(c))$. The relation $j(c) \gg p_0$ implies

$F(t^*, j(c)) \gg F(t^*, p_0) = p_{t^*}$. Therefore, $p_0 \gg p_{t^*}$, which is impossible since $p_{t^*} = p_0$ in case $t^* = \omega$ and $V(p_{t^*} - p_0) = 2$ in case $t^* \in [\frac{\omega}{2}, \frac{3\omega}{2}] \setminus \{\omega\}$. Thus, $s^n \to 0$ as $n \to \infty$.

We next claim that $t^n \to \omega$ as $n \to \infty$. If not, then $(t^n)_{n=0}^{\infty}$ has a subsequence $(t^{n_k})_{k=0}^{\infty}$ converging to some $t^* \in [\frac{\omega}{2}, \frac{3\omega}{2}] \setminus \{\omega\}$ as $k \to \infty$. Then the elements $\phi^{n_k} = F(t^{n_k}, \psi^{n_k}) = F(t^{n_k}, j(s^{n_k}))$ tend to $F(t^*, j(0)) = F(t^*, p_0) = p_{t^*} \neq p_0$ as $k \to \infty$, which contradicts $\phi^n \to p_0$ as $n \to \infty$.

Proposition 12.2 and $s^n \to 0$ as $n \to \infty$ combined yield

$$\psi^n = j(s^n) \in (p_0 + C_\delta) \cap W^u(p_0, F(\omega, \cdot), N^u)$$

for all sufficiently large $n \in \mathbb{N}$.

The definition of P_{τ_H}, the facts that $\phi^n \in H$ and $\psi^n = j(s^n)$ for all $n \in \mathbb{N}$, $s^n \to 0$ and $t^n \to \omega$ as $n \to \infty$ imply that $t^n = \tau_H(\psi^n)$ for all sufficiently large $n \in \mathbb{N}$, and thus

$$\phi^n = F(t^n, \psi^n) = P_{\tau_H}(\psi^n) \in P_{\tau_H}\left((p_0 + C_\delta) \cap W^u(p_0, F(\omega, \cdot), N^u)\right)$$

for all sufficiently large $n \in \mathbb{N}$, a contradiction. □

Recall that $c_P(\phi) = \phi(0) + f'(0) \int_{-1}^{0} e^{-\lambda_0(s+1)} \phi(s)\, ds$, $\phi \in C$.

Proposition 12.4 *Assume $\mu > 0$. Then for every $r > 0$ there exist real numbers $k_- = k_-(r)$ and $k_+ = k_+(r)$ such that*

$$0 > k_- > c_P(\xi_-), \quad \{\phi \in \overline{W} : c_P(\phi) \leq k_-\} \subset \xi_- + C_r$$

and

$$0 < k_+ < c_P(\xi_+), \quad \{\phi \in \overline{W} : c_P(\phi) \geq k_+\} \subset \xi_+ + C_r.$$

Proof We show only the existence of k_+ since the proof for k_- is analogous. Let $r > 0$ be given. Fix $M > r$ so that

$$f'(0)e^{-\lambda_0}\frac{r^2}{2M} < c_P(\xi_+), \quad \text{and} \quad |-\mu\phi(0) + f(\phi(-1))| \leq M \quad \text{for all } \phi \in \overline{W}.$$

Choose $k_+ = k_+(r)$ so that

$$c_P(\xi_+) - f'(0)e^{-\lambda_0}\frac{r^2}{2M} < k_+ < c_P(\xi_+).$$

Then $0 < k_+ < c_P(\xi_+)$ holds and only $\{\psi \in \overline{W} : c_P(\psi) \geq k_+\} \subset \xi_+ + C_r$ remains to be verified.

Suppose there exists $\phi \in \overline{W} \setminus (\xi_+ + C_r)$ with $c_P(\phi) \geq k_+$. Then there exists $s^* \in [-1, 0]$ with $\phi(s^*) \leq \xi^+ - r$, since $\phi \leq \xi_+$ for all $\phi \in \overline{W}$ by Proposition 4.1. Define $\eta_{s^*} \in C$ by

$$\begin{aligned}
\eta_{s^*}(s) &= \xi^+ && \text{on } [-1,0] \setminus \left[s^* - \frac{r}{M}, s^* + \frac{r}{M}\right], \\
\eta_{s^*}(s^*) &= \xi^+ - r, && \\
\dot{\eta}_{s^*}(s) &= -M && \text{on } [-1,0] \cap \left(s^* - \frac{r}{M}, s^*\right), \\
\dot{\eta}_{s^*}(s) &= M && \text{on } [-1,0] \cap \left(s^*, s^* + \frac{r}{M}\right).
\end{aligned}$$

Using $\phi(s^*) \le \xi^+ - r$ and $|\dot\phi(s)| \le M$ on $[-1,0]$ we infer

$$\phi \le \eta_{s^*}.$$

Then, using the monotonicity of $s \mapsto e^{-\lambda_0(s+1)}$, we obtain

$$\begin{aligned}
k_+ &\le c_P(\phi) \le c_P(\eta_{s^*}) \\
&\le \xi^+ + f'(0)\int_{-1}^{0} e^{-\lambda_0(s+1)}\eta_0(s)ds \\
&= \xi^+ + f'(0)\xi^+\int_{-1}^{-r/M} e^{-\lambda_0(s+1)}ds + f'(0)\int_{-r/M}^{0} e^{-\lambda_0(s+1)}(-Ms+\xi^+-r)ds \\
&\le \xi^+ + f'(0)\xi^+\int_{-1}^{0} e^{-\lambda_0(s+1)}ds - f'(0)e^{-\lambda_0}\int_{-r/M}^{0}(Ms+r)ds \\
&= c_P(\xi_+) - f'(0)e^{-\lambda_0}\frac{r^2}{2M} \\
&< k_+,
\end{aligned}$$

a contradiction. □

Proposition 12.5 *For every $\beta \in (0,\beta_0)$ there exist sequences $(s_n^-)_0^\infty$ in $(-\beta,0)$ and $(s_n^+)_0^\infty$ in $(0,\beta)$ such that $s_n^- \to 0$ and $s_n^+ \to 0$ as $n \to \infty$, and*

$$\begin{aligned}
&p_0 + s_n^-\psi_H + w_H^u(s_n^-\psi_H) \in \operatorname{bd} W \cap W_\beta^u(p_0, P_H) && \textit{for all } n \in \mathbb{N}, \\
&p_0 + s_n^+\psi_H + w_H^u(s_n^+\psi_H) \in \operatorname{bd} W \cap W_\beta^u(p_0, P_H) && \textit{for all } n \in \mathbb{N}.
\end{aligned}$$

In case $\mu = 0$, for every $n \in \mathbb{N}$,

$$\begin{aligned}
&F(t, p_0 + s_n^-\psi_H + w_H^u(s_n^-\psi_H))(0) \to -\infty && \textit{as } t \to \infty, \\
&F(t, p_0 + s_n^+\psi_H + w_H^u(s_n^+\psi_H))(0) \to \infty && \textit{as } t \to \infty.
\end{aligned}$$

In case $\mu > 0$, for every $n \in \mathbb{N}$,

$$\omega(p_0 + s_n^-\psi_H + w_H^u(s_n^-\psi_H)) = \{\xi_-\} \quad \textit{and} \quad \omega(p_0 + s_n^+\psi_H + w_H^u(s_n^+\psi_H)) = \{\xi_+\}.$$

Proof 1. We prove that in case $\mu > 0$ there exists $\chi \in \operatorname{bd} W \setminus \{\xi_-, \xi_+\}$ with $\omega(\chi) = \{\xi_+\}$, and that in case $\mu = 0$ there exists $\chi \in \operatorname{bd} W$ with $x^\chi(t) \to \infty$ as $t \to \infty$.

1.1. The case $\mu > 0$. Choose $r > 0$ such that $\phi \gg 0$ for every $\phi \in \xi_+ + C_r$. Using Proposition 12.4 we find $k_+ \in (0, c_P(\xi^+))$ with

$$\phi \gg 0 \qquad \text{for all } \phi \in \overline{W} \cap c_P^{-1}(k_+). \tag{12.2}$$

The set $\Pi_3 c_P^{-1}(k_+)$ is the plane $\{y \in \mathbb{R}^3 : y_3 = k_+\}$. As $\overline{W}$ is compact, it follows that $\Pi_3\overline{W}$ is compact and $\Pi_3\overline{W} \cap \Pi_3 c_P^{-1}(k_+)$ is a compact subset of $\Pi_3 c_P^{-1}(k_+)$. $\Pi_3 W$ is open in $\mathbb{R}^3$ by Proposition 6.6. It follows that $\Pi_3 W \cap \Pi_3 c_P^{-1}(k_+)$ is an open subset of $\Pi_3 c_P^{-1}(k_+)$. We claim that $\Pi_3 W \cap \Pi_3 c_P^{-1}(k_+) \neq \emptyset$. Indeed, by Proposition

4.1, there exists a solution $x:\mathbb{R}\to\mathbb{R}$ such that $x_t\in W$ for all $t\in\mathbb{R}$, $x(t)\to 0$ as $t\to-\infty$ and $x(t)\to\xi^+$ as $t\to\infty$. From $0=c_P(0)<k_+<c_P(\xi_+)$ and from the continuity of the function $\mathbb{R}\ni t\mapsto c_P(x_t)\in\mathbb{R}$, it follows that $c_P(x_t)=k_+$ for some $t\in\mathbb{R}$. So, $\Pi_3W\cap\Pi_3c_P^{-1}(k_+)\neq\emptyset$. Hence we conclude

$$\Pi_3\,\mathrm{bd}\,W\cap\Pi_3c_P^{-1}(k_+)=\left(\Pi_3\overline{W}\cap\Pi_3c_P^{-1}(k_+)\right)\setminus\left(\Pi_3W\cap\Pi_3c_P^{-1}(k_+)\right)\neq\emptyset,$$

since in the plane $\{y\in\mathbb{R}^3: y_3=k_+\}$ a nonempty bounded open set contained in a closed set cannot coincide with the closed set. Thus, there exists $\chi\in\mathrm{bd}\,W$ with $c_P(\chi)=k_+$. From $k_+<c_P(\xi_+)$ it follows that $\chi\neq\xi^+$. By (12.2), we have $\chi\gg 0$. Then $\omega(\chi)=\{\xi_+\}$ follows from Proposition 4.5.

1.2. The case $\mu=0$. Choose $B>0$ according to Proposition 2.3(ii). Let $c>B$. Consider the plane $q=\{y\in\mathbb{R}^3: y_2=c\}$. If $\phi\in\overline{W}$ and $\Pi_3\phi\in q$ then $\phi(-1)=c>B$. Proposition 2.3(ii) yields $\phi\gg 0$. In particular, $\Pi_3W\cap q\neq q$. Proposition 4.1 implies $\Pi_3W\cap q\neq\emptyset$. As Π_3W is open in $\mathbb{R}^3$ by Proposition 6.6, it follows that $\Pi_3W\cap q$ is an open subset of q. $\Pi_3\overline{W}\cap q$ is closed in q provided $\Pi_3\overline{W}$ is closed in $\mathbb{R}^3$. In order to show the closedness of $\Pi_3\overline{W}$ in $\mathbb{R}^3$, let $(u_n)_0^\infty$ be a sequence in $\Pi_3\overline{W}$ and $u\in\mathbb{R}^3$ so that $u^n\to u$ as $n\to\infty$. Then $(u_n)_0^\infty$ is bounded. As Π_3^{-1} is Lipschitz continuous on bounded sets by Proposition 6.2, it follows that $(\Pi_3^{-1}(u_n))_0^\infty$ is a Cauchy sequence in $\overline{W}$. Its limit ϕ is in $\overline{W}$, and we get $u_n\to\Pi_3\phi$ as $n\to\infty$. Therefore, $u=\Pi_3\phi\in\Pi_3\overline{W}$ and it follows that $\Pi_3\overline{W}$ is closed in $\mathbb{R}^3$. Hence, $\Pi_3\overline{W}\cap q$ is closed in q. We have $\Pi_3\,\mathrm{bd}\,W\cap q=(\Pi_3\overline{W}\cap q)\setminus(\Pi_3W\cap q)$. As $\Pi_3W\cap q$ is a nonempty open proper subset of q contained in the closed $\Pi_3\overline{W}\cap q$, we conclude $\Pi_3\,\mathrm{bd}\,W\cap q\neq\emptyset$. Thus, there is $\chi\in\mathrm{bd}\,W$ such that $\Pi_3\chi\in q$. Then $\chi\gg 0$. Proposition 3.3 implies that $x^\chi(t)\to\infty$ as $t\to\infty$.

2. As in part 1 one shows that in case $\mu>0$ there exists $\eta\in\mathrm{bd}\,W\setminus\{\xi_-,\xi_+\}$ with $\omega(\eta)=\{\xi_-\}$, and that in case $\mu=0$ there exists $\eta\in\mathrm{bd}\,W$ with $x^\eta(t)\to-\infty$ as $t\to\infty$.

Now let $\beta\in(0,\beta_0)$. Proposition 12.3 implies the existence of $\delta=\delta(\beta)\in(0,\delta_0)$ with

$$P_{\tau_H}(W^u_\delta(p_0,F(\omega,\cdot)))\subset W^u_\beta(p_0,P_H).$$

It follows that for each $s\in(-\delta,\delta)$ there is a unique $\hat{s}\in(-\beta,\beta)$ such that

$$P_{\tau_H}(p_0+s\psi_u+w^u(s\psi_u))=p_0+\hat{s}\psi_H+w^u_H(\hat{s}\psi_H).$$

Consider the functions

$$\begin{aligned}
a&:(-\delta,\delta)\ni s\mapsto\hat{s}\in\mathbb{R},\\
i_{\psi_H}&:H_{1<}\ni u\psi_H\mapsto u\in\mathbb{R},\\
i_{\psi_u}&:\mathbb{R}\ni u\mapsto u\psi_u\in C_{1<},\\
b_{p_0}&:C\ni\phi\mapsto\phi+p_0\in C,\\
c_{p_0}&:C\ni\phi\mapsto\phi-p_0\in C.
\end{aligned}$$

Clearly, $a(0)=0$, $a((-\delta,\delta))\subset(-\beta,\beta)$, and

$$a(s)=i_{\psi_H}\left(\mathrm{Pr}_{H_{1<}}\left(c_{p_0}\left(P_{\tau_H}\left(b_{p_0}\left(i_{\psi_u}(s)+w^u(i_{\psi_u}(s))\right)\right)\right)\right)\right)\qquad\text{for all } s\in(-\delta,\delta).$$

It is then easy to see that a is C^1 and

$$\begin{aligned} a'(0) &= i_{\psi_H} \circ \Pr_{H_{1<}} \circ DP_{\tau_H}(p_0) \circ i_{\psi_u}(1) \\ &= i_{\psi_H} \circ \Pr_{H_{1<}} (\lambda_u \Pr_{H,\dot{p}_0} \psi_u) \\ &= \lambda_u || \Pr_{H,\dot{p}_0} \psi_u ||_{H,u} \\ &> 0. \end{aligned}$$

It follows that there is $\tilde{\delta} \in (0, \min\{\delta(\beta), \delta_2, \delta_3\})$ such that a is negative on $(-\tilde{\delta}, 0)$ and positive on $(0, \tilde{\delta})$.

3. We notice that there exists a positive integer k with

$$(F(\omega, \cdot)_u)^{-k}(W^u(p_0, F(\omega, \cdot), N^u)) \subset W^u_{\tilde{\delta}}(p_0, F(\omega, \cdot)).$$

Hence we obtain $W^u(\mathcal{O}) = \cup_{t \geq 0} F(t, W^u_{\tilde{\delta}}(p_0, F(\omega, \cdot)))$. Consider χ, η in bd W as obtained in part 1 and at the beginning of part 2.

Using Proposition 11.1 and the last equality we find t_χ, t_η in $\mathbb{R}^+$ and $\tilde{\chi}, \tilde{\eta}$ in $W^u_{\tilde{\delta}}(p_0, F(\omega, \cdot))$ so that $\chi = F(t_\chi, \tilde{\chi})$, $\eta = F(t_\eta, \tilde{\eta})$. For every $n \in \mathbb{N}$ we have

$$\begin{aligned} F_W(-nk\omega, \tilde{\chi}) &= (F(\omega, \cdot)_u)^{-nk}(\tilde{\chi}) \in W^u_{\tilde{\delta}}(p_0, F(\omega, \cdot)) \quad \text{and} \\ F_W(-nk\omega, \tilde{\eta}) &= (F(\omega, \cdot)_u)^{-nk}(\tilde{\eta}) \in W^u_{\tilde{\delta}}(p_0, F(\omega, \cdot)), \end{aligned}$$

and $F_W(-nk\omega, \tilde{\chi}) \to p_0$, $F_W(-nk\omega, \tilde{\eta}) \to p_0$ as $n \to \infty$. The sequences $(u_n)_0^\infty$, $(v_n)_0^\infty$ in $(-\tilde{\delta}, \tilde{\delta})$ defined by

$$\begin{aligned} F_W(-t_\chi - nk\omega, \chi) &= p_0 + u_n \psi_u + w^u(u_n \psi_u) \qquad \text{for all } n \in \mathbb{N}, \\ F_W(-t_\eta - nk\omega, \eta) &= p_0 + v_n \psi_u + w^u(v_n \psi_u) \qquad \text{for all } n \in \mathbb{N}, \end{aligned}$$

converge to 0 as $n \to \infty$. Proposition 12.2, $\xi_- \ll p_0 \ll \xi_+$, Proposition 2.2, and the behaviour of x^χ,x^η as $t \to \infty$ combined yield

$$u_n \in (0, \tilde{\delta}), \ v_n \in (-\tilde{\delta}, 0) \qquad \text{for all } n \in \mathbb{N}.$$

By the definition of $\tilde{\delta}$ and the invariance of bd W, we have

$$\begin{aligned} P_{\tau_H}(F_W(-t_\chi - nk\omega, \chi)) &= p_0 + a(u_n)\psi_H + w^u_H(a(u_n)\psi_H) \in \text{bd}\, W \cap W^u_\beta(p_0, P_H), \\ P_{\tau_H}(F_W(-t_\eta - nk\omega, \eta)) &= p_0 + a(v_n)\psi_H + w^u_H(a(v_n)\psi_H) \in \text{bd}\, W \cap W^u_\beta(p_0, P_H), \end{aligned}$$

and $-\beta < a(v_n) < 0 < a(u_n) < \beta$ for all $n \in \mathbb{N}$. Defining $s_n^+ = a(u_n)$ and $s_n^- = a(v_n)$ for $n \in \mathbb{N}$, we get the desired sequences. □

Recall from Proposition 7.3 that $\Pi_2(H \cap \overline{W} \cap S) = \{(0, v)^{tr} \in \mathbb{R}^2 : p(z_1 - 1) \leq v \leq p(-1)\}$ and $\Pi_2(H \cap \mathcal{O}) = \{(0, p(z_1 - 1))^{tr}, (0, p(-1))^{tr}\}$, where z_1 is the first positive zero of the periodic solution p and $p(z_1 - 1) < 0 < p(-1)$. Let $A = p(z_1 - 1)$, $B = p(-1)$, and consider the curve

$$h : [A, B] \ni s \mapsto \Pi_2^{-1}((0, s)^{tr}) \in C$$

with range in $H \cap \overline{W} \cap S$. Proposition 6.2 implies that h is Lipschitz continuous and injective. The next proposition is the first step towards the proof that $h|_{(A,B]}$ is C^1-smooth.

Proposition 12.6 $\Pi_2^{-1}|_{\Pi_2(W\cap S)}$ *is* C^1*-smooth.*

Proof $\Pi_2(W\cap S)$ is open in $\mathbb{R}^2$ by Proposition 7.3. $P_2(W\cap S)$ is open in G_2 by Corollary 7.4. Recall the isomorphism $J_2:\mathbb{R}^2\to G_2$ from Chapter 6, and the equation $P_2\phi = J_2\circ\Pi_2(\phi)$ for all $\phi\in C$. Let $x\in\Pi_2(W\cap S)$ and $\phi=\Pi_2^{-1}(x)\in W\cap S$. By the graph representation of $\overline{W}\cap S$,

$$\begin{aligned}\phi &= P_2\phi + w_S(P_2\phi)\\ &= J_2(\Pi_2(\Pi_2^{-1}(x))) + w_S(J_2(\Pi_2(\Pi_2^{-1}(x))))\\ &= J_2(x)+w_S(J_2(x)).\end{aligned}$$

Consequently,

$$\Pi_2^{-1}(x) = J_2(x)+w_S(J_2(x)) \qquad \text{for all } x\in\Pi_2(W\cap S).$$

Now Corollary 9.6 and Proposition 10.4 imply that $\Pi_2^{-1}|_{\Pi_2(W\cap S)}$ is C^1-smooth.

□

By Proposition 7.3, $\{(0,v)^{tr}\in\mathbb{R}^2 : A<v<B\}\subset\Pi_2(W\cap S)$. Using Proposition 12.6 we obtain that $h|_{(A,B)}$ is C^1-smooth. The next result shows the smoothness of h at B.

Proposition 12.7

(i) There exists a unique continuously differentiable function $z:\mathbb{R}\to\mathbb{R}$ *satisfying*

$$\begin{cases}\dot z(t) = -\mu z(t) + f'(p(t-1))z(t-1) & \textit{for all } t\in\mathbb{R},\\ z(-1)=1,\ z(0)=0,\\ V(z_t)=2 \qquad \textit{for all } t\in\mathbb{R}.\end{cases} \tag{12.3}$$

(ii) For every $\epsilon>0$ *there exists* $\delta>0$ *so that for all* $\chi\in[B-\delta,B]$, $\nu\in[B-\delta,B]$ *with* $\chi\neq\nu$,

$$\left\|\frac{h(\chi)-h(\nu)}{\chi-\nu} - z_0\right\| < \epsilon.$$

(iii) z_0 *and* $\dot p_0$ *are linearly independent, and* $z_0\in C_{r_M<}\cap C_{\le 1}$.

Proof 1. We prove that for all sequences $(\chi^n)_0^\infty, (\nu^n)_0^\infty$ in $[A,B]$ with $\chi^n\neq\nu^n$ for all $n\in\mathbb{N}$ and $\chi^n\to B$, $\nu^n\to B$ as $n\to\infty$, there exist a strictly increasing sequence $(n_k)_{k=0}^\infty$ in $\mathbb{N}$ and a C^1-function $z:\mathbb{R}\to\mathbb{R}$ such that z satisfies (12.3), and

$$\lim_{k\to\infty}\frac{h(\chi^{n_k})-h(\nu^{n_k})}{\chi^{n_k}-\nu^{n_k}} = z_0.$$

For every $n\in\mathbb{N}$ the solutions $x^n:\mathbb{R}\to\mathbb{R}$, $y^n:\mathbb{R}\to\mathbb{R}$ of equation (1.1) with $x_0^n=h(\chi^n)$ and $y_0^n=h(\nu^n)$ satisfy

$$\chi^n = x^n(-1),\ \nu^n=y^n(-1),\ x^n(0)=y^n(0)=0.$$

For each $n\in\mathbb{N}$, define

$$z^n = \frac{1}{\chi^n-\nu^n}(x^n-y^n).$$

As x_t^n, y_t^n are in $\overline{W \cap S}$ for all $t \in \mathbb{R}$ and $x_0^n \neq y_0^n$, we have from Proposition 5.7 that

$$V(z_t^n) = 2 \qquad \text{for all } t \in \mathbb{R} \text{ and for all } n \in \mathbb{N}.$$

Moreover,

$$z^n(-1) = 1, \; z^n(0) = 0 \qquad \text{for all } n \in \mathbb{N}.$$

By the Lipschitz continuity of h, there is $N_1 > 0$ such that $||z_0^n|| \leq N_1$ for all $n \in \mathbb{N}$. Consider the functions $b^n : \mathbb{R} \to \mathbb{R}$, $n \in \mathbb{N}$, given by

$$b^n(t) = \int_0^1 f'(sx^n(t-1) + (1-s)y^n(t-1))\, ds.$$

For every $n \in \mathbb{N}$, z^n is a solution of the equation

$$\dot{z}(t) = -\mu z(t) + b^n(t) z(t-1).$$

$\chi^n \to B$ and $\nu^n \to B$ imply $x_0^n \to p_0$ and $y_0^n \to p_0$ as $n \to \infty$, and thus the functions b^n converge to the limit $\mathbb{R} \ni t \mapsto f'(p(t-1)) \in \mathbb{R}$ uniformly on compact subsets of $\mathbb{R}$ as $n \to \infty$. As x_t^n, y_t^n are in the compact set $\overline{W \cap S}$ for all $t \in \mathbb{R}$ and $n \in \mathbb{N}$, there are positive constants b_0 and $b_1 \geq b_0$ such that $b_0 \leq b^n(t) \leq b_1$ for all $t \in \mathbb{R}$ and $n \in \mathbb{N}$. Lemmas VI.5 and VI.6 imply that there is $c > 0$ with $|z^n(t)| \leq ce^{c|t|}$ for all $t \in \mathbb{R}$ and $n \in \mathbb{N}$. Then Lemma VI.4 gives a subsequence $(z^{n_k})_{k=0}^{\infty}$ of $(z^n)_{n=0}^{\infty}$ and a C^1-function $z : \mathbb{R} \to \mathbb{R}$ such that $z^{n_k} \to z$, $\dot{z}^{n_k} \to \dot{z}$ as $k \to \infty$, uniformly on every compact subset of $\mathbb{R}$, and z satisfies the differential equation in (12.3). Clearly, $z(-1) = 1$, $z(0) = 0$. Lemma VI.1 implies $V(z_t) \leq 2$ for all $t \in \mathbb{R}$. Assume that $V(z_s) = 0$ for some $s \in \mathbb{R}$. Then, by the monotonicity of V, $V(z_t) = 0$ for all $t \geq s$. From Lemma VI.2(iii) it follows that $z_t \in R$ for all $t \geq s+3$. The C^1-convergence of z^{n_k} to z and Lemma VI.1(ii) imply $V(z_t^{n_k}) = 0$ for all $t \geq s+3$ and for all sufficiently large $k \in \mathbb{N}$. This is a contradiction. Consequently, $V(z_t) = 2$ for all $t \in \mathbb{R}$.

2. Assume that $\hat{z} : \mathbb{R} \to \mathbb{R}$ is also a C^1-function satisfying (12.3), and $z \neq \hat{z}$. Using that $M = D_2F(\omega, p_0)$ is injective one shows $z_0 \neq \hat{z}_0$. z_0 and $\hat{z}_0$ are in $C_{r_M <} \cap C_{\leq 1}$ by Proposition 8.5. The function $d = z - \hat{z}$ satisfies

$$\begin{cases} \dot{d}(t) = -\mu d(t) + f'(p(t-1)) d(t-1) & \text{for all } t \in \mathbb{R}, \\ d(-1) = d(0) = 0, \end{cases}$$

and $d_0 = z_0 - \hat{z}_0 \in C_{r_M <} \cap C_{\leq 1} \setminus \{0\}$. Therefore, $V(d_t) = 2$ for all $t \in \mathbb{R}$ by Proposition 8.5. By Lemma VI.2(iii), $d_0 \in R$ which contradicts $d(-1) = d(0) = 0$. This implies (i) and (ii).

3. Note that $p(0) = 0$ and $p(-1) > 0$ imply $\dot{p}(0) > 0$. Consequently, $z_0 \notin \mathbb{R}\dot{p}_0$. Thus, z_0 and $\dot{p}_0$ are linearly independent. □

Corollary 12.8 *$h|_{(A,B]}$ is C^1-smooth.*

Let $\phi_H = z_0$, where z_0 is uniquely determined by (12.3). Define the continuous curve $\Upsilon : [A - B, \infty) \to C$ by

$$\Upsilon(s) = h(s+B) \quad \text{for } A - B \leq s \leq 0, \quad \Upsilon(s) = p_0 + s\phi_H \quad \text{for } s > 0.$$

Proposition 12.9
(i) Υ *is an injective curve,* $\Upsilon|_{(A-B,\infty)}$ *is* C^1*-smooth with* $\Upsilon'(0) = \phi_H$.
(ii) $H \cap \overline{W \cap S} = \Upsilon([A-B,0])$, $H \cap W \cap S = \Upsilon((A-B,0))$, $H \cap \mathcal{O} = \{\Upsilon(A-B), \Upsilon(0)\}$, $\Upsilon([A-B,\infty)) \subset H$.

Proof The C^1-smoothness of $\Upsilon|_{(A-B,\infty)}$ and $\Upsilon'(0) = \phi_H$ are obtained from the definition of Υ, Proposition 12.7 and Corollary 12.8. The definition of Υ and the facts $\phi_H(-1) = 1$, $p(0) = \phi_H(0) = 0$ imply

$$\Pi_2 \Upsilon(s) = (0, s+B)^{tr} = (0, s+p(-1))^{tr} \qquad \text{for all } s \in [A-B, \infty).$$

This relation gives the injectivity of Υ, and combined with Proposition 7.3(iv) yields assertion (ii). □

Define

$$\lambda_c = \min\{|\zeta| : \zeta \in \sigma \text{ and } r_M < |\zeta|\}.$$

Proposition 12.10
(i) $DP_H(p_0)\phi_H = \lambda_c \phi_H$.
(ii) $C_{r_M<} = \mathbb{R}\phi_H \oplus \mathbb{R}\dot{p}_0 \oplus \mathbb{R}\psi_H$ *and* $H_{r_M<} = \mathbb{R}\phi_H \oplus \mathbb{R}\psi_H$.
(iii) $\Pi_3\phi_H$ *and* $\Pi_3\psi_H$ *are linearly independent.*

Proof 1. Proof of assertion (i).

1.1. Proposition 12.7 and $\mathbb{R}\dot{p}_0 \subset G_{\mathbb{R}}(1)$ yield $2 \le \dim C_{r_M<} \cap C_{\le 1}$. Using $\dim C_{r_M<} \le 3$ and $\dim G_{\mathbb{R}}(\lambda_u) = 1$ we get

$$C_{r_M<} \cap C_{\le 1} = \mathbb{R}\phi_H \oplus \mathbb{R}\dot{p}_0, \quad \dim(C_{r_M<} \cap C_{\le 1}) = 2.$$

1.2. In case $\lambda_c = 1$ we note that $-1 \notin \sigma$ (Proposition 8.3), and that there are no Floquet multipliers on $S^1_{\mathbb{C}} \setminus \{-1, 1\}$ since otherwise $\dim(C_{r_M<} \cap C_{\le 1}) \ge \dim G_{\mathbb{R}}(1) + 2 \ge 3$, a contradiction. It follows in this case that $C_{r_M<} \cap C_{\le 1} = G_{\mathbb{R}}(1)$. Recall $\dot{p}_0 \in G_{\mathbb{R}}(1) \setminus H$. Therefore $\dim G_{\mathbb{R}}(1) \cap H \le 1$. Using $\phi_H \in (C_{r_M<} \cap C_{\le 1}) \cap H = G_{\mathbb{R}}(1) \cap H$ and Proposition I.5(iii) we arrive at

$$DP_H(p_0)\phi_H = 1\phi_H = \lambda_c \phi_H.$$

1.3. The case $\lambda_c < 1$ requires some preparations.

Recall the homeomorphism $\rho : v_+ \cap \Pi_2(\overline{W} \cap S) \to v_+ \cap \Pi_2(\overline{W} \cap S)$ from Chapter 7. We have

$$v_+ \cap \Pi_2(\overline{W} \cap S) = \{(0,v)^{tr} \in \mathbb{R}^2 : 0 < v \le p(-1)\} \text{ and } p(-1) = B.$$

Consider the maps $p_2 : \mathbb{R}^2 \ni (u,v)^{tr} \mapsto v \in \mathbb{R}$ and $\hat{\rho} : (0,B] \ni v \mapsto p_2\rho(0,v)^{tr} \in (0,B]$; we have $\hat{\rho}(B) = B$, and $v < \hat{\rho}(v)$ for all $v \in (0,B)$, by Proposition 7.2(ii). There exists $\epsilon > 0$ so that $\Upsilon((-\epsilon,\epsilon)) \subset H \cap N_{p_0}$, and for every $v \in (B-\epsilon, B]$ the second positive zero z of the solution $x : \mathbb{R} \to \mathbb{R}$ of equation (1.1) with $x_0 = h(v) = \Pi_2^{-1}((0,v)^{tr})$ belongs to the interval $(\omega - \epsilon_H, \omega + \epsilon_H)$. It follows that for all $v \in (-\epsilon, 0]$,

$$P_H(\Upsilon(v)) = x_z = \Pi_2^{-1}(\Pi_2 x_z) = h(\hat{\rho}(B+v)).$$

Using the differentiability of P_H and Υ we obtain

$$\begin{aligned} DP_H(p_0)\phi_H &= DP_H(p_0)D\Upsilon(0)1 = D(P_H \circ \Upsilon|_{(-\epsilon,\epsilon)})(0)1 \\ &= \lim_{n\to\infty} \frac{h\left(\hat{\rho}\left(B-\frac{1}{n}\right)\right) - h\left(\hat{\rho}(B)\right)}{-\frac{1}{n}} \\ &= \lim_{n\to\infty} \frac{\hat{\rho}\left(B-\frac{1}{n}\right) - \hat{\rho}(B)}{-\frac{1}{n}} \frac{h\left(\hat{\rho}\left(B-\frac{1}{n}\right)\right) - h\left(\hat{\rho}(B)\right)}{\hat{\rho}\left(B-\frac{1}{n}\right) - \hat{\rho}(B)}. \end{aligned}$$

For every integer $n > \frac{1}{\epsilon}$,

$$0 < \frac{\hat{\rho}\left(B-\frac{1}{n}\right) - \hat{\rho}(B)}{-\frac{1}{n}} < 1.$$

It follows that there are a strictly increasing sequence $(n_k)_{k=0}^{\infty}$ and $\theta \in [0,1]$ so that

$$\frac{\hat{\rho}\left(B-\frac{1}{n_k}\right) - \hat{\rho}(B)}{-\frac{1}{n_k}} \to \theta \qquad \text{as } k \to \infty.$$

Using $Dh(B)1 = \phi_H$ we infer

$$DP_H(p_0)\phi_H = \theta\phi_H.$$

ϕ_H and $\dot{p}_0$ are linearly independent and belong to $C_{r_M<} \cap C_{\le 1}$. Using $\dim C_{r_M} \le 3$ and $1 < \lambda_u$ we obtain

$$\dim C_{r_M<} = 3,\ \dim\left(C_{r_M<1} \cap C_{\le 1}\right) = 2,\ C_{r_M<} \cap C_{\le 1} = \mathbb{R}\phi_H \oplus \mathbb{R}\dot{p}_0.$$

In case $\lambda_c < 1$ there exists an eigenvalue ζ of M with $|\zeta| = \lambda_c$ and $\operatorname{Im}\zeta \ge 0$,

$$C_{r_M<} \cap C_{\le 1} = G_{\mathbb{R}}(\zeta) \oplus G_{\mathbb{R}}(1).$$

ζ is real since otherwise we would obtain a contradiction to $\dim G_{\mathbb{R}}(\zeta) = \dim(C_{r_M<} \cap C_{\le 1}) - \dim G_{\mathbb{R}}(1) \le 1$. We also obtain $G_{\mathbb{R}}(1) = \mathbb{R}\dot{p}_0$. Choose a vector $\psi \in G_{\mathbb{R}}(\zeta) \setminus \{0\}$. Then $M\psi = \zeta\psi$, and there are reals c_1, c_2 with $\phi_H = c_1\psi + c_2\dot{p}_0$. By Proposition 12.7(iii), $c_1 \ne 0$. It follows that

$$\begin{aligned} c_1\theta\psi + c_2\theta\dot{p}_0 = \theta\phi_H = DP_H(p_0)\phi_H &= M\phi_H - \frac{l(M\phi_H)}{l(\dot{p}_0)}\dot{p}_0 \\ &= c_1\zeta\psi + \left(c_2 - \frac{l(M\phi_H)}{l(\dot{p}_0)}\right)\dot{p}_0, \end{aligned}$$

and consequently,

$$\zeta = \theta \in [0,1].$$

Altogether,

$$\lambda_c = |\zeta| = \zeta = \theta,$$

and

$$DP_H(p_0)\phi_H = \lambda_c\phi_H.$$

2. We have $C_{r_M<} \cap C_{\leq 1} = \mathbb{R}\phi_H \oplus \mathbb{R}\dot{p}_0$, $\dim C_{1<} = 1$, $1 < \lambda_u \in \sigma$, $G_{\mathbb{R}}(\lambda_u) = \mathbb{R}\psi_u$. It follows that

$$C_{r_M<} = \mathbb{R}\phi_H \oplus \mathbb{R}\dot{p}_0 \oplus \mathbb{R}\psi_u.$$

Using $\psi_H = c\left(\psi_u - \frac{l(\psi_u)}{l(\dot{p}_0)}\dot{p}_0\right)$ with $c \neq 0$, we get

$$C_{r_M<} = \mathbb{R}\phi_H \oplus \mathbb{R}\dot{p}_0 \oplus \mathbb{R}\psi_H.$$

This fact and Proposition I.5 combined yield $H_{r_M<} = \mathbb{R}\phi_H \oplus \mathbb{R}\psi_H$.

3. Proof of assertion (iii). Choose $\beta \in (0, \beta_0)$. Consider the C^1-curves

$$c : (-1, 1) \ni s \mapsto p_0 + \beta s\psi_H + w_H^u(\beta s\psi_H) \in C$$

and

$$d : (-1, 1) \ni s \mapsto \Upsilon((B - A)s) \in C.$$

We have $c(0) = d(0) = p_0$, $c'(0) = \beta\psi_H$, $d'(0) = (B - A)\phi_H$ and $d((-1, 0]) \subset \overline{W}$. Proposition 12.5 guarantees the existence of sequences $(s_n^-)_0^\infty$ and $(s_n^+)_0^\infty$ such that $-1 < s_n^- < 0 < s_n^+ < 1$, and $c(s_n^-) \in \overline{W}$, $c(s_n^+) \in \overline{W}$ for all $n \in \mathbb{N}$, and $s_n^- \to 0$, $s_n^+ \to 0$ as $n \to \infty$. Then Proposition 6.5(iii) implies that $\Pi_3\psi_H$ and $\Pi_3\phi_H$ are linearly independent. □

By Proposition 12.10 the vectors $\Pi_3\psi_H$ and $\Pi_3\phi_H$ span the plane $\Pi_3 H = \{y \in \mathbb{R}^3 : y_1 = 0\}$ in $\mathbb{R}^3$. Let $\mathrm{Pr}_{\mathbb{R}\Pi_3\psi_H}$ denote the projection of $\Pi_3 H$ onto $\mathbb{R}\Pi_3\psi_H$ along $\mathbb{R}\Pi_3\phi_H$. Choose $\beta \in (0, \beta_0)$. The curve

$$b : (-\beta, \beta) \ni s \mapsto \Pi_3(p_0 + s\psi_H + w_H^u(s\psi_H)) \in \Pi_3 H$$

is C^1-smooth and satisfies $b(0) = \Pi_3 p_0$, $Db(0)1 = \Pi_3\psi_H$. It follows that there exist $\beta_3 \in (0, \beta_0)$, u_1 and u_2 in $(0, \infty)$, and a C^1-map

$$h_1 : (-u_1, u_2) \to \mathbb{R}$$

with

$$|h_1'(u)| < 1 \qquad \text{for all } u \in (-u_1, u_2)$$

so that

$$b\big|_{(-\beta_3, \beta_3)} \text{ is injective,}$$
$$\Pi_3\big|_{W_{\beta_3}^u(p_0, P_H)} \text{ is injective,}$$
$$h_1(0) = 0 \text{ and } h_1'(0) = 0,$$
$$\Pi_3(W_{\beta_3}^u(p_0, P_H)) = b((-\beta_3, \beta_3)) = \{\Pi_3 p_0 + u\Pi_3\psi_H + h_1(u)\Pi_3\phi_H : -u_1 < u < u_2\}.$$

Let $r_0 = \min\{u_1, u_2\}$. For $0 < r < r_0$, define

$$\begin{aligned}
D_r &= \{\Pi_3 p_0 + u\Pi_3\psi_H + v\Pi_3\phi_H : |u| < r,\ |v| < r\},\\
D_r^- &= \{\Pi_3 p_0 + u\Pi_3\psi_H + v\Pi_3\phi_H : |u| < r,\ -r < v < h_1(u)\},\\
D_r^+ &= \{\Pi_3 p_0 + u\Pi_3\psi_H + v\Pi_3\phi_H : |u| < r,\ h_1(u) < v < r\},\\
D_r^0 &= \{\Pi_3 p_0 + u\Pi_3\psi_H + v\Pi_3\phi_H : |u| < r,\ v = h_1(u)\},\\
D_r^{-0} &= D_r^- \cup D_r^0.
\end{aligned}$$

Then $D_r = D_r^- \cup D_r^0 \cup D_r^+$ and

$$D_r^0 = D_r \cap \Pi_3\left(W_{\beta_3}^u(p_0, P_H)\right). \tag{12.4}$$

Proposition 12.11
(i) There exists $r^* > 0$ *such that, for* $0 < r < r^*$,

$$D_r \cap \Pi_3(\text{bd}\, W) = D_r^0,$$
$$D_r \cap \Pi_3 W = D_r^-,$$
$$D_r \cap \Pi_3 \overline{W} = D_r^{-0},$$
$$D_r^+ \cap \Pi_3 \overline{W} = \emptyset.$$

(ii) There exists $\beta_4 \in (0, \beta_3)$ *such that* $W^u_{\beta_4}(p_0, P_H) \subset \text{bd}\, W$.

Proof 1. By Proposition 12.9(i), $D\Upsilon(0)1 = \phi_H$, and $0 \neq \Pi_3\phi_H$, according to Proposition 12.10(iii). Arguments similar to those which led to the construction of the map h_1 permit to find $\delta \in (0, B - A)$, v_1, v_2 in $(0, \infty)$, and a C^1-map $h_2 : (-v_1, v_2) \to \mathbb{R}$ such that $h_2(0) = 0$, $h_2'(0) = 0$, $|h_2'(v)| < 1$ for all $v \in (-v_1, v_2)$, and

$$\Pi_3\{\Upsilon(s) : -\delta < s \leq 0\} = \{\Pi_3 p_0 + h_2(v)\Pi_3\psi_H + v\Pi_3\phi_H : v \in (-v_1, 0]\},$$
$$\Pi_3\{\Upsilon(s) : 0 < s < \delta\} = \{\Pi_3 p_0 + h_2(v)\Pi_3\psi_H + v\Pi_3\phi_H : v \in (0, v_2)\}.$$

The set $\Pi_3\{\Upsilon(s) : A - B \leq s \leq -\delta\}$ is compact and does not contain $\Pi_3 p_0 = \Pi_3\Upsilon(0)$ since we have $\Upsilon([A - B, 0]) \subset \overline{W}$ (see Proposition 12.9(ii)), and Υ and $\Pi_3|_{\overline{W}}$ are injective (Propositions 12.9(i) and 6.1). Thus, there exists $r_1 \in (0, \min\{r_0, v_1, v_2\})$ such that

$$D_{r_1} \cap \Pi_3\{\Upsilon(s) : A - B \leq s \leq -\delta\} = \emptyset.$$

Proposition 12.3 implies the existence of $\epsilon_3 = \epsilon_3(\beta_3) > 0$ such that

$$H \cap \text{bd}\, W \cap (p_0 + C_{\epsilon_3}) \subset W^u_{\beta_3}(p_0, P_H). \tag{12.5}$$

We can choose $r^* \in (0, r_1)$ such that

$$\Pi_3^{-1}\left(D_{r^*} \cap \Pi_3\overline{W}\right) \subset p_0 + C_{\epsilon_3} \tag{12.6}$$

because of Proposition 6.2.

Let $0 < r < r^*$. If $x \in D_r \cap \Pi_3(\text{bd}\, W)$ then, using $\Pi_3 H \cap \Pi_3(C \setminus H) = \emptyset$, there is $\phi \in H \cap \text{bd}\, W$ with $\Pi_3\phi = x$. Applying (12.6), we get $\phi \in H \cap \text{bd}\, W \cap (p_0 + C_{\epsilon_3})$. (12.5) implies $\phi \in W^u_{\beta_3}(p_0, P_H)$. Then it follows that $x = \Pi_3\phi \in D_r \cap \Pi_3(W^u_{\beta_3}(p_0, P_H))$. Thus (12.4) yields

$$D_r \cap \Pi_3(\text{bd}\, W) \subseteq D_r \cap \Pi_3(W^u_{\beta_3}(p_0, P_H)) = D_r^0. \tag{12.7}$$

From $|h_1'(u)| < 1$, $|h_2'(u)| < 1$ for $|u| < r^*$, we get

$$D_r^0 \subset \{\Pi_3 p_0 + u\Pi_3\psi_H + v\Pi_3\phi_H : |v| < |u| < r\} \cup \{\Pi_3 p_0\}$$

and

$$\{\Pi_3 p_0 + u\Pi_3\psi_H + v\Pi_3\phi_H : |v| < r,\ u = h_2(v)\}$$
$$\subset \{\Pi_3 p_0 + u\Pi_3\psi_H + v\Pi_3\phi_H : |u| < |v| < r\} \cup \{\Pi_3 p_0\}.$$

It follows that

$$D_r^0 \cap \{\Pi_3 p_0 + u\Pi_3\psi_H + v\Pi_3\phi_H : |v| < r,\ u = h_2(v)\} = \{\Pi_3 p_0\}$$

and

$$\{\Pi_3 p_0 + u\Pi_3\psi_H + v\Pi_3\phi_H : -r < v < 0,\ u = h_2(v)\} \subset D_r^-. \tag{12.8}$$

2. We claim that

$$D_r^+ \cap \Pi_3\left(H \cap \overline{W \cap S}\right) = \emptyset. \tag{12.9}$$

As $\Upsilon([A-B,0]) = H \cap \overline{W \cap S}$, it suffices to show that $D_r^+ \cap \{\Pi_3\Upsilon(s) : A-B \le s \le 0\} = \emptyset$. From the definition of r_1 and from $r < r_1$ it follows that $D_r^+ \cap \{\Pi_3\Upsilon(s) : A-B \le s \le -\delta\} = \emptyset$. Since

$$\{\Pi_3\Upsilon(s) : -\delta < s \le 0\} = \{\Pi_3 p_0 + h_2(v)\Pi_3\psi_H + v\Pi_3\phi_H : v \in (-v_1, 0]\},$$

we have

$$\begin{aligned}\{\Pi_3\Upsilon(s) : -\delta < s \le 0\} &= \{\Pi_3 p_0 + h_2(v)\Pi_3\psi_H + v\Pi_3\phi_H : v \in (-v_1, -r]\} \\ &\quad \cup \{\Pi_3 p_0 + h_2(v)\Pi_3\psi_H + v\Pi_3\phi_H : v \in (-r, 0]\}.\end{aligned}$$

By the definition of D_r and D_r^+,

$$D_r^+ \cap \{\Pi_3 p_0 + h_2(v)\Pi_3\psi_H + v\Pi_3\phi_H : v \in (-v_1, -r]\} = \emptyset.$$

Then the claim follows from $D_r^- \cap D_r^+ = \emptyset$ and the fact that

$$\{\Pi_3 p_0 + h_2(v)\Pi_3\psi_H + v\Pi_3\phi_H : v \in (-r, 0]\} \subset D_r^- \cup \{\Pi_3 p_0\}.$$

3. We prove $D_r^- \subset \Pi_3 W$. As $\Pi_3 W$ is open in $\mathbb{R}^3$ (Proposition 6.6), $\Pi_3 W \cap D_r^-$ is relatively open in D_r^-.

We claim that $\Pi_3 W \cap D_r^-$ is relatively closed in D_r^-. Suppose $(x_n)_0^\infty$ is a sequence in $\Pi_3 W \cap D_r^-$ and $x \in D_r^-$ with $x_n \to x \in D_r^-$ as $n \to \infty$. We show that $x \in \Pi_3 W$. Since $(x_n)_0^\infty$ is bounded, the Lipschitz continuity of Π_3^{-1} on bounded subsets yields that the sequence $(\phi_n)_0^\infty$ defined by $\phi_n = \Pi_3^{-1}(x_n)$ is a Cauchy sequence in W. Then, for some $\phi \in \overline{W}$, $\phi_n \to \phi$ as $n \to \infty$. It follows that $\Pi_3\phi = x$. In case $\phi \in \operatorname{bd} W$, we have $x \in \Pi_3(\operatorname{bd} W) \cap D_r^-$, which contradicts the definition of D_r^- and (12.7). Thus $\phi \in W$ and $x = \Pi_3\phi \in \Pi_3 W$.

Consequently, $\Pi_3 W \cap D_r^-$ is relatively open and closed in D_r^-. Moreover, $\Pi_3 W \cap D_r^- \neq \emptyset$ because of (12.8) and

$$\{\Pi_3 p_0 + h_2(v)\Pi_3\psi_H + v\Pi_3\phi_H : -r < v < 0\} \subset \Pi_3\left(\{\Upsilon(s) : -\delta < s < 0\}\right) \subset \Pi_3 W.$$

Therefore, $\Pi_3 W \cap D_r^- = D_r^-$. This proves $D_r^- \subset \Pi_3 W$.

4. In the same way as for D_r^-, we can show that $\Pi_3 W \cap D_r^+$ is relatively open and closed in D_r^+. Thus, either $\Pi_3 W \cap D_r^+ = \emptyset$ or $\Pi_3 W \cap D_r^+ = D_r^+$.

Assume that $\Pi_3 W \cap D_r^+ = D_r^+$. Proposition 12.5 implies that in case $\mu > 0$ there exist y, z in $D_r \cap \Pi_3(\operatorname{bd} W) \cap \Pi_3(W_{\beta_3}^u(p_0, P_H))$ such that $\omega(\Pi_3^{-1}(y)) = \{\xi_+\}$, $\omega(\Pi_3^{-1}(z)) = \{\xi_-\}$, and that in case $\mu = 0$ there exist y, z in $D_r \cap \Pi_3(\operatorname{bd} W) \cap \Pi_3(W_{\beta_3}^u(p_0, P_H))$ such that $x^{\Pi_3^{-1}y}(t) \to \infty$, $x^{\Pi_3^{-1}z}(t) \to -\infty$ as $t \to \infty$. In both cases $\mu > 0$ and $\mu = 0$, we have $y \in D_r^0$ and $z \in D_r^0$ by (12.7). There exists a continuous curve $[0,1] \ni s \mapsto c(s) \in D_r$ such that $c(0) = y$, $c(1) = z$ and $c(s) \in D_r^+$ for all $s \in (0,1)$.

The sets $S^+ = \{\phi \in C : \phi \text{ is above } S\}$ and $S^- = \{\phi \in C : \phi \text{ is below } S\}$ are open. Propositions 3.3, 4.5, 7.3 and 7.5 combined yield $\Pi_3^{-1}(y) \in S^+$ and $\Pi_3^{-1}(z) \in S^-$. We conclude that

$$\{s \in [0,1] : \Pi_3^{-1}(c(s)) \in S^+\} \text{ and } \{s \in [0,1] : \Pi_3^{-1}(c(s)) \in S^-\}$$

are disjoint, nonempty and relatively open subsets of $[0,1]$. It follows that there exists an $s^* \in (0,1)$ which does not belong to the above sets. Then $\Pi_3^{-1}(c(s^*)) \in S$. Also using $c(s^*) \in D_r^+ = \Pi_3 W \cap D_r^+$ and $\Pi_3 H \cap \Pi_3(C \setminus H) = \emptyset$, we infer $c(s^*) \in D_r^+ \cap \Pi_3(H \cap \overline{W} \cap S)$. This contradicts (12.9). Therefore, $\Pi_3 W \cap D_r^+ = \emptyset$. Also using (12.7), we get $D_r^+ \cap \Pi_3 \overline{W} = \emptyset$.

5. We show that $D_r \cap \Pi_3(W^u_{\beta_3}(p_0, P_H)) \subset \Pi_3(\operatorname{bd} W)$. This combined with (12.4), (12.7) and the results in parts 3 and 4 shall complete the proof of (i).

As $\Pi_3 W$ is open in $\mathbb{R}^3$, it is clear that $\Pi_3 W \cap D_r$ is relatively open in D_r. We assert that $(\Pi_3 \overline{W}) \cap D_r$ is relatively closed in D_r. Let $x \in D_r$ and let $(x_n)_0^\infty$ be a sequence in $(\Pi_3 \overline{W}) \cap D_r$ such that $x_n \to x$ as $n \to \infty$. We show that $x \in (\Pi_3 \overline{W}) \cap D_r$. Let $\phi_n = \Pi_3^{-1}(x_n)$. By Proposition 6.2, Π_3^{-1} is Lipschitz continuous on the bounded subset $\{x_n : n \in \mathbb{N}\}$ of $\Pi_3 \overline{W}$. It follows that $(\phi_n)_0^\infty$ is a Cauchy sequence in $\overline{W}$. There exists $\phi \in \overline{W}$ such that $\phi_n \to \phi$ as $n \to \infty$. By the continuity of Π_3, $x = \Pi_3 \phi$. Therefore, $x \in (\Pi_3 \overline{W}) \cap D_r$. A consequence is that $\operatorname{cl}_{D_r}(\Pi_3 W \cap D_r) \subseteq (\Pi_3 \overline{W}) \cap D_r$, where cl_{D_r} denotes the closure relative to D_r.

Using this consequence, the definition of D_r^-, the fact that $D_r^- = \Pi_3 W \cap D_r^-$ and (12.4), we obtain

$$\begin{aligned} D_r \cap \Pi_3(W^u_{\beta_3}(p_0, P_H)) &= D_r^0 \\ &\subset \operatorname{cl}_{D_r} D_r^- \\ &= \operatorname{cl}_{D_r}(\Pi_3 W \cap D_r^-) \\ &\subseteq \operatorname{cl}_{D_r}(\Pi_3 W \cap D_r) \\ &\subseteq (\Pi_3 \overline{W}) \cap D_r \\ &\subseteq \Pi_3 \overline{W}. \end{aligned}$$

Suppose that $D_r \cap \Pi_3(W^u_{\beta_3}(p_0, P_H))$ contains a point of $\Pi_3 W$. Then using (12.4) and that $\Pi_3 W \cap D_r$ is open relatively to D_r, it follows that $\Pi_3 W \cap D_r^+ \neq \emptyset$, a contradiction. Therefore,

$$D_r \cap \Pi_3(W^u_{\beta_3}(p_0, P_H)) \subset \Pi_3 \overline{W} \setminus \Pi_3 W = \Pi_3(\operatorname{bd} W).$$

6. Proof of assertion (ii). Let $r \in (0, r^*)$ be fixed. Choose $\beta_4 \in (0, \beta_3)$ so small that $\Pi_3(W^u_{\beta_4}(p_0, P_H)) \subset D_r$. Then, from (12.4) and from the first equality in assertion (i),

$$\Pi_3(W^u_{\beta_4}(p_0, P_H)) \subset D_r \cap \Pi_3(W^u_{\beta_3}(p_0, P_H)) = D_r^0 = D_r \cap \Pi_3(\operatorname{bd} W).$$

Let $\phi \in W^u_{\beta_4}(p_0, P_H)$. Also using $D_r \subset \Pi_3 H$ and $\Pi_3 H \cap \Pi_3(C \setminus H) = \emptyset$, there is $\psi \in H \cap \operatorname{bd} W$ with $\Pi_3 \phi = \Pi_3 \psi$. From $\Pi_3 \psi \in D_r \cap \Pi_3 \overline{W}$ and (12.6) it follows that $\psi \in H \cap \operatorname{bd} W \cap (p_0 + C_{\epsilon_3})$. Hence (12.5) implies $\psi \in W^u_{\beta_3}(p_0, P_H)$. Recall that Π_3 restricted to $W^u_{\beta_3}(p_0, P_H)$ is injective. Then $\phi \in W^u_{\beta_3}(p_0, P_H)$, $\psi \in W^u_{\beta_3}(p_0, P_H)$ and $\Pi_3 \phi = \Pi_3 \psi$ imply $\phi = \psi$. Consequently, $\phi \in \operatorname{bd} W$. □

Proposition 12.12 $W^u(\mathcal{O}) \subseteq \mathrm{bd}\, W$.

Proof Using Theorem V.3, the relation

$$W^u(p_0, P_H, N_H^u) \subset \bigcup_{t\geq 0} F(t, W_{\beta_4}^u(p_0, P_H, N_H^u)),$$

and Proposition 12.11, we obtain

$$\begin{aligned} W^u(\mathcal{O}) &= \bigcup_{t\geq 0} F(t, W^u(p_0, P_H, N_H^u)) \\ &\subset \bigcup_{t\geq 0} F(t, W_{\beta_4}^u(p_0, P_H)) \\ &\subset \bigcup_{t\geq 0} F(t, \mathrm{bd}\, W) \\ &= \mathrm{bd}\, W. \end{aligned}$$

□

In case $\mu > 0$, ξ_- and ξ_+ do not belong to $W^u(\mathcal{O})$. Therefore Propositions 11.1 and 12.12 yield the following relation between the unstable set of $\mathcal{O}$ and bd W.

Corollary 12.13 *If* $\mu = 0$ *then* $W^u(\mathcal{O}) = \mathrm{bd}\, W$*; and if* $\mu > 0$ *then* $W^u(\mathcal{O}) = \mathrm{bd}\, W \setminus \{\xi_-, \xi_+\}$.

CHAPTER 13

$H \cap \overline{W}$ is smooth near p_0

The aim of this chapter is to show that $H \cap \overline{W}$ is smooth near p_0. Important information on $H \cap \overline{W}$ near p_0 is the result $D_r^{-0} = D_r \cap \Pi_3 \overline{W}$, $0 < r < r^*$, in Proposition 12.11. In the sequel we first obtain a diffeomorphism Φ from an open neighbourhood of p_0 in H onto an open neighbourhood of 0 in H, and prove that $\Phi \circ \Pi_3^{-1}(D_{r_1}^{-0})$, for some $r_1 > 0$, is given by a Lipschitz continuous function Σ mapping a subset of a 2-dimensional subspace of H into a complementary subspace. Then we show that Σ can be extended to a C^1-smooth function defined in an open neighbourhood of 0 in this 2-dimensional subspace of H. This fact easily implies a description of $H \cap \overline{W}$ in a neighbourhood of p_0 in H which is the starting point of the proof of the smoothness results in the next chapter.

Recall that $0 < r_M < \lambda_c \leq 1 < \lambda_u$, $H_{r_M<} = \mathbb{R}\phi_H \oplus \mathbb{R}\psi_H$; $\mathbb{R}\phi_H$ and $\mathbb{R}\psi_H$ are the realified generalized eigenspaces of $DP_H(p_0)$ associated with the eigenvalues λ_c and λ_u, respectively. Define

$$X_1 = \mathbb{R}\psi_H,\ X_2 = \mathbb{R}\phi_H,\ X_3 = H_{\leq r_M},\ X_{12} = X_1 \oplus X_2.$$

Then $H = X_1 \oplus X_2 \oplus X_3$.

We apply Theorem III.1 to find open neighbourhoods N_{12} of 0 in X_{12}, N_3 of 0 in X_3, N of p_0 in $N_{p_0} \cap H$ and a C^1-map $w_{12} : N_{12} \to X_3$ with $w_{12}(N_{12}) \subset N_3$, $w_{12}(0) = 0$, $Dw_{12}(0) = 0$ so that the graph

$$W_{12} = \{\chi + w_{12}(\chi) : \chi \in N_{12}\}$$

satisfies

$$P_H((p_0 + W_{12}) \cap N) \subset p_0 + W_{12}$$

and

$$\{\psi \in H : \text{There exists a trajectory } (\psi^n)_{-\infty}^0 \text{ of } P_H \text{ in } p_0 + N_{12} + N_3 \text{ with } \psi^0 = \psi\} \subset p_0 + W_{12}.$$

Recall $X_1 = T_{p_0} W^u(p_0, P_H, N_H^u)$ and $X_2 = T_{p_0} \Upsilon((A - B, \infty))$. Let

$$W^1 = W^u(p_0, P_H, N_H^u), \quad W^{12} = p_0 + W_{12}.$$

These are C^1-manifolds tangent to X_1 and X_{12} at p_0, respectively. Observe that due to Propositions V.1 and 12.12 we have

$$W^1 \subset W^u(\mathcal{O}) \subset \operatorname{bd} W \subset \overline{W}.$$

In particular,

$$W_\beta^u(p_0, P_H) \subset \operatorname{bd} W \subset \overline{W} \qquad \text{for all } \beta \in (0, \beta_0).$$

The following result gives a diffeomorphism from a neighbourhood of p_0 in H onto a neighbourhood of 0 in H mapping the manifolds W^1 and W^{12} into their tangent spaces at p_0.

Proposition 13.1 *There exist an open bounded neighbourhood $O^1_{p_0} \subset N_{p_0} \cap H$ of p_0 in H, an open neighbourhood O^1_0 of 0 in H and a C^1-diffeomorphism $\Phi : O^1_{p_0} \to O^1_0$ such that Φ and Φ^{-1} are Lipschitz continuous,*

$$\Phi(p_0) = 0, \quad D\Phi(p_0) = \mathrm{id}$$

and

$$\Phi(O^1_{p_0} \cap W^j)) = O^1_0 \cap X_j \qquad \textit{for } j \in \{1, 12\}.$$

Proof Recall from Chapter 12 that $(P_H)_u^{-1}(W^u_\beta(p_0, P_H)) \subset W^u_\beta(p_0, P_H)$ for every $\beta \in (0, \beta_0)$. Choose $\beta^* \in (0, \beta_0)$ and an open neighbourhood N_2 of 0 in X_2 so small that

$$\begin{gathered}\{s\psi_H : |s| < \beta^*\} + N_2 \subset N_{12} \quad \text{and} \\ W^u_{\beta^*}(p_0, P_H) \subset p_0 + N_{12} + N_3.\end{gathered}$$

Then $W^u_{\beta^*}(p_0, P_H) \subset p_0 + W_{12}$ follows. Thus, for $x_1 = s\psi_H$ with $|s| < \beta^*$, we have

$$\begin{gathered}x_1 + \mathrm{Pr}_{X_2}(w^u_H(x_1)) = \mathrm{Pr}_{X_{12}}(x_1 + \mathrm{Pr}_{X_2}(w^u_H(x_1))) \in N_{12} \quad \text{and} \\ x_1 + w^u_H(x_1) = x_1 + \mathrm{Pr}_{X_2}(w^u_H(x_1)) + w_{12}(x_1 + \mathrm{Pr}_{X_2}(w^u_H(x_1))).\end{gathered}$$

In particular,

$$\mathrm{Pr}_{X_3}\, w^u_H(x_1) = w_{12}(x_1 + \mathrm{Pr}_{X_2}(w^u_H(x_1))). \tag{13.1}$$

Consider the map $\Phi^* : \{s\psi_H : |s| < \beta^*\} + N_2 + X_3 \to H$ given by

$$\Phi^*(x_1 + x_2 + x_3) = x_1 + [x_2 - Pr_{X_2} w^u_H(x_1)] + [x_3 - w_{12}(x_1 + x_2)]$$

for $x_1 = s\psi_H$, $|s| < \beta^*$, $x_2 \in N_2$, $x_3 \in X_3$. For all $x_1 = s\psi_H$ with $|s| < \beta^*$, $x_2 \in N_2$, $x_3 \in X_3$ we obtain

$$\mathrm{Pr}_{X_3}\, \Phi^*(x_1 + x_2 + x_3) = 0 \quad \text{if and only if} \quad x_3 = w_{12}(x_1 + x_2).$$

Furthermore, $\mathrm{Pr}_{X_2 \oplus X_3}\, \Phi^*(x_1 + x_2 + x_3) = 0$ if and only if $x_2 = \mathrm{Pr}_{X_2}\, w^u_H(x_1)$ and $x_3 = w_{12}(x_1 + x_2)$. Consequently, using (13.1), we conclude that for all $x_1 = s\psi_H$ with $|s| < \beta^*$, $x_2 \in N_2$, $x_3 \in X_3$ we have $\mathrm{Pr}_{X_2 \oplus X_3}\, \Phi^*(x_1 + x_2 + x_3) = 0$ if and only if $x_2 + x_3 = w^u_H(x_1)$. Note that Φ^* is C^1-smooth, $\Phi^*(0) = 0$ and $\Phi^*(0) = \mathrm{id}$. We apply the Inverse Function Theorem to the composition $\Phi^{**} : p_0 + \{s\psi_H : |s| < \beta^*\} + N_2 + X_3 \ni y \mapsto \Phi^*(y - p_0) \in H$ and obtain the desired neighbourhoods $O^1_{p_0}$ of p_0, O^1_0 of 0 and the desired map Φ as a restriction of Φ^{**} to $O^1_{p_0}$. □

Choose real constants c_1 and c_2 such that

$$r_M < c_1 < c_2 < \lambda_c.$$

By Theorem I.1 there is an equivalent norm $|\cdot|_H$ on H such that

$$|D_{12} P_H(p_0) x_{12}|_H \geq \frac{\lambda_c + c_2}{2} |x_{12}|_H \qquad \text{for all } x_{12} \in X_{12}, \tag{13.2}$$

$$|D_3 P_H(p_0) x_3|_H \le \frac{r_M + c_1}{2} |x_3|_H \qquad \text{for all } x_3 \in X_3, \tag{13.3}$$

$$|x|_H = |\operatorname{Pr}_{X_{12}}|_H + |\operatorname{Pr}_{X_3}|_H \quad \text{for all } x \in H, \quad |\operatorname{Pr}_{X_{12}}|_H = |\operatorname{Pr}_{X_3}|_H = 1.$$

Recall the subsets D_r, D_r^-, D_r^0, D_r^{-0}, $0 < r$, of the plane $\Pi_3 H = \{y \in \mathbb{R}^3 : y_1 = 0\}$ of $\mathbb{R}^3$ defined in Chapter 12. Recall from Proposition 12.11 that there exists $r^* > 0$ with

$$D_r \cap \Pi_3 W = D_r^-, \; D_r \cap \Pi_3(\operatorname{bd} W) = D_r^0 \qquad \text{for all } 0 < r < r^*.$$

By the continuity of Π_3^{-1} on $\Pi_3 \overline{W}$ we find $r_0 \in (0, r^*)$ with $\Pi_3^{-1}(D_{r_0}^{-0}) \subset O_{p_0}^1$. For $r \in (0, r_0)$ define the following subsets of X_{12}:

$$\begin{aligned} \Delta_r^0 &= \operatorname{Pr}_{X_{12}}(\Phi(\Pi_3^{-1}(D_r^0))), \\ \Delta_r^- &= \operatorname{Pr}_{X_{12}}(\Phi(\Pi_3^{-1}(D_r^-))), \\ \Delta_r^{-0} &= \Delta_r^- \cup \Delta_r^0. \end{aligned}$$

The next result shows that, for small $r > 0$, the set $\Phi(\Pi_3^{-1}(D_r^{-0}))$ is the graph of a Lipschitz continuous function.

Proposition 13.2 *There exist $r_1 \in (0, r_0)$, a constant $K > 1$ and a function $\Sigma : \Delta_{r_1}^{-0} \to X_3$ such that*
(i) $|\Sigma(x_{12}) - \Sigma(y_{12})|_H \le K |x_{12} - y_{12}|_H$ *for all x_{12}, y_{12} in $\Delta_{r_1}^{-0}$;*
(ii) $\Phi(\Pi_3^{-1}(D_{r_1}^{-0})) = \{x + \Sigma(x) : x \in \Delta_{r_1}^{-0}\}$;
(iii) $\Sigma(x_1) = 0$ *for all $x_1 \in \Delta_{r_1}^{-0} \cap X_1$.*

Proof 1. We first show that there exist $r_1^* \in (0, r_0)$ and a constant $K > 1$ such that for all ϕ, ψ in $\Phi(\Pi_3^{-1}(D_{r_1^*}^{-0}))$ we have

$$|\operatorname{Pr}_{X_3} \phi - \operatorname{Pr}_{X_3} \psi|_H \le K |\operatorname{Pr}_{X_{12}} \phi - \operatorname{Pr}_{X_{12}} \psi|_H. \tag{13.4}$$

If this statement is not true then there exist sequences $(r_n)_0^\infty$ in $(0, r_0)$, $(x_{12}^n)_0^\infty$ and $(y_{12}^n)_0^\infty$ in X_{12}, $(x_3^n)_0^\infty$ and $(y_3^n)_0^\infty$ in X_3 with $r_n \to 0$ as $n \to \infty$, $x_3^n \ne y_3^n$ for all $n \in \mathbb{N}$, and

$$\phi^n = x_{12}^n + x_3^n, \; \psi^n = y_{12}^n + y_3^n \text{ in } \Phi(\Pi_3^{-1}(D_{r_n}^{-0})) \qquad \text{for all } n \in \mathbb{N},$$

$$\frac{|x_{12}^n - y_{12}^n|_H}{|x_3^n - y_3^n|_H} \to 0 \qquad \text{as } n \to \infty.$$

It follows that

$$\phi^n \to 0 \text{ and } \psi^n \to 0 \qquad \text{as } n \to \infty.$$

Let

$$\phi_*^n = \Phi^{-1}(\phi^n), \quad \psi_*^n = \Phi^{-1}(\psi^n) \qquad \text{for every } n \in \mathbb{N}.$$

Then ϕ_*^n, ψ_*^n are in $H \cap \overline{W}$ for all $n \in \mathbb{N}$, and $\phi_*^n \to p_0$, $\psi_*^n \to p_0$ as $n \to \infty$. Consider the solutions $x^{\phi_*^n} : \mathbb{R} \to \mathbb{R}$ and $x^{\psi_*^n} : \mathbb{R} \to \mathbb{R}$ of equation (1.1). For each $n \in \mathbb{N}$ set

$$w^n = \frac{x^{\phi_*^n} - x^{\psi_*^n}}{||\phi_*^n - \psi_*^n||}.$$

Clearly, $w_0^n \in H$ for all $n \in \mathbb{N}$. We claim that $\mathrm{Pr}_{X_{12}} w_0^n \to 0$ as $n \to \infty$. To show this, let $\epsilon > 0$ and $\epsilon^* \in (0,1)$ with $||\mathrm{Pr}_{X_{12}}||\epsilon^*/(1-\epsilon^*) < \epsilon$ be given. Due to the C^1-smoothness of Φ^{-1}, there exists $\delta^* > 0$ so that $H_{\delta^*} \subset O_0^1$ and $||D\Phi^{-1}(\psi) - D\Phi^{-1}(0)|| < \epsilon^*$ for all $\psi \in H_{\delta^*}$. Let $\epsilon_n = ||x_3^n - y_3^n||^{-1}||x_{12}^n - y_{12}^n||$, $n \in \mathbb{N}$. Choose a positive integer N such that $||\phi^n|| < \delta^*$, $||\psi^n|| < \delta^*$, $\epsilon_n < 1$, and

$$\frac{||\mathrm{Pr}_{X_{12}}||\epsilon^*(\epsilon_n + 1) + \epsilon_n}{(1-\epsilon^*)(1-\epsilon_n)} < \epsilon$$

for all integers $n \geq N$. Recall $D\Phi^{-1}(0) = \mathrm{id}$. For every integer $n \geq N$, we obtain

$$\begin{aligned}
||\mathrm{Pr}_{X_{12}} w_0^n|| &= \frac{||\mathrm{Pr}_{X_{12}}(\Phi^{-1}(\phi^n) - \Phi^{-1}(\psi^n))||}{||\Phi^{-1}(\phi^n) - \Phi^{-1}(\psi^n)||} \\
&= \frac{||\mathrm{Pr}_{X_{12}} \int_0^1 D\Phi^{-1}(s\phi^n + (1-s)\psi^n)\,ds(\phi^n - \psi^n)||}{||\int_0^1 D\Phi^{-1}(s\phi^n + (1-s)\psi^n)\,ds(\phi^n - \psi^n)||} \\
&\leq \frac{||x_{12}^n - y_{12}^n|| + \epsilon^*||\phi^n - \psi^n||\ ||\mathrm{Pr}_{X_{12}}||}{(1-\epsilon^*)||\phi^n - \psi^n||} \\
&\leq \frac{\epsilon_n + \epsilon^*||\mathrm{Pr}_{X_{12}}||(\epsilon_n + 1)}{(1-\epsilon^*)(1-\epsilon_n)} < \epsilon
\end{aligned}$$

which justifies the claim.

Consider the maps

$$b^n : \mathbb{R} \ni t \mapsto \int_0^1 f'(sx^{\phi_*^n}(t-1) + (1-s)x^{\psi_*^n}(t-1))\,ds \in \mathbb{R}, \qquad n \in \mathbb{N}.$$

For every $n \in \mathbb{N}$ the function w^n is a solution of the differential equation

$$\dot{w}(t) = -\mu w(t) + b^n(t) w(t-1).$$

Recall that all ϕ_*^n, ψ_*^n are in $\overline{W}$ and $\phi_*^n \to p_0$, $\psi_*^n \to p_0$ as $n \to \infty$. It follows that the functions b^n converge to $\mathbb{R} \ni t \mapsto f'(p(t-1)) \in \mathbb{R}$ as $n \to \infty$ uniformly on compact subsets of $\mathbb{R}$. By Propositions 4.1 and 4.2 there are constants $b_0 > 0$ and $b_1 > b_0$ such that $b_0 \leq b^n(t) \leq b_1$ for all $t \leq 0$ and $n \in \mathbb{N}$. As $||w_0^n|| = 1$ and $V(w_t^n) \leq 2$ for all $t \in \mathbb{R}$ and $n \in \mathbb{N}$, Lemma VI.5 implies the existence of $c > 0$ such that $|w^n(t)| \leq ce^{c|t|}$ for all $t \leq 0$ and $n \in \mathbb{N}$. By Lemma VI.4 there exist a subsequence $(w^{n_k})_{k=0}^{\infty}$ of $(w^n)_0^{\infty}$ and a differentiable function $w : (-\infty, 0] \to \mathbb{R}$ so that $w^{n_k}|_{(-\infty,0]} \to w$ and $\dot{w}^{n_k}|_{(-\infty,0]} \to \dot{w}$ as $k \to \infty$ uniformly on every compact subset of $(-\infty, 0]$, and w satisfies

$$\dot{w}(t) = -\mu w(t) + f'(p(t-1))w(t-1) \qquad \text{for all } t \leq 0.$$

In addition, $w_0 \in H$, $||w_0|| = 1$, and $\mathrm{Pr}_{X_{12}} w_0 = 0$ by the definition of w^n and the claim above. Using Lemma VI.1(i) we infer from $V(w_t^n) \leq 2$ for all $n \in \mathbb{N}$ and $t \leq 0$ that $V(w_t) \leq 2$ for all $t \leq 0$. Proposition 8.5(i) implies $w_0 \in C_{r_M<}$. From Proposition 12.10(ii) it follows that $\mathrm{Pr}_{H,\dot{p}_0} C_{r_M<} = X_{12}$. Therefore, $w_0 \in C_{r_M<}$ and $w_0 \in H$ imply $w_0 \in X_{12}$. This contradicts $\mathrm{Pr}_{X_{12}} w_0 = 0$ and $||w_0|| = 1$. Thus there exist $r_1^* \in (0, r_0)$ and $K > 1$ so that (13.4) holds for all ϕ and ψ in $\Phi(\Pi_3^{-1}(D_{r_1^*}^{-0}))$.

2. By Proposition 12.11 and by the definition of the sets $W^u_\beta(p_0, P_H)$ it is not difficult to see that we can choose $\hat\beta \in (0, \beta_4)$ so small that Π_3 maps $W^u_{\hat\beta}(p_0, P_H)$ into $D^{-0}_{r_1^*}$ and $W^u_{\hat\beta}(p_0, P_H) \subset O^1_{p_0}$. Proposition 13.1 implies $\Phi(W^u_{\hat\beta}(p_0, P_H)) \subset X_1$. As $W^u_{\hat\beta}(p_0, P_H) = \{p_0 + s\psi_H + w^u_H(s\psi_H) : |s| < \hat\beta\}$ and Φ is a diffeomorphism with $\Phi(W^u_{\hat\beta}(p_0, P_H)) \subset X_1$, it follows that $\Phi(W^u_{\hat\beta}(p_0, P_H))$ is open in X_1 in the relative topology. Using this and $\Phi(p_0) = 0$ we obtain that there exists $r_1 \in (0, r_1^*)$ with $\Delta^{-0}_{r_1} \cap X_1 \subset \Phi(W^u_{\hat\beta}(p_0, P_H))$.

By part 1 and $r_1 < r_1^*$, for every $x_{12} \in \Delta^{-0}_{r_1}$ there exists a unique $x_3 \in X_3$ such that $x_{12} + x_3 \in \Phi(\Pi_3^{-1}(D^{-0}_{r_1}))$. Set $\Sigma(x_{12}) = x_3$. Then (i) and (ii) are satisfied. In order to show (iii) let $x_1 \in \Delta^{-0}_{r_1} \cap X_1$. Then $\Sigma(x_1) \in X_3$ and $x_1 + \Sigma(x_1) \in \Phi(\Pi_3^{-1}(D^{-0}_{r_1})) \subset \Phi(\Pi_3^{-1}(D^{-0}_{r_1^*}))$. By the choice of $\hat\beta$ and r_1 we have

$$x_1 \in \Phi(W^u_{\hat\beta}(p_0, P_H)) \subset \Phi(\Pi_3^{-1}(D^{-0}_{r_1^*})).$$

As $x_1 + 0 + 0$ and $x_1 + 0 + \Sigma(x_1)$ are in $\Phi(\Pi_3^{-1}(D^{-0}_{r_1^*}))$, we obtain $\Sigma(x_1) = 0$ from (13.4). □

For every $\delta > 0$ set

$$\begin{aligned}
X_1^\delta &= \{u\psi_H : |u| < \delta\}, \\
X_{12}^\delta &= \{u\psi_H + v\phi_H : |u| < \delta,\ |v| < \delta\}, \\
X_{12}^{\delta,-0} &= \{u\psi_H + v\phi_H : |u| < \delta,\ -\delta < v \le 0\}, \\
X_{12}^{\delta,-} &= \{u\psi_H + v\phi_H : |u| < \delta,\ -\delta < v < 0\}, \\
X_{12}^{\delta,+} &= \{u\psi_H + v\phi_H : |u| < \delta,\ 0 < v < \delta\}.
\end{aligned}$$

The next proposition locates the sets $\Delta^{-0}_{r_1}$ and $\Delta^0_{r_1}$ in X_{12}.

Proposition 13.3
(i) For every $r \in (0, r_1]$, Δ^-_r is an open and connected subset of X_{12}.
(ii) For all u, v in $\mathbb{R}$ with $u\psi_H + v\phi_H \in \Delta^{-0}_{r_1}$,

$$\begin{aligned}
u\psi_H + v\phi_H \in \Delta^0_{r_1} \quad &\textit{if and only if} \quad v = 0, \\
u\psi_H + v\phi_H \in \Delta^-_{r_1} \quad &\textit{if and only if} \quad v < 0.
\end{aligned}$$

(iii) There exists $\delta_{00} > 0$ such that

$$X_{12}^{\delta_{00},-0} \subset \Delta^{-0}_{r_1}.$$

Proof 1. The map

$$a : D^{-0}_{r_1} \ni \xi \mapsto \mathrm{Pr}_{X_{12}}(\Phi(\Pi_3^{-1}(\xi))) \in X_{12}$$

is Lipschitz continuous due to Proposition 13.1, and we have $a(D^{-0}_{r_1}) = \Delta^{-0}_{r_1}$. By Proposition 13.2 its inverse

$$b : \Delta^{-0}_{r_1} \ni x \mapsto \Pi_3(\Phi^{-1}(x + \Sigma(x))) \in \mathbb{R}^3$$

is also Lipschitz continuous; there exists $K_1 > 0$ with $|b(x) - b(y)|_{\mathbb{R}^3} \le K_1 |x - y|_H$ for all x, y in $\Delta_{r_1}^{-0}$. For every $r \in (0, r_1]$ the image $\Delta_r^- = a(D_r^-)$ of the connected open subset D_r^- of the plane $\{x \in \mathbb{R}^3 : x_1 = 0\}$ under the injective continuous map a is connected and open in the 2-dimensional space X_{12}.

2. Proof of (ii). Let u, v be given with $x = u\psi_H + v\phi_H \in \Delta_{r_1}^{-0}$. Suppose $x \in \Delta_{r_1}^0$. Then there is $\xi \in D_{r_1}^0$ such that $x = \mathrm{Pr}_{X_{12}}(\Phi(\Pi_3^{-1}(\xi)))$. $r_1 < r_0$ and the choice of r_0 imply $\Pi_3^{-1}(\xi) \in O_{p_0}^1$. From $D_{r_1}^0 = D_{r_1} \cap \Pi_3(W_{\beta_3}^u(p_0, P_H)) \subset \Pi_3(W_{\beta_3}^u(p_0, P_H))$ and $W_{\beta_3}^u(p_0, P_H) \subset \overline{W}$ it follows that $\phi = \Pi_3^{-1}(\xi) \in W_{\beta_3}^u(p_0, P_H) \subset W^u(p_0, P_H, N_H^u) = W^1$. Thus $\phi \in O_{p_0}^1 \cap W^1$. Then $x = \Phi(\phi) \in X_1$ by Proposition 13.1, and thus $v = 0$ follows.

Suppose $v = 0$. Then there is $\phi \in \Pi_3^{-1}(D_{r_1}^{-0})$ such that $u\psi_H = x = \mathrm{Pr}_{X_{12}} \Phi(\phi)$. By Proposition 13.2 (ii), we have

$$\Phi(\phi) = \mathrm{Pr}_{X_{12}}(\Phi(\phi)) + \Sigma(\mathrm{Pr}_{X_{12}}(\Phi(\phi))).$$

Using $\Sigma(x_1) = 0$ for $x_1 \in X_1 \cap \Delta_{r_1}^{-0}$ and $\mathrm{Pr}_{X_{12}} \Phi(\phi) = u\psi_H \in X_1$, we obtain $\Phi(\phi) = u\psi_H$. By the choice of r_0, $\Phi(\phi) \in O_0^1$. So $\Phi(\phi) \in O_0^1 \cap X_1$. Hence $\phi \in W^1 \cap O_{p_0}^1$ follows by Proposition 13.1. By $W^1 \subset \mathrm{bd}\, W$, $\phi \in \mathrm{bd}\, W$. Proposition 12.11 shows $\Pi_3(\phi) \in D_{r_1}^0$. Consequently, $x = u\psi_H \in \mathrm{Pr}_{X_{12}}(\Phi(\Pi_3^{-1}(D_{r_1}^0))) = \Delta_{r_1}^0$. This verifies the first equivalence in assertion (ii).

Next we show that there exist reals a and $b < 0$ so that $a\psi_H + b\phi_H \in \Delta_{r_1}^-$. Consider the linear map $k : X_2 \to \mathbb{R}$ given by $k(\phi_H) = 1$. Choose $\epsilon > 0$ with $\Upsilon((-\epsilon, \epsilon)) \subset O_{p_0}^1$. Using $\Upsilon(0) = p_0$, $\Upsilon'(0) = \phi_H$, and $D\Phi(p_0) = \mathrm{id}$ we obtain

$$\frac{d}{ds} \left[(-\epsilon, \epsilon) \ni s \mapsto k\, \mathrm{Pr}_{X_2} \mathrm{Pr}_{X_{12}} \Phi(\Upsilon(s)) \right](0) = 1.$$

Hence it is easy to see that there is $s^* > 0$ such that $k\, \mathrm{Pr}_{X_2} \mathrm{Pr}_{X_{12}} \Phi(\Upsilon(s)) < 0$ for $-s^* < s < 0$. By Proposition 12.9, $\Upsilon(s) \in H \cap W$ for all $s \in (A - B, 0)$, and $\Upsilon(0) = p_0$. Consequently, if $s \in (-s^*, 0)$ is small enough then $\Upsilon(s) \in \Pi_3^{-1}(D_{r_1}^-)$, $\mathrm{Pr}_{X_{12}} \Phi(\Upsilon(s)) \in \Delta_{r_1}^-$, and $\mathrm{Pr}_{X_{12}} \Phi(\Upsilon(s)) = a\psi_H + b\phi_H$ with $a \in \mathbb{R}$ and with $b < 0$.

Now we use the last result, the fact that $\Delta_{r_1}^-$ is arcwise connected, and the first equivalence in (ii) to show that for all reals u, v with $u\psi_H + v\phi_H \in \Delta_{r_1}^-$, necessarily $v < 0$. The remaining part of the second equivalence in assertion (ii) is obvious from the first equivalence in assertion (ii).

3. Proof of (iii). By Proposition 12.11 we can choose $\hat{\beta} > 0$ so that

$$\Pi_3(W_{\hat{\beta}}^u(p_0, P_H)) \subset D_{r_1}^0 \text{ and } W_{\hat{\beta}}^u(p_0, P_H) \subset \mathrm{bd}\, W \subset \overline{W}.$$

Then assertion (ii) implies $\mathrm{Pr}_{X_{12}} \Phi(W_{\hat{\beta}}^u(p_0, P_H)) \subset X_1 \cap \Delta_{r_1}^0$. Consider the map

$$X_1^{\hat{\beta}} \ni \psi \mapsto \mathrm{Pr}_{X_{12}} \Phi(p_0 + \psi + w_H^u(\psi)) \in X_1.$$

Its derivative at 0 is the identity map since $Dw_H^u(0) = 0$ and $D\Phi(p_0) = \mathrm{id}$. The Inverse Function Theorem implies the existence of $\epsilon > 0$ so that $X_1^\epsilon \subset \mathrm{Pr}_{X_{12}} \Phi(W_{\hat{\beta}}^u(p_0, P_H))$. It follows that $X_1^\epsilon \subset \Delta_{r_1}^0$.

Choose $\delta_{00} \in (0, \epsilon)$ with

$$\sup\{|u\psi_H + v\phi_H|_H : |u| < \delta_{00},\ |v| < \delta_{00}\} \leq \frac{r_1}{2K_1}.$$

Assertion (ii), the relations $\Pi_3 p_0 \in D_{r_1}^{-0} \subset \overline{D_{r_1}^-}$, $\mathrm{Pr}_{X_{12}} \Phi(\Pi_3^{-1}(\Pi_3 p_0)) = 0$, $\Delta_{r_1}^- = \mathrm{Pr}_{X_{12}} \Phi(\Pi_3^{-1}(D_{r_1}^-))$, and the continuity of the map $D_{r_1}^{-0} \ni \chi \mapsto \mathrm{Pr}_{X_{12}} \Phi(\Pi_3^{-1}(\chi)) \in X_{12}$ at $\Pi_3 p_0$ combined imply that there exists $x = u\psi_H + v\phi_H \in \Delta_{r_1}^{-0}$ with $|u| < \delta_{00}$ and $-\delta_{00} < v < 0$. Assume that there is $\hat{x} = \hat{u}\psi_H + \hat{v}\phi_H \in X_{12} \setminus \Delta_{r_1}^{-0}$ with $|\hat{u}| < \delta_{00}$, $-\delta_{00} < \hat{v} < 0$. Consider the curve $c : [0,1] \ni s \mapsto (1-s)x + s\hat{x}$. Clearly, $|c| \cap X_1 = \emptyset$. Thus, by (ii), $|c| \cap \Delta_{r_1}^0 = \emptyset$. Let $s_0 = \inf\{s \in [0,1] :\ c(s) \notin \Delta_{r_1}^{-0}\}$. $s_0 > 0$ since $\Delta_{r_1}^-$ is open in X_{12} and $c(0) = x \in \Delta_{r_1}^-$. By the definition of s_0 and $|c| \cap \Delta_{r_1}^0 = \emptyset$, we have $c(s) \in \Delta_{r_1}^-$ for $0 \leq s < s_0$. Consider the curve

$$d : [0, s_0) \ni s \mapsto \Pi_3 \Phi^{-1}(c(s) + \Sigma(c(s))) \in D_{r_1}^-.$$

As $|c(s) - 0|_H \leq \frac{r_1}{2K_1}$ and as b is Lipschitz continuous with Lipschitz constant K_1,

$$|d(s) - \Pi_3 p_0|_{\mathbb{R}^3} \leq K_1 \frac{r_1}{2K_1} = \frac{r_1}{2} \quad \text{for } 0 \leq s < s_0$$

and

$$|d(s') - d(s'')|_{\mathbb{R}^3} \leq K_1 |c(s') - c(s'')|_H = K_1 |\hat{x} - x|_H |s' - s''| \quad \text{for } 0 \leq s' \leq s'' < s_0.$$

We conclude that d has a limit d_0 at s_0 and $d_0 \in \overline{\{d(s) : 0 \leq s < s_0\}} \subset \overline{D_{r_1/2}^{-0}} \subset D_{r_1}^{-0}$. It follows that $c(s_0) = \mathrm{Pr}_{X_{12}} \Phi(\Pi_3^{-1}(d_0))$. So, $c(s_0) \in \Delta_{r_1}^{-0}$. From $|c| \cap \Delta_{r_1}^0 = \emptyset$ we infer $c(s_0) \in \Delta_{r_1}^-$. As $\Delta_{r_1}^-$ is open in X_{12}, this contradicts the definition of s_0. Therefore, $X_{12}^{\delta_{00},-0} \subset \Delta_{r_1}^{-0}$. □

Fix $\delta_{00} > 0$ such that $X_{12}^{\delta_{00},-0} \subset \Delta_{r_1}^{-0}$. Define the function $\Sigma^* : X_{12}^{\delta_{00}} \to X_3$ by

$$\Sigma^*(x) = \begin{cases} \Sigma(x) & \text{if } x \in X_{12}^{\delta_{00},-0}, \\ 0 & \text{if } x \in X_{12}^{\delta_{00},+}. \end{cases}$$

From Proposition 13.2 it follows that Σ^* is Lipschitz continuous. Our aim is to show that there is $\delta_{01} \in (0, \delta_{00})$ such that $\Sigma^*|_{X_{12}^{\delta_{01}}}$ is C^1-smooth. Recall that $\Delta_{r_1}^-$ is an open subset of X_{12}.

Proposition 13.4 $\Sigma|_{\Delta_{r_1}^-} : \Delta_{r_1}^- \to X_3$ *is a* C^1*-map.*

Proof The maps

$$b : \Delta_{r_1}^- \ni x_{12} \mapsto \Pi_3(\Phi^{-1}(x_{12} + \Sigma(x_{12}))) \in \Pi_3 H$$

and

$$a : D_{r_1}^- \ni \xi \mapsto \mathrm{Pr}_{X_{12}}(\Phi(\Pi_3^{-1}(\xi))) \in X_{12}$$

are inverse to each other; $D_{r_1}^-$ is the image of the map b above, and $a(D_{r_1}^-) = \Delta_{r_1}^-$. Propositions 13.1 and 6.6 combined yield that a is C^1-smooth.

Let $\xi \in D^-_{r_1}$. We claim that

$$Da(\xi) = \mathrm{Pr}_{X_{12}} \circ D\Phi(\Pi_3^{-1}(\xi)) \circ D\Pi_3^{-1}(\xi)|_{\Pi_3 H}$$

is injective. Let $\gamma : (-1,1) \to \mathbb{R}^3$ be a C^1-curve with $|\gamma| \subset D^-_{r_1}$, $\gamma(0) = \xi$ and $\gamma'(0) \neq 0$. Proposition 13.2 implies that the curves $\tilde{\gamma} : (-1,1) \ni s \mapsto \Phi(\Pi_3^{-1}(\gamma(s))) \in H$ and $\hat{\gamma} : (-1,1) \ni s \mapsto \mathrm{Pr}_{X_{12}} \Phi(\Pi_3^{-1}(\gamma(s))) \in X_{12}$ satisfy $\tilde{\gamma}(s) = \hat{\gamma}(s) + \Sigma(\hat{\gamma}(s))$ for all $s \in (-1,1)$. By the choice of r_0 and $r_1 < r_0$, $\Pi_3^{-1}(\xi) \in O^1_{p_0}$. The maps $D\Phi(\Pi_3^{-1}(\xi))$ and $D\Pi_3^{-1}(\xi)$ are injective because of Propositions 13.1 and 6.6, respectively. Consequently, $\tilde{\gamma}'(0) \neq 0$. Proposition 13.2(i) and $\tilde{\gamma}(s) = \hat{\gamma}(s) + \Sigma(\hat{\gamma}(s))$ for $|s| < 1$ imply

$$|\tilde{\gamma}(s) - \tilde{\gamma}(0)|_H \leq (1+K)|\hat{\gamma}(s) - \hat{\gamma}(0)|_H \qquad \text{for all } s \in (-1,1).$$

Dividing both sides by $|s| \neq 0$ and then letting $s \to 0$, we get $|\tilde{\gamma}'(0)|_H \leq (1+K)|\hat{\gamma}'(0)|_H$. Hence $\hat{\gamma}'(0) \neq 0$ follows. Thus $Da(\xi)$ is injective. The Inverse Function Theorem implies that there are open neighbourhoods U_ξ of ξ in D^-_r and $U_{a(\xi)}$ of $a(\xi)$ in $\Delta^-_{r_1}$ so that a maps U_ξ onto $U_{a(\xi)}$, and the inverse $a^{-1} : U_{a(\xi)} \to \Pi_3 H$ of $a|_{U_\xi}$ is C^1-smooth. Using Proposition 13.2(ii), it is easy to see that

$$\Sigma(x) = (\mathrm{id} - \mathrm{Pr}_{X_{12}})(\Phi(\Pi_3^{-1}(a^{-1}(x)))) \qquad \text{for all } x \in U_{a(\xi)}.$$

This implies that the restriction of Σ to $U_{a(\xi)}$ is C^1-smooth, and it follows that the restriction of Σ to $\Delta^-_{r_1}$ is C^1-smooth. □

Choose a constant c_3 such that

$$0 < c_3 < \min\left\{\frac{c_2}{K}, \frac{c_1(c_2 - c_1)}{K(c_1 + c_2)}\right\}.$$

There exists a convex open neighbourhood O^2_0 of 0 in H such that $O^2_0 \subset O^1_0$ and

$$P_H(\Phi^{-1}(O^2_0)) \subset O^1_{p_0}, \quad P_H(\Phi^{-1}(O^2_0 \cap X_{12})) \subset W^{12}. \tag{13.5}$$

The map

$$P^*_H : O^2_0 \ni x \mapsto \Phi(P_H(\Phi^{-1}(x))) \in H$$

satisfies $P^*_H(O^2_0) \subset O^1_0$. P^*_H is C^1-smooth since Φ, Φ^{-1} and P_H are C^1-smooth. Then the partial derivatives $D_i P^*_H : O^2_0 \ni x \mapsto D_i P^*_H(x) \in L(X_i, H)$, $i \in \{12, 3\}$, of P^*_H are continuous. Consequently, we may choose $c_4 > 0$ so that

$$H(c_4) = \{x \in H : |x|_H < c_4\} \subset O^2_0,$$
$$|[D_{12}P^*_H(x) - D_{12}P^*_H(0)]x_{12}|_H \leq c_4 |x_{12}|_H \quad \text{for all } x \in H(c_4) \text{ and } x_{12} \in X_{12},$$
$$|[D_3 P^*_H(x) - D_3 P^*_H(0)]x_3|_H \leq c_4 |x_3|_H \quad \text{for all } x \in H(c_4) \text{ and } x_3 \in X_3,$$
$$\frac{\lambda_c + c_2}{2} - c_4 \geq c_2,$$
$$c_4 \leq c_3,$$
$$\frac{r_M + c_1}{2} + c_4 \leq c_1.$$

Introduce the notation $x_i = \mathrm{Pr}_{X_i} x$, $y_i = \mathrm{Pr}_{X_i} y$ for $i \in \{1, 2, 3, 12\}$ and x, y in H. Define $c(x,y) = \sup_{0 \leq s \leq 1} ||\mathrm{Pr}_{X_3} \circ D_{12} P^*_H(y + s(x-y))||$ for all x, y in $H(c_4)$.

Proposition 13.5 *For every x, y in $H(c_4)$*

$$|\operatorname{Pr}_{X_{12}} P_H^*(x) - \operatorname{Pr}_{X_{12}} P_H^*(y)|_H \geq c_2|x_{12} - y_{12}|_H - c_3|x_3 - y_3|_H,$$
$$|\operatorname{Pr}_{X_3} P_H^*(x) - \operatorname{Pr}_{X_3} P_H^*(y)|_H \leq c_1|x_3 - y_3|_H + c(x,y)|x_{12} - y_{12}|_H.$$

Proof Let x, y in $H(c_4)$ be given. We have

$$\begin{aligned} P_H^*(x) - P_H^*(y) =& DP_H^*(0)(x-y) + \int_0^1 [DP_H^*(y + s(x-y)) - DP_H^*(0)]\,(x-y)\,ds \\ =& D_{12}P_H^*(0)(x_{12} - y_{12}) + D_3 P_H^*(0)(x_3 - y_3) \\ &+ \int_0^1 [D_{12}P_H^*(y + s(x-y)) - D_{12}P_H^*(0)]\,(x_{12} - y_{12})\,ds \\ &+ \int_0^1 [D_3 P_H^*(y + s(x-y)) - D_3 P_H^*(0)]\,(x_3 - y_3)\,ds. \end{aligned}$$

From $D\Phi(0) = \mathrm{id}$ it follows that $DP_H^*(0) = DP_H(p_0)$. Hence $DP_H^*(0)X_i \subset X_i$ for $i \in \{12, 3\}$, and thus

$$\operatorname{Pr}_{X_{12}} D_3 P_H^*(0)(x_3 - y_3) = \operatorname{Pr}_{X_{12}} DP_H^*(0)(x_3 - y_3) = 0,$$
$$\operatorname{Pr}_{X_3} D_{12} P_H^*(0)(x_{12} - y_{12}) = \operatorname{Pr}_{X_3} DP_H^*(0)(x_{12} - y_{12}) = 0.$$

Using (13.2) and (13.3), the above equality for $P_H^*(x) - P_H^*(y)$, the choice of c_4, and the above observation for the partial derivatives of P_H^*, we obtain

$$\begin{aligned} &|\operatorname{Pr}_{X_{12}} P_H^*(x) - \operatorname{Pr}_{X_{12}} P_H^*(y)|_H \\ &\geq \frac{\lambda_c + c_2}{2}|x_{12} - y_{12}|_H - c_4|x_{12} - y_{12}|_H - c_4|x_3 - y_3|_H \\ &\geq c_2|x_{12} - y_{12}|_H - c_3|x_3 - y_3|_H \end{aligned}$$

and

$$\begin{aligned} &|\operatorname{Pr}_{X_3} P_H^*(x) - \operatorname{Pr}_{X_3} P_H^*(y)|_H \\ &\leq \frac{r_M + c_1}{2}|x_3 - y_3|_H + c_4|x_3 - y_3|_H + c(x,y)|x_{12} - y_{12}|_H \\ &\leq c_1|x_3 - y_3|_H + c(x,y)|x_{12} - y_{12}|_H. \end{aligned}$$

□

Choose an open neighbourhood $O_{p_0}^2$ of p_0 in H such that $O_{p_0}^2 \subset O_{p_0}^1$, $\Phi(O_{p_0}^2) \subset O_0^2 \cap H(c_4)$, $\Pi_3(O_{p_0}^2 \cap \overline{W}) \subset D_{r_1}^{-0}$ and $O_0^2 \subset N_{p_0}$.

Proposition 13.6 *There exists an open neighbourhood $O_{p_0}^3$ of p_0 in H with $O_{p_0}^3 \subset O_{p_0}^2$ so that the following holds.*

(i) For every $\phi \in O_{p_0}^3 \cap \overline{W}$ there is a unique element $\psi \in O_{p_0}^2$ with $P_H(\psi) = \phi$.

(ii) The map

$$P_H^{-1}:\ O_{p_0}^3 \cap \overline{W} \to H$$

given by $P_H^{-1}(\phi) = \psi$, $\psi \in O_{p_0}^2$, $P_H(\psi) = \phi$, is continuous. We have $P_H^{-1}(O_{p_0}^3 \cap \overline{W}) \subset O_0^2 \cap \overline{W}$ and

$$P_H(P_H^{-1}(\phi)) = \phi \qquad \textit{for all } \phi \in O_{p_0}^3 \cap \overline{W}.$$

(iii) The continuous map

$$P_H^{*-1}: \Phi\left(O_{p_0}^3 \cap \overline{W}\right) \ni x \mapsto \Phi(P_H^{-1}(\Phi^{-1}(x))) \in H$$

satisfies $P_H^{*-1}(\Phi\left(O_{p_0}^3 \cap \overline{W}\right)) \subset \Phi(O_{p_0}^2 \cap \overline{W})$ *and*

$$P_H^*(P_H^{*-1}(x)) = x \qquad \textit{for all } x \in \Phi\left(O_{p_0}^3 \cap \overline{W}\right).$$

Proof 1. The injectivity of $P_H|_{O_{p_0}^2}$. Suppose ϕ, ψ are in $O_{p_0}^2$ and $P_H(\phi) = P_H(\psi)$. Then $F(\tau_H(\phi), \phi) = F(\tau_H(\psi), \psi)$. In the case $\tau_H(\phi) = \tau_H(\psi)$ the injectivity of $F(t, \cdot)$ implies $\phi = \psi$. Suppose $\tau_H(\phi) > \tau_H(\psi)$. Then

$$F(\tau_H(\psi), \psi) = F(\tau_H(\phi), \phi) = F(\tau_H(\psi), F(\tau_H(\phi) - \tau_H(\psi), \phi))$$

and the injectivity of $F(\tau_H(\psi), \cdot)$ yield $\psi = F(\tau_H(\phi) - \tau_H(\psi), \phi)$. As $O_{p_0}^2 \subset N_{p_0} \subset p_0 + C_{\tilde{\delta}}$, it follows that $\dot{x}^\phi(t) > 0$ for all $t \in (0, \tilde{\epsilon}]$. By $\phi \in H$, $x^\phi(0) = 0$. So, $F(t, \phi) \notin H$ for all $t \in (0, \tilde{\epsilon}]$. Then $0 < \tau_H(\phi) - \tau_H(\psi) < 2\epsilon_H < \tilde{\epsilon}$ implies $\psi \notin H$, a contradiction.

2. Let $C^1([-\omega - 2, 0], \mathbb{R})$ denote the Banach space of real-valued C^1-functions χ on $[-\omega - 2, 0]$, with the norm given by

$$||\chi|| = \max_{t \in [-\omega-2,0]} |\chi(t)| + \max_{t \in [-\omega-2,0]} |\dot{\chi}(t)|.$$

Proposition 4.3 implies that the map

$$a : \overline{W} \ni \phi \mapsto x^\phi|_{[-\omega-2,0]} \in C^1([-\omega - 2, 0], \mathbb{R})$$

is continuous. As the zeros of p are simple and $\omega \in (1, 2)$, there exists $\epsilon > 0$ such that for every $\phi \in H \cap (p_0 + C_\epsilon) \cap \overline{W}$ the numbers of zeros of x^ϕ and p in $[-\omega - 2, 0]$ are the same, and all zeros of x^ϕ in $[-\omega - 2, 0]$ are simple. In particular, for every $\phi \in H \cap (p_0 + C_\epsilon) \cap \overline{W}$ there exist a largest zero $z_{-1}(\phi)$ of x^ϕ in $(-\omega - 2, 0)$, and a largest zero $z_{-2}(\phi)$ of x^ϕ in $[-\omega - 2, z_{-1}(\phi))$. There exists $\check{\epsilon} \in (0, \epsilon)$ so that for every $\phi \in H \cap (p_0 + C_{\check{\epsilon}}) \cap \overline{W}$, $|\omega + z_{-2}(\phi)| < \epsilon_H$ and $x^\phi_{z_{-2}(\phi)} \in O_{p_0}^2$. Choose the open neighbourhood $O_{p_0}^3$ of p_0 in H so that $O_{p_0}^3 \subset O_{p_0}^2$ and $O_{p_0}^3 \subset H \cap (p_0 + C_{\check{\epsilon}})$. Then for every $\phi \in O_{p_0}^3 \cap \overline{W}$ we have $z_{-2}(\phi) \in (-\omega - \epsilon_H, -\omega + \epsilon_H)$ and $x^\phi_{z_{-2}(\phi)} \in O_{p_0}^2 \subset N_{p_0}$. Consequently, $P_H(x^\phi_{z_{-2}(\phi)}) = \phi$. Part 1 shows that $x^\phi_{z_{-2}(\phi)}$ is the unique $\psi \in O_{p_0}^2$ with $P_H(\psi) = \phi$.

3. Let ϕ and a sequence $(\phi^n)_0^\infty$ in $O_{p_0}^3 \cap \overline{W}$ be given with $\phi^n \to \phi$ as $n \to \infty$. Observe that due to the continuity of a we obtain $z_{-2}(\phi^n) \to z_{-2}(\phi)$ as $n \to \infty$. Recall that the flow F_W is continuous. It follows that

$$F_W(z_{-2}(\phi^n), \phi^n) \to F_W(z_{-2}(\phi), (\phi)) \qquad \text{as } n \to \infty.$$

We conclude that the map $P_H^{-1} : O_{p_0}^3 \cap \overline{W} \to H$ given by $P_H^{-1}(\phi) = \psi$, $\psi \in O_{p_0}^2$, $P_H(\psi) = \phi$ is continuous. The inclusion $P_H^{-1}(O_{p_0}^3 \cap \overline{W}) \subset O_0^2 \cap \overline{W}$ and the equality $P_H(P_H^{-1}(\phi)) = \phi$ for all $\phi \in O_{p_0}^3 \cap \overline{W}$ are obvious from the definition of P_H^{-1}. Finally, (iii) follows from (ii) and the continuity of Φ, Φ^{-1}. □

We claim that there is $\delta_{01}^* \in (0, \delta_{00})$ such that

$$\left\{x + \Sigma(x) : x \in \overline{X_{12}^{\delta_{01}^*, -0}}\right\} \subset \Phi(O_{p_0}^3 \cap \overline{W}). \tag{13.6}$$

From Propositions 12.11, 13.2 and 13.3(iii) it follows that, for $0 < \delta_{01}^* < \delta_{00}$,

$$\left\{x + \Sigma(x) :\ x \in \overline{X_{12}^{\delta_{01}^*, -0}}\right\} \subset \{x + \Sigma(x) :\ x \in \Delta_{r_1}^{-0}\} \subset \Phi(\overline{W} \cap O_{p_0}^1).$$

As the function $\Delta_{r_1}^{-0} \ni x \mapsto \Phi^{-1}(x + \Sigma(x)) \in H$ is continuous at 0 and $\Sigma(0) = 0$, and $\Phi^{-1}(0) = p_0$, it follows that there is $\delta_{01}^* \in (0, \delta_{00})$ so that

$$\Phi^{-1}\left(\left\{x + \Sigma(x) :\ x \in \overline{X_{12}^{\delta_{01}^*, -0}}\right\}\right) \subset O_{p_0}^3.$$

Hence $\left\{x + \Sigma(x) :\ x \in \overline{X_{12}^{\delta_{01}^*, -0}}\right\} \subset \Phi(O_{p_0}^3)$ follows. By the injectivity of Φ, we have $\Phi(O_{p_0}^3 \cap \overline{W}) = \Phi(O_{p_0}^3) \cap \Phi(\overline{W} \cap O_{p_0}^1)$. This shows the existence of $\delta_{01}^* \in (0, \delta_{00})$ with property (13.6).

Observe that (13.6) implies

$$\overline{X_1^{\delta_{01}^*}} \subset H(c_4) \cap O_0^2 \tag{13.7}$$

since $\Sigma(x) = 0$ for $x \in \overline{X_1^{\delta_{01}^*}} \subset X_1^{\delta_{00}}$ and $\Phi(O_{p_0}^3) \subset \Phi(O_{p_0}^2) \subset O_0^2 \cap H(c_4)$ by the choice of $O_{p_0}^3$ and $O_{p_0}^2$.

Choose $\bar{\beta} \in (0, \beta_0)$ so that

$$W_{\bar{\beta}}^u(p_0, P_H) \subset O_{p_0}^3, \quad \Phi(W_{\bar{\beta}}^u(p_0, P_H)) \subset X_1^{\delta_{01}^*}.$$

Then $W_{\bar{\beta}}^u(p_0, P_H) \subset O_{p_0}^3 \cap \overline{W}$ and $\Phi(W_{\bar{\beta}}^u(p_0, P_H)) \subset \Phi(O_{p_0}^3 \cap \overline{W})$. It follows from Proposition 13.1 that one can select $\delta_{01} \in (0, \delta_{01}^*)$ such that

$$\overline{X_1^{\delta_{01}}} \subset \Phi(W_{\bar{\beta}}^u(p_0, P_H)). \tag{13.8}$$

Proposition 13.7 *$\Sigma^*|_{X_{12}^{\delta_{01}}}$ is C^1-smooth and $D\Sigma^*(x) = 0$ for all $x \in X_1^{\delta_{01}}$.*

Proof For $\delta > 0$ and $c > 0$, set

$$X_{12}^{\delta, c} = \{u\psi_H + v\phi_H : |u| < \delta,\ -c < v \leq 0\}$$

and

$$X_1^{\delta}(c) = \{x_1 + x_2 + x_3 \in H : x_1 \in \overline{X_1^{\delta}},\ x_2 \in X_2,\ x_3 \in X_3,\ |x_2|_H < c,\ |x_3|_H < c\}.$$

As $\Sigma^*|_{X_{12}^{\delta_{01}, -}}$ is C^1-smooth by Proposition 13.4 and $\Sigma^*(x) = 0$ for $x \in X_{12}^{\delta_{01}, +} \cup X_1^{\delta_{01}}$, it suffices to show that for every $\epsilon \in (0, 1)$ there exists $c = c(\epsilon) \in (0, \delta_{01})$ such that

$$\sup\left\{\frac{|\Sigma(x_{12}) - \Sigma(y_{12})|_H}{|x_{12} - y_{12}|_H} : x_{12} \in X_{12}^{\delta_{01}, c},\ y_{12} \in X_{12}^{\delta_{01}, c},\ x_{12} \neq y_{12}\right\} \leq \epsilon. \tag{13.9}$$

Let $\epsilon \in (0,1)$ be given. Then

$$\epsilon_1 = \epsilon \left[\frac{c_2 - c_1}{2} - \frac{K(c_1+c_2)c_3}{2c_1} \right]$$

is positive because of the choice of c_3.

From (13.5) it follows that

$$P_H^*(O_0^2 \cap X_{12}) = \Phi(P_H(\Phi^{-1}(O_0^2 \cap X_{12}))) \subset \Phi(O_{p_0}^1 \cap W^{12}) = O_0^1 \cap X_{12}.$$

Consequently,

$$\Pr_{X_3} D_{12} P_H^*(x_{12}) = 0 \qquad \text{for all } x_{12} \in O_0^2 \cap X_{12}.$$

In particular,

$$\Pr_{X_3} D_{12} P_H^*(x_1) = 0 \qquad \text{for all } x_1 \in \overline{X_1^{\delta_{01}^*}}$$

because of (13.7). Then also using the facts that $\overline{X_1^{\delta_{01}^*}}$ is a compact subset of the open subset $H(c_4) \cap O_0^2$ of H and $\Pr_{X_3} D_{12} P_H^*$ is continuous on $H(c_4) \cap O_0^2$, it follows that there exists $c_5 \in (0, \delta_{01}^*)$ with

$$X_1^{\delta_{01}^*}(c_5) \subset H(c_4) \tag{13.10}$$

and

$$|| \Pr_{X_3} D_{12} P_H^*(x) || < \epsilon_1 \qquad \text{for all } x \in X_1^{\delta_{01}^*}(c_5). \tag{13.11}$$

Let m be the nonnegative integer such that

$$\epsilon \left(\frac{\frac{c_2}{c_1}+1}{2} \right)^{m+1} > K \geq \epsilon \left(\frac{\frac{c_2}{c_1}+1}{2} \right)^{m}. \tag{13.12}$$

Recall from Chapter 12 that we have

$$P_H^{-1}(W_{\bar\beta}^u(p_0, P_H)) \subset W_{\bar\beta}^u(p_0, P_H).$$

Hence

$$P_H^{*-1}(\Phi(W_{\bar\beta}^u(p_0, P_H))) \subset \Phi(W_{\bar\beta}^u(p_0, P_H)) \subset X_1^{\delta_{01}^*} \tag{13.13}$$

follows because of the choice of $\bar\beta$.

Let $x_1 \in \overline{X_1^{\delta_{01}}}$. Then, by (13.8) and (13.13), there exist $x_1^0, x_1^{-1}, \ldots, x_1^{-m-1}$ in $\Phi(W_{\bar\beta}^u(p_0, P_H))$ such that $x_1^0 = x_1$ and $P_H^{*-1}(x_1^{-k}) = x_1^{-k-1}$ for all $k \in \{0, 1, \ldots, m\}$. Using the continuity of P_H^{*-1} at the points $x_1^0, x_1^{-1}, \ldots, x_1^{-m}$ and the fact that $x_1^0, x_1^{-1}, \ldots, x_1^{-m-1}$ all belong to $\Phi(O_{p_0}^3 \cap \overline{W}) \cap X_1^{\delta_{01}^*}$, we can find open neighbourhoods $U_0, U_{-1}, \ldots, U_{-m-1}$ of the points $x_1^0, x_1^{-1}, \ldots, x_1^{-m-1}$ in H so that

$$U_{-k} \subset \Phi(O_{p_0}^3) \cap X_1^{\delta_{01}^*}(c_5) \qquad \text{for all } k \in \{0, 1, \ldots, m+1\}$$

and

$$P_H^{*-1}(U_{-k} \cap \Phi(O_{p_0}^3 \cap \overline{W})) \subset U_{-k-1} \qquad \text{for all } k \in \{0, 1, \ldots, m\}.$$

It follows that x_1 has a neighbourhood $U_0\subset\Phi(O^3_{p_0})\cap X_1^{\delta^*_{01}}(c_5)$ in H such that for every $x\in U_0\cap\Phi(O^3_{p_0}\cap\overline{W})$ there exist $x^0,x^{-1},\dots,x^{-m-1}$ in $X_1^{\delta^*_{01}}(c_5)\cap\Phi(O^3_{p_0}\cap\overline{W})$ with $x^0=x$ and $P^*_H(x^{-k-1})=x^{-k}$ for all $k\in\{0,1,\dots,m\}$.

Therefore, as $\overline{X_1^{\delta_{01}}}$ is compact, we can choose $c_6>0$ so that

$$\begin{cases}\text{for every } x\in X_1^{\delta_{01}}(c_6)\cap\Phi(O^3_{p_0}\cap\overline{W}) \text{ there exist } x^0,x^{-1},\dots,x^{-m-1}\\ \text{in } X_1^{\delta^*_{01}}(c_5)\cap\Phi(O^3_{p_0}\cap\overline{W}) \text{ with } x^0=x \text{ and } P^*_H(x^{-k-1})=x^{-k}\\ \text{for all } k\in\{0,1,\dots,m\}.\end{cases}\tag{13.14}$$

Choose $c=c(\epsilon)\in\mathbb{R}$ so that

$$0<c<\min\left\{\delta_{01},\frac{c_6}{K|\phi_H|_H}\right\}.$$

We claim that (13.9) holds. Let x_{12},y_{12} in $X_{12}^{\delta_{01},c}$ be given with $x_{12}\neq y_{12}$. Define

$$x=x_{12}+\Sigma(x_{12}),\qquad y=y_{12}+\Sigma(y_{12}).$$

Then (13.6) implies that x,y belong to $\Phi(O^3_{p_0}\cap\overline{W})$. Note that for every $w_{12}=w_1+w_2\in X_{12}^{\delta_{01},c}$ with $w_1\in X_1$ and $w_2\in X_2$ we have $w_1\in X_1^{\delta_{01}}$ and $w_2=v\phi_H$ with $-c<v\le 0$. Hence

$$|w_2|_H=|v||\phi_H|_H<c|\phi_H|_H<\frac{c_6}{K|\phi_H|_H}|\phi_H|_H=\frac{c_6}{K}<c_6.$$

Using $\Sigma(w_1)=0$ and the fact that Σ is Lipschitz continuous with Lipschitz constant K, we obtain

$$|\Sigma(w_1+w_2)|_H=|\Sigma(w_1+w_2)-\Sigma(w_1)|_H\le K|w_2|_H<K\frac{c_6}{K}=c_6.$$

Consequently,

$$x\in X_1^{\delta_{01}}(c_6)\cap\Phi(O^3_{p_0}\cap\overline{W}),\quad y\in X_1^{\delta_{01}}(c_6)\cap\Phi(O^3_{p_0}\cap\overline{W}).$$

Then, by (13.14), there exist

$$x^0,x^{-1},\dots,x^{-m-1},y^0,y^{-1},\dots,y^{-m-1} \text{ in } X_1^{\delta^*_{01}}(c_5)\cap\Phi(O^3_{p_0}\cap\overline{W})$$

with $x^0=x$, $y^0=y$ and

$$P^*_H(x^{-k-1})=x^{-k} \text{ and } P^*_H(y^{-k-1})=y^{-k}\qquad\text{for all } k\in\{0,1,\dots,m\}.$$

By the choice of $O^2_{p_0}$ we have $\Pi_3(O^2_{p_0}\cap\overline{W})\subset D^{-0}_{r_1}$. Hence

$$\Phi(O^3_{p_0}\cap\overline{W})\subset\Phi(O^2_{p_0}\cap\overline{W})\subset\Phi(\Pi_3^{-1}(D^{-0}_{r_1}))=\{x+\Sigma(x):\ x\in\Delta^{-0}_{r_1}\}$$

follows. Consequently, we obtain for the points $x_{12}^{-k}=\mathrm{Pr}_{X_{12}}x^{-k}$, $y_{12}^{-k}=\mathrm{Pr}_{X_{12}}y^{-k}$, $k\in\{0,1,\dots,m+1\}$,

$$x^{-k}=x_{12}^{-k}+\Sigma(x_{12}^{-k}),\quad y^{-k}=y_{12}^{-k}+\Sigma(y_{12}^{-k}),$$

and

$$\Sigma(x_{12}^{-k}) = \mathrm{Pr}_{X_3}\, x^{-k}, \quad \Sigma(y_{12}^{-k}) = \mathrm{Pr}_{X_3}\, y^{-k}.$$

If $x_{12}^{-k} = y_{12}^{-k}$ for some $k \in \{1, 2, \dots, m+1\}$, then $x^{-k} = x_{12}^{-k} + \Sigma(x_{12}^{-k}) = y_{12}^{-k} + \Sigma(y_{12}^{-k}) = y^{-k}$ follows. This implies $x^{-k+1} = P_H^*(x^{-k}) = P_H^*(y^{-k}) = y^{-k+1}$ and, by induction $x^0 = y^0$ which contradicts $x_{12} \neq y_{12}$. Thus, $x_{12}^{-k} \neq y_{12}^{-k}$ for all $k \in \{0, 1, \dots, m+1\}$.

Define

$$a_k = \frac{|\Sigma(x_{12}^{-k}) - \Sigma(y_{12}^{-k})|_H}{|x_{12}^{-k} - y_{12}^{-k}|_H} \qquad \text{for } k \in \{0, 1, \dots, m+1\}.$$

As $x^0, x^{-1}, \dots, x^{-m-1}, y^0, y^{-1}, \dots, y^{-m-1}$ belong to $X_1^{\delta_{01}^*}(c_5)$, $x^{-k} = P_H^*(x^{-k-1})$, and $y^{-k} = P_H^*(y^{-k-1})$ for $k \in \{0, 1, \dots, m\}$, applying (13.10), (13.11) and Proposition 13.5, it follows that

$$\begin{aligned} a_k &= \frac{|\mathrm{Pr}_{X_3} P_H^*(x^{-k-1}) - \mathrm{Pr}_{X_3} P_H^*(y^{-k-1})|_H}{|\mathrm{Pr}_{X_{12}} P_H^*(x^{-k-1}) - \mathrm{Pr}_{X_{12}} P_H^*(y^{-k-1})|_H} \\ &\leq \frac{c_1|x_3^{-k-1} - y_3^{-k-1}|_H + \epsilon_1|x_{12}^{-k-1} - y_{12}^{-k-1}|_H}{c_2|x_{12}^{-k-1} - y_{12}^{-k-1}|_H - c_3|x_3^{-k-1} - y_3^{-k-1}|_H} \\ &= \frac{c_1|\Sigma(x_{12}^{-k-1}) - \Sigma(y_{12}^{-k-1})|_H + \epsilon_1|x_{12}^{-k-1} - y_{12}^{-k-1}|_H}{c_2|x_{12}^{-k-1} - y_{12}^{-k-1}|_H - c_3|\Sigma(x_{12}^{-k-1}) - \Sigma(y_{12}^{-k-1})|_H} \\ &= \frac{c_1 a_{k+1} + \epsilon_1}{c_2 - c_3 a_{k+1}} \qquad \text{for all } k \in \{0, 1, \dots, m\}, \end{aligned} \tag{13.15}$$

where the denominators are positive since Σ is Lipschitz continuous with Lipschitz constant K and $c_3 K < c_2$.

We want to prove that

$$a_k < \epsilon \left(\frac{\frac{c_2}{c_1} + 1}{2}\right)^k \qquad \text{for all } k \in \{0, 1, \dots, m+1\}. \tag{13.16}$$

The inequality $0 < c_1 < c_2$ and (13.12) combined imply

$$\frac{\frac{c_2}{c_1} + 1}{2} > 1 \text{ and } \epsilon \left(\frac{\frac{c_2}{c_1} + 1}{2}\right)^k \leq K \qquad \text{for all } k \in \{0, 1, \dots, m\}.$$

Hence

$$\begin{aligned} \epsilon_1 &= \epsilon \left[\frac{c_2 - c_1}{2} - K \frac{\frac{c_2}{c_1} + 1}{2} c_3\right] \\ &\leq \epsilon \left(\frac{\frac{c_2}{c_1} + 1}{2}\right)^k \left[\frac{c_2 - c_1}{2} - \epsilon \left(\frac{\frac{c_2}{c_1} + 1}{2}\right)^{k+1} c_3\right] \qquad \text{for all } k \in \{0, 1, \dots, m\}. \end{aligned} \tag{13.17}$$

We also have

$$0 < c_2 - \epsilon \left(\frac{\frac{c_2}{c_1} + 1}{2}\right)^{k+1} c_3 \qquad \text{for all } k \in \{0, 1, \dots, m\}$$

since

$$\epsilon \left(\frac{\frac{c_2}{c_1} + 1}{2} \right)^k \leq K \qquad \text{for such } k$$

and

$$0 < c_2 - K \frac{\frac{c_2}{c_1} + 1}{2} c_3$$

because of

$$c_3 < \frac{1}{K} \frac{c_1 c_2}{c_2 + c_1} < \frac{1}{K} \frac{2}{\frac{c_2}{c_1} + 1} c_2.$$

We now prove (13.16) by induction. From (13.12) and the Lipschitz continuity of Σ we obtain

$$a_{m+1} \leq K < \epsilon \left(\frac{\frac{c_2}{c_1} + 1}{2} \right)^{m+1}.$$

Assume

$$a_{k+1} < \epsilon \left(\frac{\frac{c_2}{c_1} + 1}{2} \right)^{k+1} \tag{13.18}$$

for some $k \in \{0, 1, \ldots, m\}$. Then (13.15), the strict monotonicity of the function $\left[0, \frac{c_2}{c_3}\right) \ni s \mapsto \frac{c_1 s + \epsilon_1}{c_2 - c_3 s} \in \mathbb{R}$, the inequalities

$$a_{k+1} \leq K < \frac{c_2}{c_3}$$

and

$$\epsilon \left(\frac{\frac{c_2}{c_1} + 1}{2} \right)^{k+1} < \frac{c_2}{c_3},$$

and (13.18) combined yield

$$a_k < \frac{c_1 \epsilon \left(\frac{\frac{c_2}{c_1}+1}{2} \right)^{k+1} + \epsilon_1}{c_2 - \epsilon \left(\frac{\frac{c_2}{c_1}+1}{2} \right)^{k+1} c_3}.$$

Using (13.17) we get

$$\begin{aligned} a_k &< \frac{c_1 \epsilon \left(\frac{\frac{c_2}{c_1}+1}{2} \right)^{k+1} + \epsilon \left(\frac{\frac{c_2}{c_1}+1}{2} \right)^{k} \left[\frac{c_2 - c_1}{2} - \epsilon \left(\frac{\frac{c_2}{c_1}+1}{2} \right)^{k+1} c_3 \right]}{c_2 - \epsilon \left(\frac{\frac{c_2}{c_1}+1}{2} \right)^{k+1} c_3} \\ &= \epsilon \left(\frac{\frac{c_2}{c_1} + 1}{2} \right)^k. \end{aligned}$$

Therefore (13.16) holds. In particular, $a_0 < \epsilon$, which implies (13.9). □

For $\delta > 0$ set $k(\delta) = \sup\{|x_{12}|_H : x_{12} \in X_{12}^{\delta}\}$. Choose $\delta_{02} \in (0, \delta_{01})$ so small that for the open neighbourhood

$$O_0^3 = \left\{x_{12} + x_3 \in H : x_{12} \in X_{12}^{\delta_{02}},\ x_3 \in X_3,\ |x_3|_H < (K+1)k(\delta_{02})\right\}$$

of 0 in H,

$$O_0^3 \subset O_0^1 \quad \text{and} \quad \Phi^{-1}(O_0^3) \subset O_{p_0}^2$$

hold. Then

$$O_{p_0}^4 = \Phi^{-1}(O_0^3)$$

is an open neighbourhood of p_0 in H, and

$$\left\{x + \Sigma^*(x) : x \in X_{12}^{\delta_{02}}\right\} \subset O_0^3$$

since Σ^* is Lipschitz continuous with Lipschitz constant K. Define

$$\Psi : X_{12}^{\delta_{02}} \ni x \mapsto \Phi^{-1}(x + \Sigma^*(x)) \in H.$$

Observe that $\Psi(X_{12}^{\delta_{02}}) \subset O_{p_0}^2$.

The last result of this chapter proves some properties of Ψ which are used in the next chapter, and shows the smoothness of $H \cap \overline{W}$ near p_0 in the sense that the C^1-diffeomorphism Φ maps the open neighbourhood $O_{p_0}^4$ of p_0 in H onto the open neighbourhood O_0^3 of 0 in H so that $\Phi(O_{p_0}^4 \cap \overline{W})$ is the graph of the C^1-map Σ^* restricted to $X_{12}^{\delta_{02},-0}$.

Proposition 13.8

(i) *Ψ is C^1-smooth, $\Psi(0) = p_0$ and $D\Psi(0)x_{12} = x_{12}$ for all $x_{12} \in X_{12}$.*

(ii) *For all u, v in $(-\delta_{02}, \delta_{02})$,*

$$\Psi(u\psi_H + v\phi_H) \in W \qquad \textit{if and only if} \quad v < 0,$$

and

$$\Psi(u\psi_H + v\phi_H) \in \operatorname{bd} W \qquad \textit{if and only if} \qquad v = 0.$$

(iii) $\Phi(O_{p_0}^4 \cap \overline{W}) = \left\{x + \Sigma^*(x) : x \in X_{12}^{\delta_{12},-0}\right\}$.

Proof 1. Assertion (i) follows from Propositions 13.1 and 13.7.

2. Proof of the first equivalence in assertion (ii). Let u, v in $(-\delta_{02}, \delta_{02})$ be given. Assume $\phi = \Psi(u\psi_H + v\phi_H) \in W$. Recall that, by the choice of $O_{p_0}^2$, we have $\Pi_3(O_{p_0}^2 \cap \overline{W}) \subset D_{r_1}^{-0}$. This fact and $\Psi(X_{12}^{\delta_{02}}) \subset O_{p_0}^2$ imply $\Pi_3\phi \in D_{r_1}^{-0}$. Proposition 12.11 gives $\Pi_3\phi \in D_{r_1}^-$. It follows that $\Pr_{X_{12}} \circ \Phi(\phi) \in \Delta_{r_1}^-$. On the other hand, $u\psi_H + v\phi_H = \Pr_{X_{12}} \circ \Phi(\phi)$. Hence, Proposition 13.3(ii) implies $v < 0$.

If $v < 0$ then the inequality $\delta_{02} < \delta_{00}$, Proposition 13.3 and the definition of $\Delta_{r_1}^-$ show that there exists $x \in D_{r_1}^-$ such that $u\psi_H + v\phi_H = \Pr_{X_{12}} \circ \Phi(\Pi_3^{-1}(x))$. From Proposition 12.11 we obtain $\phi = \Pi_3^{-1}(x) \in W$. By Proposition 13.2 $\Phi(\phi) = \Pr_{X_{12}} \Phi(\phi) + \Sigma(\Pr_{X_{12}} \Phi(\phi))$. So,

$$\begin{aligned}
\Psi(u\psi_H + v\phi_H) &= \Phi^{-1}\left(u\psi_H + v\phi_H + \Sigma^*(u\psi_H + v\phi_H)\right) && \text{(the definition of } \Psi)\\
&= \Phi^{-1}\left(u\psi_H + v\phi_H + \Sigma(u\psi_H + v\phi_H)\right) && (\text{using } v < 0)\\
&= \Phi^{-1}\left(\Pr_{X_{12}} \Phi(\phi) + \Sigma(\Pr_{X_{12}} \Phi(\phi))\right)\\
&= \Phi^{-1}(\Phi(\phi))\\
&= \phi \in W.
\end{aligned}$$

3. The second equivalence of assertion (ii) follows in the same way if we replace W, $D_{r_1}^-$, $v < 0$ in the proof of the first equivalence above by $\operatorname{bd} W$, $D_{r_1}^0$, $v = 0$, respectively.

4. Proof of assertion (iii). From the choice of δ_{02} and O_0^3, and from assertion (ii) we obtain

$$\Phi\big(O_{p_0}^4 \cap \overline{W}\big) \supseteq \Big\{x + \Sigma^*(x) : x \in X_{12}^{\delta_{12},-0}\Big\}.$$

To prove the other inclusion let $\phi \in O_{p_0}^4 \cap \overline{W}$ be given. By the choice of $O_{p_0}^2$ we have $\Pi_3\big(O_{p_0}^2 \cap \overline{W}\big) \subset D_{r_1}^{-0}$. Thus $\Pi_3\phi \in D_{r_1}^{-0}$. By the definition of $O_{p_0}^4$, $\Phi(\phi) \in \Phi(O_{p_0}^4) = O_0^3$. These facts and Proposition 13.3(ii) combined yield

$$\operatorname{Pr}_{X_{12}} \Phi(\phi) = \operatorname{Pr}_{X_{12}} \Phi\big(\Pi_3^{-1}(\Pi_3\phi)\big) \in X_{12}^{\delta_{02},-0}.$$

The relation $\phi \in \Pi_3^{-1}(D_{r_1}^{-0})$ permits to apply Proposition 13.2(ii) to get

$$\Phi(\phi) = \operatorname{Pr}_{X_{12}} \Phi(\phi) + \Sigma\big(\operatorname{Pr}_{X_{12}} \Phi(\phi)\big).$$

Using $\operatorname{Pr}_{X_{12}} \Phi(\phi) \in X_{12}^{\delta_{02},-0}$ and the definition of Σ^* we conclude $\Phi(\phi) \in \big\{x + \Sigma^*(x) : x \in X_{12}^{\delta_{12},-0}\big\}$. □

CHAPTER 14

Smoothness of $\overline{W}$, $\operatorname{bd} W$ and $\overline{W} \cap S$

Recall that G_i, $i \in \{1,2,3\}$, are i-dimensional subspaces of C with $G_3 = G_1 \oplus G_2$, E is a closed complementary space of G_3 in C, $P_2 : C \to C$ and $P_3 : C \to C$ are projections onto G_2 and G_3, respectively, and there are maps

$$w : P_3\overline{W} \to E, \qquad w_S : P_2(\overline{W} \cap S) \to G_1 \oplus E$$

such that

$$\overline{W} = \{\chi + w(\chi) : \ \chi \in P_3\overline{W}\}, \qquad \overline{W} \cap S = \{\chi + w_S(\chi) : \ \chi \in P_2(\overline{W} \cap S)\}.$$

The map w is locally Lipschitz continuous, and w_S is Lipschitz continuous. The sets P_3W and $P_2(W \cap S)$ are open subsets of G_3 and G_2, respectively. The restrictions $w|_{P_3W}$ and $w_S|_{P_2(W\cap S)}$ are C^1-smooth. The boundaries of $P_3\overline{W}$ and $P_2(\overline{W} \cap S)$ in G_3 and G_2 are $\partial(P_3\overline{W}) = P_3\overline{W} \setminus P_3W = P_3(\operatorname{bd} W)$ and $\partial(P_2(\overline{W} \cap S)) = P_2(\overline{W} \cap S) \setminus P_2(W \cap S) = P_2\mathcal{O}$, respectively.

In this chapter we show that w in the case $\mu = 0$ and w_S are also smooth at the boundary points of their domains in the sense that close to boundary points of $P_3\overline{W}$ and $P_2(\overline{W} \cap S)$, w and w_S can be extended to C^1-functions on open sets. If $\mu > 0$ then the smoothness of w in the above sense is proved at all boundary points of its domain with the exception of $P_3\xi_-$ and $P_3\xi_+$. We also obtain in the case $\mu = 0$ that $P_3(\operatorname{bd} W)$ is a 2-dimensional C^1-submanifold of G_3, and that $\operatorname{bd} W$ is a 2-dimensional C^1-submanifold of C. If $\mu > 0$ then the same holds for the sets $P_3(\operatorname{bd} W) \setminus \{P_3\xi_-, P_3\xi_+\}$ and $\operatorname{bd} W \setminus \{\xi_-, \xi_+\}$. The precise statement for w is contained in the following theorem.

Theorem 14.1 *Let $\phi \in \operatorname{bd} W$ if $\mu = 0$, and $\phi \in \operatorname{bd} W \setminus \{\xi_-, \xi_+\}$ if $\mu > 0$. Then there exist*

(i) a 2-dimensional subspace G_3^2 of G_3 and $\xi \in G_3 \setminus G_3^2$,

(ii) open neighbourhoods V_1 of 0 in $G_3^1 = \mathbb{R}\xi$ and V_2 of 0 in G_3^2, respectively,

(iii) a C^1-function $d : V_2 \to V_1$, with $d(0) = 0$, so that the neighbourhood $U = P_3\phi + V_1 + V_2$ of $P_3\phi$ in G_3 is the disjoint union of the sets

$$\begin{aligned} V^- &= \{P_3\phi + \chi_1 + \chi_2 : \ \chi_1 \in V_1, \ \chi_2 \in V_2, \ d(\chi_2) - \chi_1 \in (-\infty, 0)\xi\}, \\ V^0 &= \{P_3\phi + \chi_1 + \chi_2 : \ \chi_1 \in V_1, \ \chi_2 \in V_2, \ \chi_1 = d(\chi_2)\}, \\ V^+ &= \{P_3\phi + \chi_1 + \chi_2 : \ \chi_1 \in V_1, \ \chi_2 \in V_2, \ d(\chi_2) - \chi_1 \in (0, \infty)\xi\}, \end{aligned}$$

(iv) and a C^1-function $w^e : U \to E$ with the following properties:

$$U \cap P_3(\operatorname{bd} W) = V^0. \tag{14.1}$$

$$U \cap P_3 W = V^* \quad \textit{with} \quad V^* \in \{V^-, V^+\}. \tag{14.2}$$

$$w^e\big|_{U \cap P_3 \overline{W}} = w\big|_{U \cap P_3 \overline{W}}. \tag{14.3}$$

$$(U+E) \cap \operatorname{bd} W = \{P_3\phi + \chi_2 + d(\chi_2) + w\left(P_3\phi + \chi_2 + d(\chi_2)\right) : \ \chi_2 \in V_2\}. \tag{14.4}$$

$$(U+E) \cap \overline{W} = \{\chi + w(\chi) : \ \chi \in V^0 \cup V^*\}. \tag{14.5}$$

Proof 1. Preparatory definitions. As Ψ is C^1 in $X_{12}^{\delta_{02}}$ and $D\Psi(0)$ is injective, we can choose $\delta_{03} \in (0, \delta_{02})$ such that $D\Psi(x)$ is injective for all $x \in X_{12}^{\delta_{03}}$. In addition, we may assume that δ_{03} is so small that

$$\frac{d}{dt}\Psi(u\psi_H) \notin H \qquad \text{for all } \ u \in (-\delta_{03}, \delta_{03}),$$

since $\frac{d}{dt}\Psi(0) = \dot{p}_0 \notin H$ and the function $\overline{W} \ni \phi \mapsto \dot{\phi} \in C$ is continuous by Proposition 4.3.

Assume $\mu > 0$ and consider a point $\phi \in \operatorname{bd} W \setminus \{\xi_-, \xi_+\}$. The proof for $\mu = 0$ and $\phi \in \operatorname{bd} W$ is analogous.

Recall β_4 from Proposition 12.11. Choose $\tilde{\beta} \in (0, \beta_4)$ such that $W^u_{\tilde{\beta}}(p_0, P_H) \subset O^1_{p_0}$, $\Pi_3(W^u_{\tilde{\beta}}(p_0, P_H)) \subset D_{r_1}$ and $\operatorname{Pr}_{X_{12}} \circ \Phi(W^u_{\tilde{\beta}}(p_0, P_H)) \subset X_{12}^{\delta_{03}}$. Corollary 12.13 and Theorem V.3 combined imply

$$\phi \in \cup_{t \geq 0} F(t, W^u(p_0, P_H, N^u_H)) = \cup_{t \geq 0} F(t, W^u_{\tilde{\beta}}(p_0, P_H)).$$

Thus there exist $\psi \in W^u_{\tilde{\beta}}(p_0, P_H)$ and $t_0 > 1$ with $\phi = F(t_0, \psi)$. By the choice of $\tilde{\beta}$, Proposition 12.11 implies $\psi \in \operatorname{bd} W$ and $\Pi_3 \psi \in D^0_{r_1}$. Hence $\operatorname{Pr}_{X_{12}} \circ \Phi(\psi) \in \Delta^0_{r_1} \cap X_{12}^{\delta_{03}}$ follows. Consequently, by Proposition 13.3, we obtain $\psi = \Psi(u^0 \psi_H)$ with $|u^0| < \delta_{03}$.

Choose $\epsilon_1 > 0$ such that

$$0 < \epsilon_1 < \min\{t_0 - 1, \delta_{03} - |u^0|\}.$$

For $0 < \epsilon \leq \epsilon_1$, set

$$\begin{aligned}
N_\epsilon &= \{(u + u^0)\psi_H + v\phi_H : |u| < \epsilon, |v| < \epsilon\}, \\
N^-_\epsilon &= \{(u + u^0)\psi_H + v\phi_H : |u| < \epsilon, -\epsilon < v < 0\}, \\
N^+_\epsilon &= \{(u + u^0)\psi_H + v\phi_H : |u| < \epsilon, 0 < v < \epsilon\}, \\
N^0_\epsilon &= \{(u + u^0)\psi_H : |u| < \epsilon\} \\
N^{-0}_\epsilon &= N^-_\epsilon \cup N^0_\epsilon, \\
M_\epsilon &= (-\epsilon, \epsilon) \times N_\epsilon, \\
M^i_\epsilon &= (-\epsilon, \epsilon) \times N^i_\epsilon \qquad \text{for } i \in \{-, 0, +, -0\}.
\end{aligned}$$

The functions

$$\begin{aligned}
a &: M_{\epsilon_1} \ni (t, x) \mapsto F(t_0 + t, \Psi(x)) \in C, \\
\hat{a} &: M^0_{\epsilon_1} \ni (t, x) \mapsto F(t_0 + t, \Psi(x)) \in C
\end{aligned}$$

are C^1-smooth since $N_{\epsilon_1} \subset X_{12}^{\delta_{01}}$, $N_{\epsilon_1}^0 \subset X_{12}^{\delta_{01}}$ and $\Psi : X_{12}^{\delta_{01}} \to C$, $F : (1,\infty) \times C \to C$ are C^1-smooth.

The invariance of W, $\overline{W}$, bd W, and Proposition 13.8 combined imply

$$a(M_{\epsilon_1}^-) \subset W,\ a(M_{\epsilon_1}^{-0}) \subset \overline{W},\ a(M_{\epsilon_1}^0) = \hat{a}(M_{\epsilon_1}^0) \subset \operatorname{bd} W,\ a(M_{\epsilon_1}^+) \cap \overline{W} = \emptyset. \quad (14.6)$$

2. Curves and tangent vectors. The curves

$$\begin{aligned}
\gamma_1 :&(-\epsilon_1, \epsilon_1) \ni s \mapsto (0, (s + u^0)\psi_H) \in M_{\epsilon_1},\\
\gamma_2 :&(-\epsilon_1, \epsilon_1) \ni s \mapsto (0, u^0\psi_H + s\phi_H) \in M_{\epsilon_1},\\
\gamma_3 :&(-\epsilon_1, \epsilon_1) \ni s \mapsto (s, u^0\psi_H) \in M_{\epsilon_1}
\end{aligned}$$

and

$$\gamma^i : (-\epsilon_1, \epsilon_1) \ni s \mapsto a(\gamma_i(s)) \in C, \qquad i \in \{1,2,3\},$$

are C^1-smooth, with

$$\begin{aligned}
(\gamma^1)'(0) &= D_2F(t_0, \psi)D\Psi(u^0\psi_H)\psi_H,\\
(\gamma^2)'(0) &= D_2F(t_0, \psi)D\Psi(u^0\psi_H)\phi_H,\\
(\gamma^3)'(0) &= \dot{\phi} = D_2F(t_0, \psi)\dot{\psi}.
\end{aligned}$$

We claim that the vectors $(\gamma^i)'(0)$, $i \in \{1,2,3\}$, are linearly independent. Since $D_2F(t_0,\psi)$ is injective, it suffices to prove that $D\Psi(u^0\psi_H)\psi_H$, $D\Psi(u^0\psi_H)\phi_H$ and $\dot{\psi}$ are linearly independent. We have $\dot{\psi} \notin H$ by the choice of δ_{03}. The vectors $D\Psi(u^0\psi_H)\psi_H$ and $D\Psi(u^0\psi_H)\phi_H$ in H are linearly independent since $D\Psi(u^0\psi_H)$ is injective and ψ_H and ϕ_H are linearly independent. This shows the claim.

Define

$$E_1 = \mathbb{R}(\gamma^1)'(0) \oplus \mathbb{R}(\gamma^2)'(0) \oplus \mathbb{R}\dot{\phi}, \quad \hat{E}_1 = \mathbb{R}(\gamma^1)'(0) \oplus \mathbb{R}\dot{\phi}.$$

Let E_2 be a closed complement of E_1 in C. Let $\hat{E}_2 = E_2 \oplus \mathbb{R}(\gamma^2)'(0)$. Then

$$C = E_1 \oplus E_2 = \hat{E}_1 \oplus \hat{E}_2.$$

Let Pr_{E_1} and $\mathrm{Pr}_{\hat{E}_1}$ denote the projections of C onto E_1 and $\hat{E}_1$ along E_2 and $\hat{E}_2$, respectively. Let $j : C \to C$ be defined by $j(\chi) = \chi - \phi$ for $\chi \in C$.

3. Graph representation for $a(M_\epsilon)$ and $\hat{a}(M_\epsilon^0)$. Consider the maps

$$\begin{aligned}
\mathrm{Pr}_{E_1} \circ j \circ a &: M_{\epsilon_1} \to E_1,\\
\mathrm{Pr}_{\hat{E}_1} \circ j \circ \hat{a} &: M_{\epsilon_1}^0 \to \hat{E}_1.
\end{aligned}$$

It is easy to see that

$$\begin{aligned}
\frac{d}{ds}\left[\mathrm{Pr}_{E_1} \circ j \circ a \circ \gamma_i\right](0) &= (\gamma^i)'(0), \qquad i \in \{1,2,3\},\\
\frac{d}{ds}\left[\mathrm{Pr}_{\hat{E}_1} \circ j \circ a \circ \gamma_i\right](0) &= (\gamma^i)'(0), \qquad i \in \{1,3\}.
\end{aligned}$$

Consequently,

$$D(\mathrm{Pr}_{E_1}\circ j\circ a)(0,u^0\psi_H)(\mathbb{R}\times X_{12})=E_1,$$
$$D(\mathrm{Pr}_{\hat{E}_1}\circ j\circ \hat{a})(0,u^0\psi_H)(\mathbb{R}\times X_1)=\hat{E}_1.$$

Thus, by the Inverse Function Theorem, $\mathrm{Pr}_{E_1}\circ j\circ a$ and $\mathrm{Pr}_{\hat{E}_1}\circ j\circ \hat{a}$ are local C^1-diffeomorphisms at $(0,u^0\psi_H)$. Therefore there exists $\epsilon_2\in(0,\epsilon_1)$ such that for every $\epsilon\in(0,\epsilon_2]$, $\mathrm{Pr}_{E_1}\circ j\circ a$ maps M_ϵ injectively onto an open neighbourhood $\mathcal{N}_\epsilon$ of 0 in E_1, the inverse $\left(\mathrm{Pr}_{E_1}\circ j\circ a|_{M_\epsilon}\right)^{-1}:\mathcal{N}_\epsilon\to\mathbb{R}\times X_{12}$ is C^1-smooth, $\mathrm{Pr}_{\hat{E}_1}\circ j\circ\hat{a}$ maps M_ϵ^0 injectively onto an open neighbourhood $\hat{\mathcal{N}}_\epsilon$ of 0 in $\hat{E}_1$ and the inverse $\left(\mathrm{Pr}_{\hat{E}_1}\circ j\circ\hat{a}|_{M_\epsilon^0}\right)^{-1}:\hat{\mathcal{N}}_\epsilon\to(-\epsilon,\epsilon)\times X_1$ is C^1-smooth. It follows that Pr_{E_1} maps $j\circ a(M_{\epsilon_2})$ onto $\mathcal{N}_{\epsilon_2}$ injectively. Therefore there is a map $b:\mathcal{N}_{\epsilon_2}\to E_2$ such that $b(0)=0$ and

$$j\circ a(M_\epsilon)=\{\chi+b(\chi):\chi\in\mathcal{N}_\epsilon\},\qquad 0<\epsilon\le\epsilon_2.$$

It is easy to see that

$$b=(\mathrm{id}-\mathrm{Pr}_{E_1})\circ j\circ a\circ(\mathrm{Pr}_{E_1}\circ j\circ a|_{M_{\epsilon_2}})^{-1}$$

and hence b is C^1-smooth. We have $Db(0)=0$ since $Da(0,0)(\mathbb{R}\times X_{12})\subset E_1$. Similarly, there is a C^1-map $\hat{b}:\hat{\mathcal{N}}_{\epsilon_2}\to\hat{E}_2$ with $\hat{b}(0)=0$ and $D\hat{b}(0)=0$ such that

$$j\circ\hat{a}(M_\epsilon^0)=\{\chi+\hat{b}(\chi):\chi\in\hat{\mathcal{N}}_\epsilon\},\qquad 0<\epsilon\le\epsilon_2.$$

Consequently, for $0<\epsilon\le\epsilon_2$,

$$a(M_\epsilon)=\{\phi+\chi+b(\chi):\chi\in\mathcal{N}_\epsilon\},$$
$$\hat{a}(M_\epsilon^0)=\{\phi+\chi+\hat{b}(\chi):\chi\in\hat{\mathcal{N}}_\epsilon\}.$$

4. We claim that for every $\epsilon\in(0,\epsilon_2]$ there exists $\delta=\delta(\epsilon)>0$ with

$$(\phi+C_\delta)\cap a(M_\epsilon^{-0})=(\phi+C_\delta)\cap\overline{W}. \tag{14.7}$$

Suppose the last statement is false. Then there exist $\epsilon\in(0,\epsilon_2]$ and a sequence $(\phi^n)_0^\infty$ in $\overline{W}$ such that $\phi^n\to\phi$ as $n\to\infty$ and $\phi^n\notin a(M_\epsilon^{-0})$ for all $n\in\mathbb{N}$. Then $\psi^n=F_W(-t_0,\phi^n)\to F_W(-t_0,\phi)=\psi\in\overline{W}$ as $n\to\infty$ and $\psi^n\in\overline{W}$, $\psi^n\notin\{F_W(t,\Psi((u+u^0)\psi_H+v\phi_H)):|t|<\epsilon,|u|<\epsilon,-\epsilon<v\le 0\}$ for all $n\in\mathbb{N}$ follow. Moreover, $F_W(-2,\psi^n)\to F_W(-2,\psi)$ as $n\to\infty$. Using $\psi\in H$ and $\dot{\psi}\notin H$, we can construct an intersection map as in Appendix I. Thus, there exist an open neighbourhood N of $F_W(-2,\psi)$ in C and a C^1-map $\tau:N\to(0,\infty)$ such that $\tau(F_W(-2,\psi))=2$ and $F_W(\tau(\chi),\chi)\in H$ for all $\chi\in N$. It follows that $\tau(F_W(-2,\psi^n))\to 2$ as $n\to\infty$. Thus the sequence $(t^n)_0^\infty$ given by $t^n=\tau(F_W(-2,\psi^n))-2$ satisfies $t^n\to 0$ as $n\to\infty$, $F_W(t^n,\psi^n)\in H\cap\overline{W}$ for all $n\in\mathbb{N}$, and $F_W(t^n,\psi^n)\to\psi$ as $n\to\infty$. Then $\Pi_3\psi=\lim_{n\to\infty}\Pi_3F_W(t^n,\psi^n)$ and Proposition 12.11 combined imply $\Pi_3F_W(t^n,\psi^n)\in D_{r_1}^{-0}$ for all sufficiently large $n\in\mathbb{N}$. Hence $\mathrm{Pr}_{X_{12}}\circ\Phi(F_W(t^n,\psi^n))\in\Delta_{r_1}^{-0}$ for sufficiently large n and $\mathrm{Pr}_{X_{12}}\circ\Phi(F_W(t^n,\psi^n))\to\mathrm{Pr}_{X_{12}}\circ\Phi(\psi)=u^0\psi_H$ as $n\to\infty$. Using Proposition

13.3(ii) we find $n \in \mathbb{N}$ with $|t^n| < \epsilon$ and $\mathrm{Pr}_{X_{12}} \circ \Phi(F_W(t^n, \psi^n)) \in N_\epsilon^{-0} \subset X_{12}^{\delta_{03}}$. By $F_W(t^n, \psi^n) \in H \cap \overline{W}$ and $\Pi_3 F_W(t^n, \psi^n) \in D_{r_1}^{-0}$,

$$\Phi(F_W(t^n, \psi^n)) = \mathrm{Pr}_{X_{12}} \circ \Phi(F_W(t^n, \psi^n)) + \Sigma(\mathrm{Pr}_{X_{12}} \circ \Phi(F_W(t^n, \psi^n))),$$

and consequently

$$F_W(t^n, \psi^n) \in \{\Psi((u + u^0)\psi_H + v\phi_H) : |u| < \epsilon, -\epsilon < v \leq 0\},$$

and thereby

$$\begin{aligned}\psi^n &= F_W(-t^n, F_W(t^n, \psi^n)) \\ &\in \{F_W(t, \Psi((u + u^0)\psi_H + v\phi_H)) : |t| < \epsilon, |u| < \epsilon, -\epsilon < v \leq 0\},\end{aligned}$$

a contradiction. Therefore, (14.7) holds.

5. Set $B_r = P_3\phi + G_{3,r}$, for every $r > 0$. We show that for every $\epsilon \in (0, \epsilon_2]$ there exists $r = r(\epsilon) > 0$ with

$$B_r \cap P_3(a(M_\epsilon^{-0})) = B_r \cap P_3\overline{W}. \tag{14.8}$$

The inclusion $a(M_\epsilon^{-0}) \subset \overline{W}$ yields $B_r \cap P_3(a(M_\epsilon^{-0})) \subset B_r \cap P_3\overline{W}$ for every $r > 0$. Then assume that for a given $\epsilon \in (0, \epsilon_2]$ there is no $r > 0$ with (14.8). Then we can find a sequence $(x^n)_0^\infty$ in G_3 such that $x^n \in P_3\overline{W} \setminus P_3(a(M_\epsilon^{-0}))$ for all $n \in \mathbb{N}$ and $x^n \to P_3\phi$ as $n \to \infty$. The points $\xi^n = \left(P_3|_{\overline{W}}\right)^{-1}(x^n) \in \overline{W}$, $n \in \mathbb{N}$, satisfy $\xi^n \notin a(M_\epsilon^{-0})$. Choose $\delta > 0$ according to part 4 so that (14.7) holds. It follows that $\xi^n \notin \phi + C_\delta$ for all $n \in \mathbb{N}$. On the other hand, using that $\left(P_3|_{\overline{W}}\right)^{-1}$ is continuous, from $P_3\phi = \lim_{n\to\infty} x^n$ we have

$$\xi^n = \left(P_3|_{\overline{W}}\right)^{-1}(x^n) \to \left(P_3|_{\overline{W}}\right)^{-1}(P_3\phi) = \phi,$$

a contradiction. This proves the existence of $r = r(\epsilon) > 0$ so that (14.8) holds.

6. Smooth extension of w. The map

$$c : \mathcal{N}_{\epsilon_2} \ni \chi \mapsto P_3(\phi + \chi + b(\chi)) \in G_3$$

is C^1-smooth, and we have $Dc(0)\chi = P_3\chi$ for all $\chi \in E_1$ since $Db(0) = 0$.

We claim that $P_3|_{E_1}$ is injective. It suffices to show that $\Pi_3|_{E_1}$ is injective since $P_3 = J_3 \circ \Pi_3$ and $J_3 : \mathbb{R}^3 \to C$ is injective. Let $\chi \in E_1 \setminus \{0\}$. There exist $s^* > 0$ and reals c_1, c_2, c_3 so that $\sum_{j=1}^3 c_j\gamma_j(s) \in M_{\epsilon_1}$ for all $s \in (-s^*, s^*)$, and the curve

$$\Gamma : (-s^*, s^*) \ni s \mapsto a(c_1\gamma_1(s) + c_2\gamma_2(s) + c_3\gamma(s)) \in C$$

satisfies $D\Gamma(0)1 = \chi$. The equation (14.8) implies that either $\Gamma(s) \in \overline{W}$ for all $s \in (-s^*, 0]$ or $\Gamma(s) \in \overline{W}$ for all $s \in [0, s^*)$. Then from Proposition 6.5 it follows that $\Pi_3\Gamma'(0) \neq 0$ since $\Gamma'(0) = \chi \neq 0$. Thus, $\Pi_3|_{E_1}$ is injective. Consequently, $P_3|_{E_1}$ is injective.

The Inverse Function Theorem permits to choose $\epsilon_3 \in (0, \epsilon_2)$ such that $c|_{\mathcal{N}_{\epsilon_3}}$ is a C^1-diffeomorphism. It follows that the restriction of P_3 to $a(M_{\epsilon_3}) = \{\phi+\chi+b(\chi) : \chi \in \mathcal{N}_{\epsilon_3}\}$ is injective. Choose $\delta(\epsilon_3)$ and $r(\epsilon_3)$ according to the claims in parts 4 and

5, respectively. We can find open neighbourhoods U_0 of 0 in E_1 and $V_{P_3\phi}$ of $P_3\phi$ in G_3 such that $U_0 \subset \mathcal{N}_{\epsilon_3} \cap C_{\delta(\epsilon_3)}$, $V_{P_3\phi} \subset B_{r(\epsilon_3)}$, and c maps U_0 injectively onto $V_{P_3\phi}$ with an inverse $c^{-1} : V_{P_3\phi} \to E_1$ which is C^1-smooth.

Let $\eta \in V_{P_3\phi} \cap P_3\overline{W}$. The equation (14.8) implies $\eta \in V_{P_3\phi} \cap P_3(a(M_{\epsilon_3}^{-0}))$. Also, $c^{-1}(\eta) \in U_0$ and $\eta = P_3(\phi + c^{-1}(\eta) + b(c^{-1}(\eta)))$. From $c^{-1}(\eta) \in U_0 \subset \mathcal{N}_{\epsilon_3}$ we obtain that $\zeta = \phi + c^{-1}(\eta) + b(c^{-1}(\eta))$ belongs to $a(M_{\epsilon_3})$. We also have $P_3\zeta = \eta$. By the injectivity of P_3 on $a(M_{\epsilon_3})$ and by $\eta \in P_3(a(M_{\epsilon_3}^{-0}))$ we infer

$$\zeta \in a(M_{\epsilon_3}^{-0}) \subset \overline{W}.$$

Now we get

$$\phi + c^{-1}(\eta) + b(c^{-1}(\eta)) = \zeta = P_3\zeta + w(P_3\zeta) = \eta + w(\eta).$$

Hence $w(\eta) = \phi + c^{-1}(\eta) + b(c^{-1}(\eta)) - \eta$ follows. Observe also that for every $\eta \in V_{P_3\phi}$,

$$P_3(\phi + c^{-1}(\eta) + b(c^{-1}(\eta)) - \eta) = c(c^{-1}(\eta)) - \eta = 0,$$

or

$$\phi + c^{-1}(\eta) + b(c^{-1}(\eta)) - \eta \in P_3^{-1}(0) = E.$$

Therefore the C^1-map $w^* : V_{P_3\phi} \to E$ given by

$$w^*(\eta) = \phi + c^{-1}(\eta) + b(c^{-1}(\eta)) - \eta \qquad \text{for all } \eta \in V_{P_3\phi} \tag{14.9}$$

is an extension of w from $V_{P_3\phi} \cap P_3\overline{W}$ to the open neighbourhood $V_{P_3\phi}$ of $P_3\phi$ in G_3.

7. Construction of a map for the representation of $P_3 \operatorname{bd} W$ close to $P_3\phi$. The set $G_3^2 = P_3(\hat{E}_1)$ is a hyperplane in G_3 since $P_3|_{E_1}$ is injective. Choose $\xi \in G_3 \setminus G_3^2$. Then $G_3 = G_3^2 \oplus \mathbb{R}\xi$. Let $\mathrm{Pr}_{G_3^2}$ denote the projection of G_3 onto G_3^2 along $\mathbb{R}\xi$. Then the C^1-map

$$\hat{c} : \hat{\mathcal{N}}_{\epsilon_3} \ni \chi \mapsto \mathrm{Pr}_{G_3^2} \circ P_3 \circ (\mathrm{id} + \hat{b})(\chi) \in G_3^2$$

satisfies $\hat{c}(0) = 0$, and $D\hat{c}(0)$ is injective since $D\hat{c}(0)\chi = P_3\chi$ for all $\chi \in \hat{E}_1 \subset E_1$. The Inverse Function Theorem implies that there are open neighbourhoods $\hat{U}_0$ of 0 in $\hat{E}_1$ and $\hat{V}_0$ of 0 in G_3^2 with $\hat{U}_0 \subset U_0$ and $P_3\phi + \hat{V}_0 \subset V_{P_3\phi}$ such that $\hat{c}$ maps $\hat{U}_0$ injectively onto $\hat{V}_0$ and the inverse $\hat{c}^{-1} : \hat{V}_0 \to \hat{E}_1$ of $\hat{c}|_{\hat{U}_0}$ is C^1-smooth. In particular, $\mathrm{Pr}_{G_3^2}$ maps the set $\{P_3(\chi + \hat{b}(\chi)) : \chi \in \hat{U}_0\}$ injectively onto $\hat{V}_0$. Thus there is a map $d : \hat{V}_0 \to \mathbb{R}\xi$ such that

$$\{P_3(\chi + \hat{b}(\chi)) : \chi \in \hat{U}_0\} = \{\chi_2 + d(\chi_2) : \ \chi_2 \in \hat{V}_0\}. \tag{14.10}$$

It is easy to see that

$$d = (\mathrm{id} - \mathrm{Pr}_{G_3^2}) \circ P_3 \circ (\mathrm{id} + \hat{b}) \circ \hat{c}^{-1}.$$

Therefore d is C^1-smooth, and $d(0) = 0$.

8. Construction of a neighbourhood of ϕ in C. Choose $\epsilon_4 \in (0, \epsilon_3)$ such that

$$\mathcal{N}_{\epsilon_4} = \mathrm{Pr}_{E_1} \circ j \circ a(M_{\epsilon_4}) \subset U_0, \quad \hat{\mathcal{N}}_{\epsilon_4} = \mathrm{Pr}_{\hat{E}_1} \circ j \circ \hat{a}(M_{\epsilon_4}^0) \subset \hat{U}_0.$$

Set $\hat{V}_{\epsilon_4} = \hat{c}(\hat{\mathcal{N}}_{\epsilon_4})$. Using (14.10) we find

$$\begin{aligned} P_3(\hat{a}(M^0_{\epsilon_4})) &= P_3\{\phi + \chi + \hat{b}(\chi) :\ \chi \in \hat{\mathcal{N}}_{\epsilon_4}\} \\ &= \{P_3\phi + \chi_2 + d(\chi_2) :\ \chi_2 \in \hat{V}_{\epsilon_4}\}. \end{aligned}$$

Observe that we have

$$P_3(a(M_{\epsilon_4})) = c(\mathcal{N}_{\epsilon_4}) \subset V_{P_3\phi}.$$

For $\kappa > 0$ define

$$\begin{aligned} V_1(\kappa) &= \{\chi_1 \in \mathbb{R}\xi : ||\chi_1|| < \kappa(||Dd(0)|| + 1)\}, \\ V_2(\kappa) &= \{\chi_2 \in G_3^2 : ||\chi_2|| < \kappa\}. \end{aligned}$$

There exists $\kappa_1 > 0$ so that for all $\kappa \in (0, \kappa_1)$ we have $V_2(\kappa) \subset \hat{V}_0$ ans $d(V_2(\kappa)) \subset V_1(\kappa)$. For $\kappa \in (0, \kappa_1)$ define

$$\begin{aligned} V_\kappa^- &= \{P_3\phi + \chi_1 + \chi_2 :\ \chi_1 \in V_1(\kappa),\ \chi_2 \in V_2(\kappa),\ d(\chi_2) - \chi_1 \in (-\infty, 0)\xi\}, \\ V_\kappa^0 &= \{P_3\phi + \chi_1 + \chi_2 :\ \chi_1 \in V_1(\kappa),\ \chi_2 \in V_2(\kappa),\ \chi_1 = d(\chi_2)\}, \\ V_\kappa^+ &= \{P_3\phi + \chi_1 + \chi_2 :\ \chi_1 \in V_1(\kappa),\ \chi_2 \in V_2(\kappa),\ d(\chi_2) - \chi_1 \in (0, \infty)\xi\}. \end{aligned}$$

Let $V_\kappa = V_\kappa^- \cup V_\kappa^0 \cup V_\kappa^+$, $V_\kappa^{-0} = V_\kappa^- \cup V_\kappa^0$, $V_\kappa^{0+} = V_\kappa^0 \cup V_\kappa^+$. Then $V_\kappa = P_3\phi + V_1(\kappa) + V_2(\kappa)$.

Choose $r(\epsilon_4)$ according to the claim in part 5. Select $\kappa_2 \in (0, \kappa_1)$ so small that $V_{\kappa_2} \subset V_{P_3\phi} \cap B_{r(\epsilon_4)}$ and $V_2(\kappa_2) \subset \hat{V}_{\epsilon_4}$. Then it follows from $P_3(\hat{a}(M^0_{\epsilon_4})) = \{P_3\phi + \chi_2 + d(\chi_2) : \chi_2 \in \hat{V}_{\epsilon_4}\}$ and $\hat{a}(M^0_{\epsilon_4}) = a(M^0_{\epsilon_4})$ that we have $V^0_{\kappa_2} \subset P_3(a(M^0_{\epsilon_4}))$.

Select $\epsilon_5 \in (0, \epsilon_4)$ so small that $P_3(a(M_{\epsilon_5})) \subset V_{\kappa_2}$.

Recall the inclusions $\mathcal{N}_{\epsilon_5} = \mathrm{Pr}_{E_1} \circ j \circ a(M_{\epsilon_5}) \subset \mathcal{N}_{\epsilon_4} \subset U_0$ (by the choice of ϵ_4), the fact that $\mathcal{N}_{\epsilon_5}$ is an open neighbourhood of 0 in E_1, the equation $P_3(a(M_{\epsilon_5})) = c(\mathcal{N}_{\epsilon_5})$ and the fact that c is a C^1-diffeomorphism on U_0. We deduce that $P_3(a(M_{\epsilon_5}))$ is an open neighbourhood of $P_3\phi$ in G_3. Thus, there exists $\kappa_3 \in (0, \kappa_2)$ such that

$$V_{\kappa_3} \subset P_3(a(M_{\epsilon_5})).$$

The set $U_\phi = V_{\kappa_3} + E$ is an open neighbourhood of ϕ in C.

9. We claim that

$$\text{either}\ \ P_3(U_\phi \cap \overline{W}) = V_{\kappa_3}^{-0}\ \ \text{or}\ \ P_3(U_\phi \cap \overline{W}) = V_{\kappa_3}^{0+}. \tag{14.11}$$

The inclusions $V_{\kappa_3} \subset V_{\kappa_2} \subset B_{r(\epsilon_4)}$, the equation (14.8) with $\epsilon = \epsilon_4$ and $r = r(\epsilon_4)$, the inclusion $V_{\kappa_3} \subset P_3(a(M_{\epsilon_5}))$, and the equations $V_{\kappa_3} = P_3 U_\phi$, $U_\phi = P_3^{-1}(V_{\kappa_3})$ combined imply

$$\begin{aligned} P_3(U_\phi \cap \overline{W}) &= P_3(U_\phi) \cap P_3\overline{W} \\ &= V_{\kappa_3} \cap P_3\overline{W} \\ &= V_{\kappa_3} \cap P_3(a(M^{-0}_{\epsilon_4})) \\ &= V_{\kappa_3} \cap \left(P_3(a(M^{-0}_{\epsilon_4})) \cap P_3(a(M_{\epsilon_5}))\right). \end{aligned}$$

The inclusion $\mathcal{N}_{\epsilon_4} \subset U_0$ implies that P_3 maps $a(M_{\epsilon_4})$ injectively into G_3. Recall from part 2 that the restriction of a to $M_{\epsilon_4} \subset M_{\epsilon_2}$ is injective. Using the last statements and $\epsilon_5 < \epsilon_4$, we obtain

$$P_3(U_\phi \cap \overline{W}) = V_{\kappa 3} \cap P_3(a(M_{\epsilon_5}^{-0})). \tag{14.12}$$

The set $P_3(a(M_{\epsilon_5})) = c(\mathcal{N}_{\epsilon_5})$ is open in G_3 since $\mathcal{N}_{\epsilon_5} \subset U_0$ is open in E_1 and $c : U_0 \to G_3$ is continuous and injective.

From the relations $\hat{a}(M_{\epsilon_4}^0) = a(M_{\epsilon_4}^0)$, $P_3(\hat{a}(M_{\epsilon_4}^0)) = \{P_3\phi + \chi_2 + d(\chi_2) : \chi_2 \in \hat{V}_{\epsilon_4}\}$ and $V_{\kappa_2}^0 \subset P_3(a(M_{\epsilon_4}^0))$ we deduce

$$V_{\kappa_2} \cap P_3(a(M_{\epsilon_4}^0)) = V_{\kappa_2}^0.$$

The last equation, the injectivity of the restriction of a to M_{ϵ_4} and of P_3 restricted to $a(M_{\epsilon_4})$, and the inequality $\epsilon_5 < \epsilon_4$ combined imply

$$V_{\kappa_2}^0 \cap P_3(a(M_{\epsilon_5}^- \cup M_{\epsilon_5}^+)) = \emptyset.$$

Using the preceding equation and the inclusion $V_{\kappa 3} \subset P_3(a(M_{\epsilon_5}))$ we obtain

$$V_{\kappa 3}^0 = V_{\kappa 3} \cap P_3(a(M_{\epsilon_5}^0)). \tag{14.13}$$

$P_3(a(M_{\epsilon_5}^-))$ and $P_3(a(M_{\epsilon_5}^+))$ are arcwise connected subsets of V_{κ_2}. Their intersections with $V_{\kappa_2}^0$ are empty. Consequently, either $P_3(a(M_{\epsilon_5}^-)) \subset V_{\kappa_2}^-$ or $P_3(a(M_{\epsilon_5}^-)) \subset V_{\kappa_2}^+$, and either $P_3(a(M_{\epsilon_5}^+)) \subset V_{\kappa_2}^-$ or $P_3(a(M_{\epsilon_5}^+)) \subset V_{\kappa_2}^+$. Suppose $P_3(a(M_{\epsilon_5}^- \cup M_{\epsilon_5}^+)) \subset V_{\kappa_2}^-$. For each $\chi \in V_{\kappa 3} \subset P_3(a(M_{\epsilon_5}))$, either $\chi \in V_{\kappa 3} \cap P_3(a(M_{\epsilon_5}^0)) = V_{\kappa 3}^0$ or $\chi \in P_3(a(M_{\epsilon_5}^- \cup M_{\epsilon_5}^+)) \subset V_{\kappa_2}^-$. In both cases there exist $\chi_i \in V_i(\kappa_2)$, $i \in \{1,2\}$ with $\chi = P_3\phi + \chi_1 + \chi_2$ and $d(\chi_2) - \chi_1 \in (-\infty, 0]\xi$, in contradiction to the fact that V_{κ_3} contains points $P_3\phi + \chi_1 + \chi_2$, $\chi_i \in V_i(\kappa_2)$ for $i \in \{1,2\}$, with

$$d(\chi_2) - \chi_1 \in (0, \infty)\xi.$$

It follows that

$$P_3(a(M_{\epsilon_5}^- \cup M_{\epsilon_5}^+)) \not\subset V_{\kappa_2}^-.$$

Arguing as above we also get

$$P_3(a(M_{\epsilon_5}^- \cup M_{\epsilon_5}^+)) \not\subset V_{\kappa_2}^+.$$

Consequently, either

$$P_3(a(M_{\epsilon_5}^-)) \subset V_{\kappa_2}^- \quad \text{and} \quad P_3(a(M_{\epsilon_5}^+)) \subset V_{\kappa_2}^+,$$

or

$$P_3(a(M_{\epsilon_5}^-)) \subset V_{\kappa_2}^+ \quad \text{and} \quad P_3(a(M_{\epsilon_5}^+)) \subset V_{\kappa_2}^-.$$

Using the inclusion $V_{\kappa 3} \subset P_3(a(M_{\epsilon_5}))$ and the equation $V_{\kappa 3}^0 = V_{\kappa 3} \cap P_3(a(M_{\epsilon_5}^0))$ we deduce in the first of the two cases above that we have

$$V_{\kappa 3} \cap P_3(a(M_{\epsilon_5}^{-0})) = V_{\kappa 3}^{-0},$$

while in the second case

$$V_{\kappa_3} \cap P_3(a(M_{\epsilon_5}^{-0})) = V_{\kappa_3}^{0+}.$$

It follows that the claim holds.

We have $P_3(W) \cap P_3(\mathrm{bd}\, W) = \emptyset$ since $P_3|_{\overline{W}}$ is injective. Using this and (14.6), (14.12), (14.13), we obtain

$$P_3(U_\phi \cap \mathrm{bd}\, W) = V_{\kappa_3} \cap P_3(a(M_{\epsilon_5}^0)) = V_{\kappa_3}^0. \tag{14.14}$$

10. Set $V_1 = V_1(\kappa_3)$, $V_2 = V_2(\kappa_3)$, $U = V_{\kappa_3}$, $V^- = V_{\kappa_3}^-$, $V^0 = V_{\kappa_3}^0$, and $V^+ = V_{\kappa_3}^+$. Then the conditions (i), (ii) and (iii) in the theorem are satisfied. The restriction w^e of w^* to $U = V_{\kappa_3} \subset V_{P_3\phi}$ is C^1-smooth. The equations $U_\phi = V_{\kappa_3} + E$ and (14.14) combined yield

$$V_{\kappa_3} \cap P_3(\mathrm{bd}\, W) = P_3(U_\phi \cap \mathrm{bd}\, W) = V_{\kappa_3}^0$$

which proves (14.1). Using the equations $U_\phi = V_{\kappa_3} + E$, (14.11), and (14.14) we also obtain

$$V_{\kappa_3} \cap P_3 W = P_3(U_\phi \cap W) \in \{V_{\kappa_3}^-, V_{\kappa_3}^+\}$$

which gives (14.2). (14.3) is shown in part 6 since $V_{\kappa_3} \subset V_{P_3\phi}$. The graph representation of $\overline{W}$ and the equations $P_3((U+E) \cap \mathrm{bd}\, W) = U \cap P_3(\mathrm{bd}\, W) = V^0 = V_{\kappa_3}^0$ combined yield (14.4). Similarly we obtain (14.5) using the graph representation of $\overline{W}$, the relations $P_3((U+E) \cap \overline{W}) = U \cap P_3\overline{W} \in \{V_{\kappa_3}^{-0}, V_{\kappa_3}^{0+}\}$, and the definitions of $V_{\kappa_3}^{-0}$, $V_{\kappa_3}^{0+}$. □

We now consider the smoothness of w_S at the boundary $P_2\mathcal{O}$ of its domain $P_2(\overline{W} \cap S)$. Recall that $J_2 : \mathbb{R}^2 \to G_2$ is an isomorphism and that we have $P_2 = J_2 \circ \Pi_2$. Therefore Propositions 7.3 and 7.5 imply that $P_2 \circ \eta$ defines a simple closed C^1-curve η^2 in G_2 and

$$\begin{aligned} &P_2(\overline{W} \cap S) = |\eta^2| \cup \mathrm{int}(\eta^2), \\ &P_2\mathcal{O} = |\eta^2|, \quad P_2(W \cap S) = \mathrm{int}(\eta^2). \end{aligned} \tag{14.15}$$

Hence it follows that the boundary $P_2\mathcal{O}$ of the domain $P_2(\overline{W} \cap S)$ of w_S is a 1-dimensional C^1-submanifold of G_2. The next result contains the smoothness of w_S at $P_2\mathcal{O}$.

Theorem 14.2 *For every $\phi \in \mathcal{O}$ there exist an open neighbourhood U of $P_2\phi$ in G_2 and a C^1-map*

$$w_S^e : U \to G_1 \oplus E$$

such that

$$w_S^e\big|_{U \cap P_2(\overline{W} \cap S)} = w_S\big|_{U \cap P_2(\overline{W} \cap S)},$$

and $U \setminus |\eta^2|$ is the union of open connected disjoint subsets U^+ and U^- with $U^+ \cap P_2(\overline{W} \cap S) = \emptyset$ and $U^- \subset P_2(W \cap S)$.

Sketch of the proof We only sketch certain steps of the proof since it is similar to that of Theorem 14.1.

Let $\phi \in \mathcal{O}$. Choose $t_0 > 1$ such that $\phi = F(t_0, p_0)$. If $0 < \epsilon < \min\{t_0 - 1, B - A\}$, then the map

$$a : (-\epsilon, \epsilon) \times (-\epsilon, \epsilon) \ni (t, s) \mapsto F(t_0 + t, \Upsilon(s)) \in C$$

is C^1-smooth. Using Proposition 12.9, we obtain

$$Da(0,0)\mathbb{R}^2 = \mathbb{R}\dot{\phi} \oplus \mathbb{R}D_2F(t_0, p_0)\phi_H.$$

The vectors $\dot{\phi} = D_2F(t_0, p_0)\dot{p}_0$ and $D_2F(t_0, p_0)\phi_H$ are linearly independent since $D_2F(t_0, p_0)$ is injective, and $\dot{p}_0 \notin H$ and $\phi_H \in H$ are linearly independent. Lemma I.8 implies that for small $\epsilon > 0$ the set $a((-\epsilon, \epsilon) \times (-\epsilon, \epsilon))$ is a 2-dimensional C^1-submanifold of C with

$$T_\phi a((-\epsilon, \epsilon) \times (-\epsilon, \epsilon)) = Da(0,0)\mathbb{R}^2.$$

Set $E_1 = \mathbb{R}\dot{\phi} \oplus \mathbb{R}D_2F(t_0, p_0)\phi_H$. It follows that the sets $a((-\epsilon, \epsilon) \times (-\epsilon, 0))$ and $a((-\epsilon, \epsilon) \times (0, \epsilon))$ are also 2-dimensional C^1-submanifolds of C. Let E_2 be a closed complement of E_1 in C. Then for small $\epsilon > 0$ the set $a((-\epsilon, \epsilon) \times (-\epsilon, \epsilon))$ can be written as the shifted graph of a C^1-function $b : \mathcal{N}_\epsilon \to E_2$, where $\mathcal{N}_\epsilon$ is an open neighbourhood of 0 in E_1, $b(0) = 0$ and $Db(0) = 0$;

$$a((-\epsilon, \epsilon) \times (-\epsilon, \epsilon)) = \phi + \{\chi + b(\chi) : \chi \in \mathcal{N}_\epsilon\}.$$

The C^1-map

$$c : E_1 \supset \mathcal{N}_\epsilon \ni \chi \mapsto P_2(\phi + \chi + b(\chi)) \in G_2$$

satisfies $Dc(0)\chi = P_2\chi$ for all $\chi \in E_1$. We claim that $P_2|_{E_1}$ is injective. E_1 is spanned by the derivatives $D\Gamma(0)1$ of the curves

$$\Gamma : (-s^*, s^*) \ni s \mapsto a(c_1 s, c_2 s) \in C,$$

with $s^* > 0$, $(c_1, c_2)^{tr} \in \mathbb{R}^2$. From the invariance of $\overline{W} \cap S$ and Proposition 12.9 it follows that in case $c_2 s \le 0$, $\Gamma(s) \in \overline{W} \cap S$. Proposition 6.10 guarantees $\Pi_2\Gamma'(0) \neq 0$ provided $\Gamma'(0) \neq 0$. So, $\Pi_2|_{E_1}$ is injective. Then $P_2|_{E_1}$ is also injective, as we have $P_2 = J_2 \circ \Pi_2$, and $J_2 : \mathbb{R}^2 \to G_2$ is an isomorphism. It follows that there exists $\epsilon_0 > 0$ such that P_2 is injective on $a((-\epsilon_0, \epsilon_0) \times (-\epsilon_0, \epsilon_0)) = \phi + \{\chi + b(\chi) : \chi \in \mathcal{N}_{\epsilon_0}\}$ and c defines a C^1-diffeomorphism from $\mathcal{N}_{\epsilon_0}$ onto an open neighbourhood $V_{P_2\phi}$ of $P_2\phi$ in G_2. Define $c^{-1} : V_{P_2\phi} \to E_1$ by the inverse of the diffeomorphism.

Set

$$\begin{aligned}
U &= P_2 a((-\epsilon_0, \epsilon_0) \times (-\epsilon_0, \epsilon_0)) = V_{P_2\phi}, \\
U^- &= P_2 a((-\epsilon_0, \epsilon_0) \times (-\epsilon_0, 0)), \\
U^0 &= P_2 a((-\epsilon_0, \epsilon_0) \times \{0\}), \\
U^+ &= P_2 a((-\epsilon_0, \epsilon_0) \times (0, \epsilon_0)).
\end{aligned}$$

The arcwise connected sets U^- and U^+ are open in G_2, since $a((-\epsilon_0, \epsilon_0) \times (-\epsilon_0, 0))$ and $a((-\epsilon_0, \epsilon_0) \times (0, \epsilon_0))$ are 2-dimensional C^1-submanifolds of C and the restrictions P_2 to the latter are injective.

The relation $\Upsilon(0) = p_0 \in \mathcal{O}$ implies $U^0 \subset P_2\mathcal{O}$. We have $U^- \subset P_2(W \cap S)$ because of Proposition 12.9 and the invariance of $W \cap S$.

We claim that U^+ and $P_2(\overline{W} \cap S)$ are disjoint. Using that $U^0 \subset P_2\mathcal{O}$ belongs to the boundary of both connected components of $G_2 \setminus |\eta^2|$ by the Jordan Curve Theorem, we show that U^- and U^+ belong to different connected components of $G_2 \setminus |\eta^2|$. As

$$U^- \subset P_2(W \cap S) = \operatorname{int}(\eta^2),$$

we infer

$$U^+ \subset \operatorname{ext}(\eta^2).$$

This fact and (14.15) imply $U^+ \cap P_2(\overline{W} \cap S) = \emptyset$.

Consider the C^1-map

$$w_S^* : U \ni \eta \mapsto \phi + c^{-1}(\eta) + b(c^{-1}(\eta)) - \eta \in C.$$

For $\eta \in U$ we have $c^{-1}(\eta) \in \mathcal{N}_{\epsilon_0}$ and thus

$$\begin{aligned} P_2(\phi + c^{-1}(\eta) + b(c^{-1}(\eta)) - \eta) &= P_2(\phi + c^{-1}(\eta) + b(c^{-1}(\eta))) - P_2\eta \\ &= c(c^{-1}(\eta)) - \eta = 0. \end{aligned}$$

Therefore w_S^* maps U into $P_2^{-1}(0) = G_1 \oplus E$ and defines a C^1-map $w_S^e : U \to G_1 \oplus E$.

Finally, we claim that w_S^e and w_S coincide on $U \cap P_2(\overline{W} \cap S)$. Let $\eta \in U \cap P_2(\overline{W} \cap S)$. From $U \cap P_2(\overline{W} \cap S) = (U \cap P_2\mathcal{O}) \cup (U \cap P_2(W \cap S)) = U^0 \cup U^-$ and from the definition of U^0, U^- it follows that $\eta = P_2 a(t,s)$ for some $t \in (-\epsilon_0, \epsilon_0)$ and $s \in (-\epsilon_0, 0]$. We have $a(t,s) = F(t_0 + t, \Upsilon(s)) \in \overline{W} \cap S$, due to Proposition 12.9 and to the invariance of $\overline{W} \cap S$. As P_2 is injective on $\overline{W} \cap S$, we obtain

$$\eta + w_S(\eta) = a(t,s).$$

On the other hand, the definition of w_S^e implies

$$\eta + w_S^e(\eta) = \phi + c^{-1}(\eta) + b(c^{-1}(\eta)) \in a((-\epsilon_0, \epsilon_0) \times (-\epsilon_0, \epsilon_0)).$$

Using the injectivity of P_2 on $a((-\epsilon_0, \epsilon_0) \times (-\epsilon_0, \epsilon_0))$ we infer

$$\eta + w_S^e(\eta) = a(t,s).$$

Consequently, $w_S(\eta) = w_S^e(\eta)$. □

CHAPTER 15

Homeomorphisms from bd W onto the sphere and the cylinder

The aim of this chapter is to obtain a homeomorphism from bd W onto S^2 in case $\mu > 0$ and a homeomorphism from bd W onto $\mathbb{R} \times S^1_{\mathbb{C}}$ in case $\mu = 0$. Our approach is constructive and based on the important relation between bd W and $W^u(\mathcal{O})$ described in Corollary 12.13.

Choose δ_2 and δ_3 according to Proposition 12.1. Choose $\epsilon \in (0, \min\{\delta_2, \delta_3\})$ and consider the set

$$W^u_\epsilon(p_0, F(\omega, \cdot)) = \{p_0 + s\psi_u + w^u(s\psi_u) : |s| < \epsilon\}.$$

Choose $s_0 \in (0, \epsilon)$. Proposition 12.1 shows that $s_1 = g(s_0) \in (0, \delta_1)$ satisfies

$$F(\omega, p_0 + s_0\psi_u + w^u(s_0\psi_u)) = p_0 + s_1\psi_u + w^u(s_1\psi_u).$$

Using Proposition 12.1 and the flow on bd $W \supset W^u(\mathcal{O}) \supset W^u_\epsilon(p_0, F(\omega, \cdot))$ we obtain that the backward iterates $s_{-n} = g^{-n}(s_0) \in (0, \epsilon)$, $1 \le n \in \mathbb{N}$, satisfy

$$F_W(-n\omega, p_0 + s_0\psi_u + w^u(s_0\psi_u)) = p_0 + s_{-n}\psi_u + w^u(s_{-n}\psi_u) \quad \text{for all integers } n \ge 1.$$

Also by Proposition 12.1, we have

$$p_0 + \alpha_1\psi_u + w^u(\alpha_1\psi_u) \ll p_0 + \alpha_2\psi_u + w^u(\alpha_2\psi_u) \quad \text{if } -\epsilon < \alpha_1 < \alpha_2 < \epsilon \tag{15.1}$$

and

$$s_{-n} > s_{-n-1} \quad \text{for all integers } n \ge 1; \quad s_{-n} \to 0 \quad \text{as } n \to \infty. \tag{15.2}$$

The curve

$$\gamma : [s_0, s_1] \ni s \mapsto p_0 + s\psi_u + w^u(s\psi_u) \in C$$

defines a homeomorphism onto its image. Set

$$\Gamma_F = \{\gamma(s) : s_0 \le s < s_1\}.$$

Analogously we obtain that $s_1^- = g(-s_0) \in (-\delta_1, 0)$ satisfies

$$F(\omega, p_0 + (-s_0)\psi_u + w^u(-s_0\psi_u)) = p_0 + s_1^-\psi_u + w^u(s_1^-\psi_u),$$

and that the backward iterates $s_{-n}^- = g^{-n}(-s_0) \in (-\epsilon, 0)$, $1 \le n \in \mathbb{N}$, of $s_0^- = -s_0$ satisfy

$$F(-n\omega, p_0 + (-s_0)\psi_u + w^u(-s_0\psi_u)) = p_0 + s_{-n}^-\psi_u + w^u(s_{-n}^-\psi_u)$$

and

$$s^-_{-n} < s^-_{-n-1} \qquad \text{for all integers } n \geq 1,$$

and

$$s^-_{-n} \to 0 \qquad \text{as } n \to \infty.$$

The curve

$$\gamma^- : [s^-_1, s^-_0] \ni s \mapsto p_0 + s\psi_u + w^u(s\psi_u) \in C$$

defines a homeomorphism onto its image. Set

$$\Gamma^-_F = \{\gamma^-(s) : s^-_1 < s \leq s^-_0\}.$$

In case $\mu > 0$, set

$$\text{bd}^+ W = \{\phi \in \text{bd}\, W \setminus \{\xi_+, \xi_-\} : x^\phi_t \to \xi_+ \text{ as } t \to \infty\},$$
$$\text{bd}^- W = \{\phi \in \text{bd}\, W \setminus \{\xi_+, \xi_-\} : x^\phi_t \to \xi_- \text{ as } t \to \infty\}.$$

In case $\mu = 0$, set

$$\text{bd}^+ W = \{\phi \in \text{bd}\, W : x^\phi(t) \to \infty \text{ as } t \to \infty\},$$
$$\text{bd}^- W = \{\phi \in \text{bd}\, W : x^\phi(t) \to -\infty \text{ as } t \to \infty\}.$$

Then in case $\mu > 0$,

$$\text{bd}\, W = \{\xi_-\} \cup \text{bd}^- W \cup \mathcal{O} \cup \text{bd}^+ W \cup \{\xi_+\},$$

and in case $\mu = 0$,

$$\text{bd}\, W = \text{bd}^- W \cup \mathcal{O} \cup \text{bd}^+ W.$$

Proposition 15.1

(i) For every $\phi \in \text{bd}^+ W$ there is a unique $t \in \mathbb{R}$ such that $F_W(t, \phi) \in \Gamma_F$. For every $\phi \in \text{bd}^- W$ there is a unique $t \in \mathbb{R}$ such that $F_W(t, \phi) \in \Gamma^-_F$.

(ii) $\text{bd}^+ W = F_W(\mathbb{R} \times \Gamma_F)$, $\text{bd}^- W = F(\mathbb{R} \times \Gamma^-_F)$.

Proof 1. Let $\phi \in \text{bd}^+ W$. The definition of $W^u(\mathcal{O})$ and Corollary 12.13 permit to find $t^* \geq 0$ such that $F_W(-t^*, \phi) \in W^u(p_0, F(\omega, \cdot), N^u)$. Then we can find a nonnegative integer n_1 such that $F_W(-t^* - n_1\omega, \phi) = p_0 + s\psi_u + w^u(s\psi_u)$ for some $s \in (-s_0, s_0)$. We have $0 < s_0$ since otherwise Proposition 12.1(ii) would give $F_W(-t^* - n_1\omega, \phi) \leq p_0$ which leads to $F_W(n\omega - t^* - n_1\omega, \phi) \leq p_0$ for all integers $n \geq 1$, in contradiction to the limiting behaviour of $\mathbb{R} \ni t \mapsto F_W(t, \phi) \in C$ as $t \to \infty$. Since $(0, s_0)$ is the union of the disjoint intervals $[s_{-n}, s_{-n+1})$, $1 \leq n \in \mathbb{N}$, there exists a nonnegative integer n_2 such that $s \in [s_{-n_2}, s_{-n_2+1})$. Then $F_W(-t^* + n_2\omega - n_1\omega, \phi) = p_0 + g^{n_2}(s)\psi_u + w^u(g^{n_2}(s)\psi_u) \in \Gamma_F$. This establishes the existence of t such that $F_W(t, \phi) \in \Gamma_F$.

Assume now that t_1, t_2 in $\mathbb{R}$ are given so that $F_W(t_1, \phi)$, $F_W(t_2, \phi)$ belong to Γ_F. Then $F_W(t_1 - n\omega, \phi) \to p_0$ and $F_W(t_2 - n\omega, \phi) \to p_0$ as $n \to \infty$. But $F_W(t_2 - n\omega, \phi) = F_W(t_2 - t_1, F_W(t_1 - n\omega, \phi))$ for all $n \in \mathbb{N}$. By continuity of F_W, $p_0 = p_{t_2 - t_1}$ from which it follows that $t_2 - t_1 = m\omega$ for some integer m. Without loss of generality, assume $t_2 \geq t_1$ and let $F_W(t_2, \phi) = \gamma(s)$ for some $s \in [s_0, s_1)$. Then Proposition 12.1 implies that

$$\begin{aligned} F_W(t_1,\phi) &= F_W(-m\omega, F_W(t_2,\phi)) = F_W(-m\omega, \gamma(s)) \\ &= p_0 + g^{-m}(s)\psi_u + w^u(g^{-m}(s)\psi_u). \end{aligned}$$

Using the relations $g^{-m}(s) \in [s_{-m}, s_{-m+1})$, $F_W(t_1,\phi) \in \Gamma_F$, and the injectivity of γ we obtain $m = 0$. That is, $t_2 = t_1$. This shows the uniqueness of $t \in \mathbb{R}$ such that $F_W(t,\phi) \in \Gamma_F$. The second part of assertion (i) is proved analogously.

2. We show $F_W(\mathbb{R} \times \Gamma_F) \subset \mathrm{bd}^+ W$. Using Corollary 12.13 and the invariance of bd W we obtain $\Gamma_F \subset W^u_\epsilon(p_0, F(\omega,\cdot)) \subset \mathrm{bd}\, W$, and $F_W(\mathbb{R} \times \Gamma_F) \subset \mathrm{bd}\, W$, respectively. Let $\psi \in F_W(\mathbb{R} \times \Gamma_F)$ be given. Then there exist $t_0 \in \mathbb{R}$, $\phi \in \Gamma_F$, $s \in [s_0, s_1)$ with $\psi = F_W(t_0, \phi)$ and $\phi = \gamma(s) = p_0 + s\psi_u + w^u(s\psi_u)$. There is a backward trajectory $(s^n)^0_{-\infty}$ of g in $(0, s_0]$ with $s^0 = s$ which is strictly increasing and satisfies

$$F_W(n\omega, \phi) = p_0 + s^n\psi_u + w^u(s^n\psi_u) \qquad \text{for all integers } n \leq 0.$$

Consequently $\phi \notin \mathcal{O}$. We have $s^n \to 0$ as $n \to -\infty$, and conclude from this in case $\mu > 0$ that $\phi \notin \{\xi_-, \xi_+\}$. In both cases $\mu = 0$ and $\mu > 0$ we deduce $\psi \in \mathrm{bd}^- W \cup \mathrm{bd}^+ W$. Note that, due to $0 < s$, $p_0 \ll \phi$, and consequently

$$p_0 \ll F_W(j\omega, \phi) \qquad \text{for all } j \in \mathbb{N},$$

which excludes in case $\mu = 0$ that x^ψ tends to $-\infty$ as $t \to \infty$, and in case $\mu > 0$ that the segments $F_W(t,\psi)$ converge to $\xi_- \in -\overset{\circ}{K}$ for $t \to \infty$. It follows that in both cases, $\psi \in \mathrm{bd}^+ W$. The proof of the second part of assertion (ii) is analogous.

□

Set $D^2 = \{x \in \mathbb{R}^2 : x_1^2 + x_2^2 \leq 1\}$. The next constructions prepare the proof that in case $\mu > 0$ the sets $\mathrm{bd}^+ W \cup \mathcal{O} \cup \{\xi_+\}$, $\mathrm{bd}^- W \cup \mathcal{O} \cup \{\xi_-\}$ are homeomorphic to D^2.

The map $G : \mathbb{R} \times D^2 \to D^2$ given by

$$G(t, (0,0)^{tr}) = (0,0)^{tr} \qquad \text{for all } t \in \mathbb{R}$$

and by

$$G(t,(x,y)^{tr}) = \left(\frac{r}{(1-r)e^t + r}\cos\Big(\theta + \frac{2\pi}{\omega}t\Big), \frac{r}{(1-r)e^t + r}\sin\Big(\theta + \frac{2\pi}{\omega}t\Big)\right)^{tr},$$
$$r = \sqrt{x^2+y^2},\ 0 \leq \theta < 2\pi,\ x = r\cos\theta,\ y = r\sin\theta$$

for all $t \in \mathbb{R}$ and $(x,y)^{tr} \in D^2 \setminus \{(0,0)^{tr}\}$ is a continuous flow on D^2; polar coordinates of flowlines $t \mapsto G(t,(x,y)^{tr})$ in $D^2 \setminus \{(0,0)^{tr}\}$ satisfy the ordinary differential equations

$$\dot{r} = r(r-1), \quad \dot{\theta} = \frac{2\pi}{\omega}. \tag{15.3}$$

We have

$$G\left(\omega,\left(\frac{1}{2},0\right)^{tr}\right)=\left(\frac{1}{1+e^{\omega}},0\right)^{tr}.$$

Set

$$\Gamma_G=\left\{(x,0)^{tr}:\frac{1}{1+e^{\omega}}<x\leq\frac{1}{2}\right\}.$$

Then

$$G(\mathbb{R}\times\Gamma_G)=\{x\in\mathbb{R}^2:0<x_1^2+x_2^2<1\}.$$

The maps $i_{FG}:\Gamma_F\to\Gamma_G$ and $i_{FG}^-:\Gamma_F^-\to\Gamma_G$ given by

$$i_{FG}(\gamma(s))=\left(\frac{s-s_0}{s_1-s_0}\frac{1}{1+e^{\omega}}+\frac{s_1-s}{s_1-s_0}\frac{1}{2},0\right)^{tr},\qquad s_0\leq s<s_1,$$

and

$$i_{FG}^-(\gamma^-(s))=\left(\frac{s-s_0^-}{s_1^--s_0^-}\frac{1}{1+e^{\omega}}+\frac{s_1^--s}{s_1^--s_0^-}\frac{1}{2},0\right)^{tr},\qquad s_1^-<s\leq s_0^-$$

are homeomorphisms. In case $\mu>0$ we use Proposition 15.1 to define maps h : $\mathrm{bd}^+W\cup\mathcal{O}\cup\{\xi_+\}\to D^2$ and $h^-:\mathrm{bd}^-W\cup\mathcal{O}\cup\{\xi_-\}\to D^2$ by the relations

$$\left\{\begin{array}{ll} h(\phi)=G(-t,i_{FG}(F_W(t,\phi))), & \phi\in\mathrm{bd}^+W,\ t\in\mathbb{R},\ F_W(t,\phi)\in\Gamma_F,\\ h(\xi_+)=0, & \\ h(\phi)=\left(\cos\frac{2\pi}{\omega}t,\sin\frac{2\pi}{\omega}t\right)^{tr}, & \phi=p_t,\ t\in\mathbb{R}\end{array}\right\}\tag{15.4}$$

and

$$\left\{\begin{array}{ll} h^-(\phi)=G(-t,i_{FG}^-(F_W(t,\phi))), & \phi\in\mathrm{bd}^-W,\ t\in\mathbb{R},\ F_W(t,\phi)\in\Gamma_F^-,\\ h^-(\xi_-)=0, & \\ h^-(\phi)=\left(\cos\frac{2\pi}{\omega}t,\sin\frac{2\pi}{\omega}t\right)^{tr}, & \phi=p_t,\ t\in\mathbb{R}.\end{array}\right\}$$

Proposition 15.2 *Let $\mu>0$. The maps h and h^- are homeomorphisms, and $h|_{\mathcal{O}}=h^-|_{\mathcal{O}}$, $h(\mathcal{O})=S^1$.*

Proof 1. We claim that h is surjective. Obviously $\{0\}\cup S^1\subset h(\mathrm{bd}^+W\cup\mathcal{O}\cup\{\xi_+\})$. Observe $\{x\in\mathbb{R}^2:0<x_1^2+x_2^2<1\}=G(\mathbb{R}\times\Gamma_G)=G(\mathbb{R}\times i_{FG}(\Gamma_F))$. Use that, according to Proposition 15.1(ii), for every $t\in\mathbb{R}$ and $\phi\in\Gamma_F$, the segment $\psi=F_W(-t,\phi)$ belongs to bd^+W. This yields $G(t,i_{FG}(\phi))=h(\psi)$.

2. We claim that h is injective. To show this, let ϕ^1,ϕ^2 in $\mathrm{bd}^+W\cup\mathcal{O}\cup\{\xi_+\}$ be given with $\phi^1\neq\phi^2$. Clearly, $h(\phi^1)\neq h(\phi^2)$ if ϕ^1 or ϕ^2 belong to $\mathcal{O}\cup\{\xi_+\}$. So, we only need to consider the case where $\phi^1\in\mathrm{bd}^+W$, $\phi^2\in\mathrm{bd}^+W$. There exist t^1,t^2 in $\mathbb{R}$ and ψ^1,ψ^2 in Γ_F with $\phi^i=F(t^i,\psi^i)$ for $i\in\{1,2\}$. In case $t^1=t^2$, necessarily $\psi^1\neq\psi^2$ since otherwise we obtain a contradiction to $\phi^1\neq\phi^2$. Then the injectivity of i_{FG} and of the map $G(-t^1,\cdot)=G(-t^2,\cdot)$ yields

$$h(\phi^1)=G(-t^1,i_{FG}(\psi^1))\neq G(-t^2,i_{FG}(\psi^2))=h(\phi^2).$$

In case $t^1 \neq t^2$ and $\psi^1 = \psi^2$ we obtain

$$h(\phi^1) = G(-t^1, i_{FG}(\psi^1)) \neq G(-t^2, i_{FG}(\psi^2))$$

since there are no periodic flowlines of G through points in $\{x \in \mathbb{R}^2 : 0 < x_1^2 + x_2^2 < 1\}$. In case $t^1 \neq t^2$ and $\psi^1 \neq \psi^2$ we also obtain the last inequality since the points $i_{FG}(\psi^1)$, $i_{FG}(\psi^2)$ in Γ_G are different, each orbit of G has at most one point in Γ_G, and orbits are either equal or disjoint.

3. We now verify the continuity of h. Let $\phi \in \mathrm{bd}^+ W \cup \mathcal{O} \cup \{\xi_+\}$ be given. It suffices to show that every sequence $(\phi^n)_0^\infty$ in $\mathrm{bd}^+ W \cup \mathcal{O} \cup \{\xi_+\}$ which converges to ϕ has a subsequence $(\phi^{n_k})_{k=0}^\infty$ so that

$$h(\phi^{n_k}) \to h(\phi) \qquad \text{as } k \to \infty.$$

Let a sequence $(\phi^n)_0^\infty$ in $\mathrm{bd}^+ W \cup \mathcal{O} \cup \{\xi_+\}$ with $\lim_{n\to\infty} \phi^n = \phi$ be given.

Case I: $\phi = \xi_+$. If there is a subsequence of $(\phi^n)_0^\infty$ with constant value ξ_+ then we obviously have a subsequence as desired. In the remaining case there is a subsequence $(\phi^{n_k})_{k=0}^\infty$ in $\mathrm{bd}^+ W$ since $\lim_{n\to\infty} \phi^n = \xi_+ \notin \mathcal{O}$. Then there is a sequence $(t^k)_0^\infty$ in $\mathbb{R}$ so that for every $k \in \mathbb{N}$,

$$F_W(t^k, \phi^{n_k}) \in \Gamma_F \quad \text{and} \quad h(\phi^{n_k}) = G(-t^k, i_{FG}(F_W(t^k, \phi^{n_k}))).$$

It is convenient to set $\psi^k = \phi^{n_k}$ for every $k \in \mathbb{N}$.

We claim

$$t^k \to -\infty \qquad \text{as } k \to \infty.$$

Suppose this is false. It follows that either there are a subsequence $(t^{k_j})_{j=0}^\infty$ and a sequence $(s^j)_0^\infty$ in $[s_0, s_1)$ with

$$F_W(t^{k_j}, \psi^{k_j}) = \gamma(s^j) \qquad \text{for all } j \in \mathbb{N}$$

which converge to reals t^* and $s^* \in [s_0, s_1] \subset (0, \delta_1)$, respectively, or there is a subsequence $(t^{k_j})_0^\infty$ which tends to $+\infty$.

In the first case,

$$\psi^{k_j} \to F_W(-t^*, \gamma(s^*)) \in F_W(\mathbb{R} \times W^u_{\delta_1}(p_0, F(\omega, \cdot))) \subset \mathrm{bd}\, W \setminus \{\xi_-, \xi_+\},$$

in contradiction to $\lim_{n\to\infty} \phi^n = \xi_+$.

In the second case, the contraction property of $(F_W(\omega, \cdot)_u)^{-1}$ and the inclusion $W^u_{\delta_1}(p_0, F(\omega, \cdot)) \subset \overline{W}$ combined show that for every $\tilde{\epsilon} > 0$ there exists $\tilde{m} \in \mathbb{Z}$ so that for all integers $m \le \tilde{m}$,

$$F_W(\{m\omega\} \times \Gamma_F) \subset p_0 + C_{\tilde{\epsilon}}.$$

Set $\epsilon^* = \frac{1}{2}\,\mathrm{dist}(\xi_+, \mathcal{O})$. The previous statement and the continuity of F_W permit to find $t^* \in \mathbb{R}$ so that for all $t \le t^*$,

$$F_W(\{t\} \times \Gamma_F) \subset \mathcal{O} + C_{\epsilon^*}.$$

It follows that for all sufficiently large $j \in \mathbb{N}$,

$$\psi^{k_j} = F_W(-t^{k_j}, F_W(t^{k_j}, \psi^{k_j})) \in \mathcal{O} + C_{\epsilon^*},$$

in contradiction to $\lim_{n\to\infty} \phi^n = \xi_+$.

Using the last claim and the definitions of G and Γ_G we obtain

$$h(\phi^{n_k}) = G(-t^k, i_{FG}(F_W(t^k, \phi^{n_k}))) \to 0 = h(\xi_+) = h(\phi)$$

as $k \to \infty$.

Case II: $\phi \in \mathcal{O}$. Then $\phi = p_s$ and $h(\phi) = \left(\cos \frac{2\pi}{\omega} s, \sin \frac{2\pi}{\omega} s\right)^{tr}$ for some $s \in \mathbb{R}$. From $\xi_+ \notin \mathcal{O}$ we infer that the set of $n \in \mathbb{N}$ with $\phi^n = \xi_+$ is empty or finite.

In case there exists a subsequence $(\phi^{n_k})_{k=0}^{\infty}$ in $\mathcal{O}$ we obtain a sequence $(t^k)_0^\infty$ in $\left[s - \frac{\omega}{2}, s + \frac{\omega}{2}\right)$ with $\phi^{n_k} = p_{t^k}$ and $h(\phi^{n_k}) = \left(\cos \frac{2\pi}{\omega} t^k, \sin \frac{2\pi}{\omega} t^k\right)^{tr}$ for all $k \in \mathbb{N}$. Using $\lim_{n\to\infty} \phi^n = \phi = p_s$ we conclude that necessarily

$$t^k \to s \qquad \text{as } k \to \infty.$$

Consequently

$$h(\phi^{n_k}) \to h(\phi) \qquad \text{as } k \to \infty.$$

In the remaining case there are a subsequence $(\phi^{n_k})_{k=0}^{\infty}$ in $\mathrm{bd}^+ W$ and a sequence $(t^k)_0^\infty$ of reals so that for every $k \in \mathbb{N}$,

$$F_W(t^k, \phi^{n_k}) \in \Gamma_F, \quad h(\phi^{n_k}) = G(-t^k, i_{FG}(F_W(t^k, \phi^{n_k}))).$$

It is convenient to set $\psi^k = \phi^{n_k}$ for all $k \in \mathbb{N}$.

Claim: $t^k \to \infty$ as $k \to \infty$.

Suppose the last statement is false. Then it follows that either there are a subsequence $(t^{k_j})_{j=0}^{\infty}$ and a sequence $(s^j)_0^\infty$ in $[s_0, s_1)$ with

$$F_W(t^{k_j}, \psi^{k_j}) = \gamma(s^j) \tag{15.5}$$

for all $j \in \mathbb{N}$, which converge to reals t^* and $s^* \in [s_0, s_1] \subset (0, \delta_1)$, respectively, or there is a subsequence $(t^{k_j})_{j=0}^{\infty}$ in $(-\infty, 0]$ which converges to $-\infty$, and the sequence $(s^j)_0^\infty$ in $[s_0, s_1)$ given by (15.5) converges to some $s^* \in [s_0, s_1] \subset (0, \delta_1)$.

In the first case,

$$\psi^{k_j} \to F_W(-t^*, \gamma(s^*)) \in F_W(\mathbb{R} \times [W^u_{\delta_1}(p_0, F(\omega, \cdot)) \setminus \{p_0\}])$$

as $j \to \infty$, and we obtain a contradiction since $\lim_{n\to\infty} \phi^n \in \mathcal{O}$ while solutions $x : \mathbb{R} \to \mathbb{R}$ of equation (1.1) with $x_0 \in W^u_{\delta_1}(p_0, F(\omega, \cdot)) \setminus \{p_0\}$ are not periodic.

In the second case the inequalities

$$F_W(\tilde{t}, \gamma(s_0)) \le F_W(\tilde{t}, \gamma(s^j)) \le \xi_+ \qquad \text{for all } j \in \mathbb{N},\ \tilde{t} \ge 0,$$

and the relation

$$\gamma(s_0) \in \mathrm{bd}^+ W$$

due to Proposition 15.1(ii) show that for $\epsilon^* = \frac{1}{2}\operatorname{dist}(\xi_+, \mathcal{O})$ there exists $j_0 \in \mathbb{N}$ so that for all integers $j \geq j_0$ and for all $s \in [-1, 0]$ we have

$$\xi^+ - \epsilon^* \leq F_W(-t^{k_j}, \gamma(s_0))(s) \leq F_W(-t^{k_j}, \gamma(s^j))(s) \leq \xi^+,$$

in contradiction to $\lim_{n\to\infty} \phi^n \in \mathcal{O}$.

Define sequences $(m_k)_0^\infty$ in $\mathbb{Z}$ and $(\delta^k)_0^\infty$ in $[0, \omega)$ by $m_k\omega \leq t^k < (m_k + 1)\omega$ and $\delta^k = t^k - m_k\omega$ for all $k \in \mathbb{N}$. Then $\lim_{k\to\infty} m_k = \infty$, and the contraction property of $(F_W(\omega, \cdot)_u)^{-1}$ implies that for the sequence $(s^k)_0^\infty$ in $[s_0, s_1)$ given by

$$F_W(t^k, \phi^{n_k}) = \gamma(s^k)$$

we have

$$F_W(-m_k\omega, \gamma(s^k)) \to p_0 \qquad \text{as } k \to \infty.$$

A subsequence $(\delta^{k_j})_{j=0}^\infty$ of $(\delta^k)_0^\infty$ converges to some $\delta \in [0, \omega]$. We obtain

$$\phi^{n_{k_j}} = F_W(-t^{k_j}, \gamma(s^{k_j})) = F_W(-\delta^{k_j}, F_W(-m_{k_j}\omega, \gamma(s^{k_j}))) \to F_W(-\delta, p_0) = p_{-\delta}.$$

Using $\lim_{n\to\infty} \phi^n = p_s$ we infer $-\delta \in s + \mathbb{Z}\omega$. Using the definition of G we obtain

$$h(\phi^{n_{k_j}}) = G(-t^{k_j}, i_{FG}(F_W(t^{k_j}, \phi^{n_{k_j}}))) \to$$

$$\left(\cos\frac{2\pi}{\omega}(-\delta), \sin\frac{2\pi}{\omega}(-\delta)\right)^{tr} = \left(\cos\frac{2\pi}{\omega}s, \sin\frac{2\pi}{\omega}s\right)^{tr} = h(p_s) = h(\phi)$$

as $j \to \infty$.

Case III: $\phi \in \mathrm{bd}^+ W$. Then $\xi_+ \neq \phi \notin \mathcal{O}$, and there exists $n_0 \in \mathbb{N}$ with $\phi^n \in \mathrm{bd}^+ W$ for all integers $n \geq n_0$. It follows that there are sequences $(t^n)_{n=n_0}^\infty$ in $\mathbb{R}$, $(s^n)_{n=n_0}^\infty$ in $[s_0, s_1)$, and reals t^ϕ and $s^\phi \in [s_0, s_1)$ with

$$F_W(t^\phi, \phi) = \gamma(s^\phi) \quad \text{and} \quad F_W(t^n, \phi^n) = \gamma(s^n) \qquad \text{for all integers } n \geq n_0.$$

Arguing as in the proofs of the claims in Case I and Case II, with $\epsilon^* = \frac{1}{2}\operatorname{dist}(\xi_+, \mathcal{O})$ replaced by $\epsilon^* = \frac{1}{2}\min\{\operatorname{dist}(\phi, \mathcal{O}), \operatorname{dist}(\phi, \xi_+)\}$, we obtain that the sequence $(t^n)_{n=n_0}^\infty$ is bounded.

Subcase III.1: $s_0 < s^\phi$. We claim $\lim_{n\to\infty} t^n = t^\phi$.

It suffices to show that every convergent subsequence of $((t^n, s^n)^{tr})_{n=n_0}^\infty$ in $\mathbb{R}^2$ tends to $(t^\phi, s^\phi)^{tr}$. Suppose a subsequence $((t^{n_k}, s^{n_k})^{tr})_{k=0}^\infty$ tends to a point $(t^*, s^*)^{tr} \in \mathbb{R}^2$, with $s^* \in [s_0, s_1]$. Then

$$\phi^{n_k} = F_W(-t^{n_k}, \gamma(s^{n_k})) \to F_W(-t^*, \gamma(s^*)) \qquad \text{as } k \to \infty,$$

and consequently

$$F_W(-t^\phi, \gamma(s^\phi)) = \phi = F_W(-t^*, \gamma(s^*)).$$

In case $s^* < s_1$ we obtain $F_W(t^*, \phi) \in \Gamma_F$; in case $s^* = s_1$ we infer $F_W(t^* - \omega, \phi) = F_W(-\omega, \gamma(s_1)) = \gamma(s_0) \in \Gamma_F$. Using Proposition 15.1(i) and $F_W(t^\phi, \phi) \in \Gamma_F$ we get $s^* < s_1$ and $t^* = t^\phi$, or $s^* = s_1$ and $t^* - \omega = t^\phi$. In the last case,

$$F_W(-t^\phi, \gamma(s^\phi)) = F_W(-t^*, \gamma(s^*)) = F_W(-t^\phi - \omega, \gamma(s_1)) = F_W(-t^\phi, \gamma(s_0)),$$

hence $s_0 = s^\phi$ in contradiction to the hypothesis $s_0 < s^\phi$. Consequently, $s^* < s_1$ and $t^* = t^\phi$.

It follows that

$$h(\phi^n) = G(-t^n, i_{FG}(F_W(t^n, \phi^n))) \to G(-t^\phi, i_{FG}(F_W(t^\phi, \phi))) = h(\phi).$$

Subcase III.2: $s_0 = s^\phi$. There exists a subsequence $((t^{n_k}, s^{n_k})^{tr})_{k=0}^\infty$ of the sequence $((t^n, s^n)^{tr})_{n=n_0}^\infty$ in $\mathbb{R}^2$ which converges to a point $(t^*, s^*)^{tr} \in \mathbb{R}^2$ with $s^* \in [s_0, s_1]$. It follows that

$$\phi^{n_k} = F_W(-t^{n_k}, \gamma(s^{n_k})) \to F_W(-t^*, \gamma(s^*)) \qquad \text{as } k \to \infty,$$

hence

$$F_W(-t^*, \gamma(s^*)) = \phi = F_W(-t^\phi, \gamma(s^\phi)) = F_W(-t^\phi, \gamma(s_0)).$$

In case $s^* < s_1$ we conclude from

$$F_W(t^*, \phi) = \gamma(s^*) \in \Gamma_F \quad \text{and} \quad F_W(t^\phi, \phi) \in \Gamma_F$$

that we have $t^* = t^\phi$, and therefore

$$h(\phi^{n_k}) = G(-t^{n_k}, i_{FG}(F_W(t^{n_k}, \phi^{n_k}))) \to G(-t^\phi, i_{FG}(F_W(t^\phi, \phi))) = h(\phi) \text{ as } k \to \infty.$$

In case $s^* = s_1$ we conclude from

$$F_W(t^* - \omega, \phi) = F_W(-\omega, \gamma(s^*)) = F_W(-\omega, \gamma(s_1)) = \gamma(s_0) \in \Gamma_F$$

and $F_W(t^\phi, \phi) \in \Gamma_F$ that $t^* - \omega = t^\phi$. The definition of i_{FG} yields

$$\begin{aligned} \lim_{k\to\infty} i_{FG}(F_W(t^{n_k}, \phi^{n_k})) &= \lim_{k\to\infty} i_{FG}(\gamma(s^{n_k})) \\ &= \frac{1}{1+e^\omega} = G\Big(\omega, \frac{1}{2}\Big) = G(\omega, i_{FG}(\gamma(s_0))). \end{aligned}$$

It follows that

$$\begin{aligned} h(\phi^{n_k}) &= G(-t^{n_k}, i_{FG}(F_W(t^{n_k}, \phi^{n_k}))) \\ &\to G(-t^\phi - \omega, G(\omega, i_{FG}(\gamma(s_0))) = G(-t^\phi, i_{FG}(\gamma(s^\phi))) = h(\phi) \end{aligned}$$

as $k \to \infty$.

4. We show that the set $\mathcal{O} \cup \mathrm{bd}^+ W \cup \{\xi_+\}$ is compact. Observe that $\mathcal{O} \cup \mathrm{bd}^+ W \cup \{\xi_+\}$ is invariant for the flow F_W, and that we have $\xi_- \notin \mathcal{O} \cup \mathrm{bd}^+ W \cup \{\xi_+\}$. For every $\phi \in \mathrm{bd}\, W \setminus (\mathcal{O} \cup \mathrm{bd}^+ W \cup \{\xi_+\}) = \mathrm{bd}^- W \cup \{\xi_-\}$,

$$F_W(t, \phi) \to \xi_- \qquad \text{as } t \to \infty,$$

and there exists $t_\phi \geq 0$ so that $F(t_\phi, \phi) \in -\overset{\circ}{K}$. It follows that for some open neighbourhood V of ϕ in C, $F(\{t_\phi\} \times V) \subset -\overset{\circ}{K}$. We have $-\overset{\circ}{K} \cap \mathcal{O} = \emptyset$, $\xi_+ \notin -\overset{\circ}{K}$, and $\mathrm{bd}^+ W \cap (-\overset{\circ}{K}) = \emptyset$ since $F(t, \psi) \leq 0 \ll \xi_+$ for all $\psi \in -\overset{\circ}{K}$ and $t \geq 0$. Therefore

$$F(\{t_\phi\} \times V) \cap (\mathcal{O} \cup \mathrm{bd}^+ W \cup \{\xi_+\}) = \emptyset.$$

By invariance,

$$(V \cap \mathrm{bd}\, W) \cap (\mathcal{O} \cup \mathrm{bd}^+ W \cup \{\xi_+\}) = \emptyset.$$

We conclude that the complement of $\mathcal{O} \cup \mathrm{bd}^+ W \cup \{\xi_+\}$ in $\mathrm{bd}\, W$ is an open subset of $\mathrm{bd}\, W$. Now use that $\mathrm{bd}\, W$ is compact.

5. It follows that the inverse of the bijective continuous map h is also continuous. The equations $h(\mathcal{O}) = S^1$ and $h|_\mathcal{O} = h^-|_\mathcal{O}$ are obvious.

6. The proof of the remaining assertions concerning h^- is analogous. □

Theorem 15.3 *In case $\mu > 0$ the set* $\operatorname{bd} W$ *is homeomorphic to S^2.*

Proof The map $h^* : \operatorname{bd} W \to S^2$ given by

$$
\begin{aligned}
h^*(\phi) &= h(\phi) = h^-(\phi) \qquad \text{for } \phi \in \mathcal{O},\\
h^*(\phi) &= \begin{pmatrix} x_1 \\ x_2 \\ \sqrt{1 - x_1^2 - x_2^2} \end{pmatrix}, \quad \begin{pmatrix} x_1 \\ x_2 \end{pmatrix} = h(\phi) \qquad \text{for } \phi \in \operatorname{bd}^+ W \cup \{\xi_+\} \setminus \mathcal{O},\\
h^*(\phi) &= \begin{pmatrix} x_1 \\ x_2 \\ -\sqrt{1 - x_1^2 - x_2^2} \end{pmatrix}, \quad \begin{pmatrix} x_1 \\ x_2 \end{pmatrix} = h^-(\phi) \qquad \text{for } \phi \in \operatorname{bd}^- W \cup \{\xi_-\} \setminus \mathcal{O}
\end{aligned}
$$

is a homeomorphism. □

Theorem 15.4 *In case $\mu = 0$ the set* $\operatorname{bd} W$ *is homeomorphic to $\mathbb{R} \times S^1_{\mathbb{C}}$.*

Sketch of the proof The map $G^0 : \mathbb{R} \times [0, \infty) \times S^1_{\mathbb{C}} \to [0, \infty) \times S^1_{\mathbb{C}}$ given by

$$
G^0(t, (x, e^{is})) = \left(e^t x, e^{i(\frac{2\pi}{\omega}t + s)}\right) \qquad \text{for } (x, e^{is}) \in [0, \infty) \times S^1_{\mathbb{C}}
$$

is a flow on $[0, \infty) \times S^1_{\mathbb{C}}$, and

$$
G^0(\omega, (1, 1)) = (e^\omega, 1).
$$

Let

$$
\Gamma_{G^0} = \{(x, 1) : 1 \le x < e^\omega\}.
$$

The maps $i_{FG^0} : \Gamma_F \to \Gamma_{G^0}$ and $i^-_{FG^0} : \Gamma^-_F \to \Gamma_{G^0}$ given by

$$
i_{FG^0}(\gamma(s)) = \left(\frac{s - s_0}{s_1 - s_0} e^\omega + \frac{s_1 - s}{s_1 - s_0}, 1\right), \qquad s_0 \le s < s_1,
$$

and

$$
i^-_{FG^0}(\gamma^-(s)) = \left(\frac{s - s_0^-}{s_1^- - s_0^-} e^\omega + \frac{s_1^- - s}{s_1^- - s_0^-}, 1\right), \qquad s_1^- < s \le s_0^-,
$$

respectively, are homeomorphisms. The maps $h^0 : \operatorname{bd}^+ W \cup \mathcal{O} \to [0, \infty) \times S^1_{\mathbb{C}}$ and $h^{0-} : \operatorname{bd}^- W \cup \mathcal{O} \to [0, \infty) \times S^1_{\mathbb{C}}$ given by

$$
\begin{aligned}
h^0(p_t) &= \left(0, e^{i\frac{2\pi}{\omega}t}\right), \qquad t \in \mathbb{R},\\
h^0(\phi) &= G^0(-t, i_{FG^0}(F_W(t, \phi))), \quad F_W(t, \phi) \in \Gamma_F, \quad t \in \mathbb{R},\ \phi \in \operatorname{bd}^+ W,
\end{aligned}
$$

and

$$
\begin{aligned}
h^{0-}(p_t) &= \left(0, e^{i\frac{2\pi}{\omega}t}\right), \qquad t \in \mathbb{R},\\
h^{0-}(\phi) &= G^0(-t, i^-_{FG^0}(F_W(t, \phi))), \quad F_W(t, \phi) \in \Gamma^-_F, \quad t \in \mathbb{R},\ \phi \in \operatorname{bd}^- W,
\end{aligned}
$$

respectively, are homeomorphisms, and

$$
h^{0-}(\phi) = h^0(\phi) \qquad \text{for all } \phi \in \mathcal{O}.
$$

Complete the proof in analogy to the proof of Theorem 15.3. □

CHAPTER 16

Homeomorphisms from $\overline{W}$ onto the closed ball and the solid cylinder

The aim of this chapter is to show that in case $\mu > 0$ the set $\overline{W}$ is homeomorphic to D^3, and that in case $\mu = 0$ the set $\overline{W}$ is homeomorphic to $\mathbb{R} \times D^2$. In this chapter we use the hypothesis (H4) formulated in Chapter 2. Observe that none of the results obtained in the preceding chapters requires this hypothesis.

Proposition 16.1 *Assume $\mu > 0$. There exist neighbourhoods U_+ of ξ_+ in C and U_- of ξ_- in C such that $\dot{\phi} \in \overset{\circ}{K}$ for every $\phi \in \overline{W} \cap U_+ \setminus \{\xi_+\}$ and $\dot{\phi} \in -\overset{\circ}{K}$ for every $\phi \in \overline{W} \cap U_- \setminus \{\xi_-\}$.*

Proof If a neighbourhood U_+ with the stated property does not exist then there are sequences $(\phi^n)_0^\infty$ in $\overline{W} \setminus \{\xi_+\}$ and $(s^n)_0^\infty$ in $[-1, 0]$ such that $\phi^n \to \xi_+$ as $n \to \infty$, and $\dot{\phi}^n(s^n) \leq 0$ for all $n \in \mathbb{N}$.

For every $n \in \mathbb{N}$, set $x^n = x^{\phi^n}$. Then the functions

$$y^n : \mathbb{R} \ni t \mapsto \frac{\xi^+ - x^n(s^n + t)}{||\xi_+ - x^n_{s^n}||} \in \mathbb{R}, \qquad n \in \mathbb{N},$$

satisfy

$$\begin{cases} \dot{y}^n(t) = -\mu y^n(t) + \displaystyle\int_0^1 f'((1-\theta)x^n(s^n + t - 1) + \theta\xi^+)\, d\theta\, y^n(t-1) \ \text{ for } t \in \mathbb{R}, \\ \dot{y}^n(0) \geq 0, \quad ||y_0^n|| = 1. \end{cases}$$

Proposition 4.1(iii) implies that $y^n(t) \geq 0$ for all $n \in \mathbb{N}$ and $t \in \mathbb{R}$. There are reals $b_0 > 0$ and $b_1 \geq b_0$ so that $b_0 \leq \int_0^1 f'((1-\theta)x^n(s^n + t - 1) + \theta\xi^+)\, d\theta \leq b_1$ for all $t \in \mathbb{R}$ and $n \in \mathbb{N}$ since $\overline{W}$ is compact. Clearly we have $V(y_t^n) = 0$ for all $n \in \mathbb{N}$ and $t \in \mathbb{R}$. Applying Lemmas VI.5 and VI.6, we get a constant $c > 0$ with $||y_t^n|| \leq ce^{c|t|}$ for all $n \in \mathbb{N}$ and $t \in \mathbb{R}$. This allows us to apply Lemma VI.4 to get a C^1-function $y : \mathbb{R} \to \mathbb{R}$ and a subsequence $(y^{n_k})_{k=0}^\infty$ of $(y^n)_0^\infty$ such that

$$y^{n_k} \to y, \quad \dot{y}^{n_k} \to \dot{y} \qquad \text{as } k \to \infty$$

uniformly on compact subsets of $\mathbb{R}$, and

$$\begin{cases} \dot{y}(t) = -\mu y(t) + f'(\xi^+)y(t-1) & \text{for all } t \in \mathbb{R}, \\ y(t) \geq 0 \quad \text{for all } t \in \mathbb{R}, \\ \dot{y}(0) \geq 0, \quad ||y_0|| = 1. \end{cases}$$

Recall that λ_0^+ is the leading real eigenvalue of the generator of the semigroup $(D_2F(t,\xi_+))_{t\geq 0}$. We have $\lambda_0^+ < 0$ under hypothesis (H4). Proposition 5.1 yields $c_0 \in \mathbb{R}$ so that $y(s) = c_0 e^{\lambda_0^+ s}$, $-1 \leq s \leq 0$. Using $y_0 \geq 0$ and $||y_0|| = 1$ we get $c_0 > 0$. Hence, $\dot{y}(0) = c_0\lambda_0^+ < 0$, a contradiction to $\dot{y}(0) \geq 0$. This proves the existence of a neighbourhood U_+ with the desired properties. A suitable neighbourhood U_- of ξ_- is obtained analogously. □

In case $\mu > 0$, choose open neighbourhoods U_+ of ξ_+ and U_- of ξ_- according to Proposition 16.1. Proposition 12.4 guarantees that for every $r > 0$ with $\xi_+ + C_r \subset U_+$, $\xi_- + C_r \subset U_-$ there exist $k_+ \in (0, c_P(\xi_+))$, $k_- \in (c_P(\xi_-), 0)$ so that

$$\begin{aligned} \{\phi \in \overline{W} : c_P(\phi) \geq k_+\} \subset \xi_+ + C_r, \\ \{\phi \in \overline{W} : c_P(\phi) \leq k_-\} \subset \xi_- + C_r. \end{aligned} \tag{16.1}$$

In the subsequent Propositions 16.2, 16.4, 16.5, 16.6 it is always assumed that in case $\mu > 0$ open neighbourhoods U_+ of ξ_+, U_- of ξ_-, and reals $r > 0$, $k_+ \in (0, c_P(\xi_+))$, $k_- \in (c_P(\xi_-), 0)$ are given so that $\xi_+ + C_r \subset U_+$, $\xi_- + C_r \subset U_-$ and (16.1) hold; we set

$$c_+ = k_+ \quad \text{and} \quad c_- = k_-.$$

In case $\mu = 0$ recall $B > 0$ from Proposition 2.3(ii), set $B_1 = B + f_B$, $B_2 = B_1 + f_{B_1}$, define $\phi_{B_2} \in C$ by $\phi_{B_2}(t) = B_2$ for all $t \in [-1, 0]$, choose

$$k_0 > \max\{c_P(\phi_{B_2}), \max_{0 \leq t \leq \omega} c_P(p_t)\} + 1,$$

and define

$$c_- = -k_0 \quad \text{and} \quad c_+ = k_0.$$

Recall $c_P^{-1}(0) = Q \oplus L$. For every $k \in \mathbb{R}$ there exists $\chi_k \in C$ with $c_P(\chi_k) = k$, and the hyperplane

$$c_P^{-1}(k) = Q \oplus L + \chi_k$$

is a C^1-submanifold of codimension 1 in C.

Proposition 16.2

(i) *The set* $\{\phi \in \overline{W} : c_- \leq c_P(\phi) \leq c_+\}$ *is bounded.*

(ii) $\dot{\phi} \in \overset{\circ}{K}$ *for all* $\phi \in \overline{W} \cap c_P^{-1}(c_+)$, *and* $\dot{\phi} \in -\overset{\circ}{K}$ *for all* $\phi \in \overline{W} \cap c_P^{-1}(c_-)$.

(iii) $\frac{d}{dt}c_P(F_W(t,\phi))|_{t=0} > 0$ *for all* $\phi \in \overline{W} \cap c_P^{-1}(c_+)$, *and* $\frac{d}{dt}c_P(F_W(t,\phi))|_{t=0} < 0$ *for all* $\phi \in \overline{W} \cap c_P^{-1}(c_-)$.

(iv) *The sets* $\{\phi \in \overline{W} : c_P(\phi) > c_+\}$ *and* $\{\phi \in \overline{W} : c_P(\phi) < c_-\}$ *are positively invariant.*

Proof 1. Assume $\mu > 0$. Assertion (i) follows from the compactness of $\overline{W}$. Proposition 16.1 and the first inclusion in (16.1) combined imply $\dot{\phi} \in \overset{\circ}{K}$ and

$$\begin{aligned} \frac{d}{dt}c_P(F_W(t,\phi))|_{t=0} &= c_P(D_1F_W(0,\phi)1) = c_P(\dot{\phi}) \\ &= \dot{\phi}(0) + f'(0)\int_{-1}^{0} e^{-\lambda_0(s+1)}\dot{\phi}(s)\,ds > 0 \end{aligned}$$

for all $\phi \in \overline{W} \cap c_P^{-1}(c_+)$. This proves the first parts of assertions (ii) and (iii) in case $\mu > 0$. Hence the positive invariance of the set $\{\phi \in \overline{W} : c_P(\phi) > c_+\}$ is easily obtained since the function $\mathbb{R} \ni t \mapsto c_P(F_W(t, \phi)) \in \mathbb{R}$ is C^1-smooth for all $\phi \in \overline{W}$. The remaining statements for the case $\mu > 0$ are proved analogously.

2. Assume $\mu = 0$. Let $\phi \in \{\psi \in \overline{W} : c_- \leq c_P(\psi) \leq c_+\}$. If there exists $s \in [-2, -1]$ with $x^\phi(s) = 0$, then Proposition 2.3(ii) yields $|x^\phi(t)| \leq B$ for all $t \in [s-1, s]$ and hence Proposition 2.1 implies $|x^\phi(t)| \leq B_1$ for all $t \in [s, s+1]$, and $|x^\phi(t)| \leq B_2$ for all $t \in [s+1, s+2]$. Consequently, $||\phi|| \leq B_2$ since $[-1, 0] \subset [s, s+2]$. If $x^\phi_{-1} \in \overset{\circ}{K} \cup (-\overset{\circ}{K})$ then Proposition 2.2 and equation (1.1) with $\mu = 0$ yield $\phi \in \overset{\circ}{K} \cup (-\overset{\circ}{K})$ and $\dot{\phi} \in \overset{\circ}{K} \cup (-\overset{\circ}{K})$. Hence $|\phi(s)| \leq |\phi(0)|$ for all $s \in [-1, 0]$, and $|\phi(0)| \leq \max\{c_+, -c_-\} = k_0$. This proves assertion (i). Let $\phi \in \overline{W} \cap c_P^{-1}(c_+)$. If x^ϕ has a zero in $[-2, -1]$, then $||\phi|| \leq B_2$ by the above proof for (i) in case $\mu = 0$. Then $c_P(\phi) \leq c_P(\phi_{B_2}) < c_+$ follows, a contradiction. So $x^\phi_{-1} \in \overset{\circ}{K} \cup (-\overset{\circ}{K})$. Using $c_P(\phi) = c_+$ we get $x^\phi_{-1} \in \overset{\circ}{K}$. Hence Proposition 2.2 and equation (1.1) with $\mu = 0$ yield $\phi \in \overset{\circ}{K}$ and $\dot{\phi} \in \overset{\circ}{K}$. Thus the first part of assertion (ii) holds. The proof of the second part of (ii) is analogous. Assertions (iii) and (iv) in case $\mu = 0$ can be proved as in the case $\mu > 0$. □

Our next goal is to construct a set in W near 0 which is homeomorphic to S^2 and has the property that for each $\phi \in W \setminus \{0\}$ the curve $\mathbb{R} \ni t \mapsto x^\phi_t \in C$ intersects the set exactly once. It is convenient to introduce a new norm on C. For $\phi \in L$, define

$$|\phi|_L = \left[\phi^2(0) + e^{\frac{\pi\mu_1}{\nu_1}} \phi^2\Big(-\frac{\pi}{2\nu_1}\Big)\right]^{1/2},$$

where $\mu_1 = \operatorname{Re} \lambda_1 > 0$, $\nu_1 = \operatorname{Im} \lambda_1 > 0$. Note that for every $\phi \in L$ there exist reals b, c with

$$\phi(s) = be^{\mu_1 s} \sin(\nu_1 s + c) \qquad \text{for all } s \in [-1, 0].$$

Using this, it is easy to verify that the map

$$L \ni \phi \mapsto \left(\phi(0), e^{\frac{\pi\mu_1}{2\nu_1}} \phi\Big(-\frac{\pi}{2\nu_1}\Big)\right)^{tr} \in \mathbb{R}^2$$

is an isomorphism. Thus, $|\cdot|_L$ is a norm on L. Then

$$|\cdot|_e : C \ni \phi \mapsto \max\{||\operatorname{Pr}_Q \phi||, |\operatorname{Pr}_L \phi|_L, ||\operatorname{Pr}_P \phi||\} \in \mathbb{R}$$

is a norm on C which is equivalent to the norm $||\cdot||$; there are constants $c_{||\cdot||} > 0$ and $c_e \geq c_{||\cdot||}$ with

$$c_{||\cdot||}||\phi|| \leq |\phi|_e \leq c_e||\phi|| \qquad \text{for all } \phi \in C.$$

Proposition 16.3 *For every $\mu^* \in (0, \mu_1)$ there exists $\gamma > 0$ such that for all $\phi \in W \setminus \{0\}$ with $|\phi|_e < \gamma$ we have*

$$\liminf_{\delta \to 0+} \frac{1}{\delta}\left(|x^\phi_\delta|_e - |\phi|_e\right) > \mu^* |\phi|_e.$$

The proof of Proposition 16.3 requires a variation-of-constants formula which involves sun duals of the strongly continuous semigroup $(T(t))_{t\ge 0}$ on the complex Banach space $C_{\mathbb{C}}$ given by the segments x_t of the complex-valued solutions of equation (1.2). Let us recall the facts we shall need from Chapters II, III, IV and VI in **[22]**.

For every $\phi \in C_{\mathbb{C}}$ the functions

$$\operatorname{Re}\phi : [-1,0] \ni t \mapsto \operatorname{Re}\phi(t) \in \mathbb{R}$$

and

$$\operatorname{Im}\phi : [-1,0] \ni t \mapsto \operatorname{Im}\phi(t) \in \mathbb{R}$$

satisfy

$$(T(t)\phi)(s) = (D_2F(t,0)\operatorname{Re}\phi)(s) + i(D_2F(t,0)\operatorname{Im}\phi)(s)$$

for all $t \ge 0$ and all $s \in [-1,0]$. The spectra of the generators A_C and A of the C_0-semigroups $(D_2F(t,0))_{t\ge 0}$ on C and $(T(t))_{t\ge 0}$ on $C_{\mathbb{C}}$ coincide (Recall that we defined spectra of linear operators in real Banach spaces via complexifications; complexifications of C are not identical with $C_{\mathbb{C}}$). Consider a bounded subset σ_0 of these spectra, with

$$\overline{\sigma_0} = \sigma_0.$$

Let Pr and $\mathrm{Pr}_{\mathbb{C}}$ denote the associated projection operators in $L(C,C)$ and $L(C_{\mathbb{C}},C_{\mathbb{C}})$ onto the realified generalized eigenspace of A_C and σ_0 and onto the generalized eigenspace of A and σ_0, respectively. For every $\phi \in C$ the element

$$\phi_{\mathbb{C}} : [-1,0] \ni t \mapsto \phi(t) \in \mathbb{C}$$

in $C_{\mathbb{C}}$ then satisfies

$$\mathrm{Pr}\,\phi = \mathrm{Pr}_{\mathbb{C}}\,\phi_{\mathbb{C}}.$$

It is convenient to denote dual spaces and adjoint operators by an asterisk in the sequel. For example,

$$C_{\mathbb{C}}^* = L(C_{\mathbb{C}}, \mathbb{C}\ ).$$

The elements $\phi^{\odot}$ (pronounced phi-sun) of $C_{\mathbb{C}}^*$ for which the curve

$$[0,\infty) \ni t \mapsto T(t)^*\phi^{\odot} \in C_{\mathbb{C}}^*$$

is continuous form a closed subspace $C_{\mathbb{C}}^{\odot}$ which is positively invariant under the adjoints $T(t)^*$, $t \ge 0$. The operators

$$T^{\odot}(t) : C_{\mathbb{C}}^{\odot} \ni \phi^{\odot} \mapsto T(t)^*\phi^{\odot} \in C_{\mathbb{C}}^{\odot}, \qquad t \ge 0,$$

constitute a strongly continuous semigroup on $C_{\mathbb{C}}^{\odot}$. Repeating this process we obtain a subspace

$$C_{\mathbb{C}}^{\odot\odot} \subset C_{\mathbb{C}}^{\odot *}.$$

The original state space $C_{\mathbb{C}}$ is sun-reflexive in the sense that there exists a norm-preserving linear map

$$j : C_{\mathbb{C}} \to C_{\mathbb{C}}^{\odot *}$$

with

$$jC_{\mathbb{C}} = C_{\mathbb{C}}^{\odot\odot};$$

$C_{\mathbb{C}}$ is embedded into the larger state space $C_{\mathbb{C}}^{\odot *}$.

We need an integral with values in $C_{\mathbb{C}}^{\odot *}$. For every continuous map $\tilde{g} : \mathbb{R} \to C_{\mathbb{C}}^{\odot *}$ and for all a and $b \geq a$ in $\mathbb{R}$, all functions $[a,b] \ni t \mapsto (T^{\odot}(b-t)^* \tilde{g}(t))\phi^{\odot} \in \mathbb{C}$, $\phi^{\odot} \in C_{\mathbb{C}}^{\odot}$, are Lebesgue-integrable, and the weak-star integral

$$\int_a^b T^{\odot}(b-t)^* \tilde{g}(t)\, dt \in C_{\mathbb{C}}^{\odot *}$$

is defined by

$$\left(\int_a^b T^{\odot}(b-t)^* \tilde{g}(t)\, dt \right)(\phi^{\odot}) = \int_a^b \left(T^{\odot}(b-t)^* \tilde{g}(t) \right)(\phi^{\odot})\, dt$$

for all $\phi^{\odot} \in C_{\mathbb{C}}^{\odot}$. One finds that all such weak-star integrals are elements of the subspace $C_{\mathbb{C}}^{\odot\odot} = jC_{\mathbb{C}}$.

There is a topological isomorphism

$$k : C_{\mathbb{C}}^{\odot *} \to \mathbb{C} \times L^{\infty}(-1,0;\mathbb{C}).$$

Set

$$r^{\odot *} = k^{-1}(1,0).$$

If $g : \mathbb{R} \to \mathbb{R}$ is continuous and if $x : [-1,\infty) \to \mathbb{R}$ is a solution of the linear inhomogeneous equation

$$\dot{x}(t) = -\mu x(t) + f'(0)x(t-1) + g(t)$$

then the curve

$$u : [0,\infty) \ni t \mapsto x_{t,\mathbb{C}} \in C_{\mathbb{C}}$$

satisfies

$$ju(t) = jT(t)u(0) + \int_0^t T^{\odot}(t-s)^*(g(s)r^{\odot *})\, ds \qquad \text{for all } t \geq 0.$$

Consider a bounded subset σ_0 of the spectra of the generators A_C and A. Also the spectra of A and of the generator $A^{\odot}$ of the semigroup $(T^{\odot}(t))_{t\geq 0}$ coincide. Let $\mathrm{Pr}_{\mathbb{C}}$ and $\mathrm{Pr}^{\odot}$ denote the projection operators in $L(C_{\mathbb{C}}, C_{\mathbb{C}})$ and $L(C_{\mathbb{C}}^{\odot}, C_{\mathbb{C}}^{\odot})$ which are associated with σ_0 and map onto the corresponding generalized eigenspaces of A and $A^{\odot}$, respectively. The adjoint operator $\mathrm{Pr}^{\odot *} \in L(C_{\mathbb{C}}^{\odot *}, C_{\mathbb{C}}^{\odot *})$ satisfies

$$\mathrm{Pr}^{\odot *} C_{\mathbb{C}}^{\odot *} \subset C_{\mathbb{C}}^{\odot\odot} = jC_{\mathbb{C}}, \quad \mathrm{Pr}^{\odot *} \circ j = j \circ \mathrm{Pr}_{\mathbb{C}},$$

and for g, x, u as before,

$$\mathrm{Pr}^{\odot *} ju(t) = \mathrm{Pr}^{\odot *} jT(t)u(0) + \int_0^t T^{\odot}(t-s)^* \,\mathrm{Pr}^{\odot *}(g(s)r^{\odot *})\, ds \qquad \text{for all } t \geq 0,$$

or

$$j\,\mathrm{Pr}_{\mathbb{C}}\,u(t) = jT(t)\,\mathrm{Pr}_{\mathbb{C}}\,u(0) + \int_0^t T^{\odot}(t-s)^*\,\mathrm{Pr}^{\odot *}(g(s)r^{\odot *})\,ds \qquad \text{for all } t \geq 0.$$

By the definition of weak-star integrals,

$$\left\|\int_0^t T^{\odot}(t-s)^*\,\mathrm{Pr}^{\odot *}(g(s)r^{\odot *})\,ds\right\| = \sup\left\{\left|\int_0^t [T^{\odot}(t-s)^*\,\mathrm{Pr}^{\odot *}(g(s)r^{\odot *})]\phi^{\odot}\,ds\right| : \phi^{\odot} \in C^{\odot}_{\mathbb{C}}\right\}.$$

For every $\phi^{\odot} \in C^{\odot}_{\mathbb{C}}$ and all $s \in [0,t]$,

$$\begin{aligned}\left|[T^{\odot}(t-s)^*\,\mathrm{Pr}^{\odot *}(g(s)r^{\odot *})]\phi^{\odot}\right| &= |g(s)|\left|r^{\odot *}(\mathrm{Pr}^{\odot}\,T^{\odot}(t-s)\phi^{\odot})\right| \\ &\leq |g(s)|\,||r^{\odot *}||\,||T^{\odot}(t-s)||\,||\,\mathrm{Pr}^{\odot}\,||.\end{aligned}$$

For every $\lambda > \lambda_0 = \max\{\mathrm{Re}\,\zeta : \zeta \text{ in the spectrum of } A^{\odot}\}$ there exists $M \geq 1$ with

$$||T^{\odot}(v)|| \leq Me^{\lambda v} \qquad \text{for all } v \geq 0.$$

It follows that for all $t \geq 0$,

$$\left\|\int_0^t T^{\odot}(t-s)^*\,\mathrm{Pr}^{\odot *}(g(s)r^{\odot *})\,ds\right\| \leq ||r^{\odot *}||\,||\,\mathrm{Pr}^{\odot}\,||\,Me^{\lambda t}\int_0^t |g(s)|\,ds.$$

Let $t \geq 0$. If we define $\psi_t \in C_{\mathbb{C}}$ to be the pre-image of

$$\int_0^t T^{\odot}(t-s)^*\,\mathrm{Pr}^{\odot *}(g(s)r^{\odot *})\,ds$$

with respect to j then we obtain

$$|\psi_t|_e \leq c_e||\psi_t|| = c_e||j\psi_t|| \leq c_e||r^{\odot *}||\,||\,\mathrm{Pr}^{\odot}\,||\,Me^{\lambda t}\int_0^t |g(s)|\,ds,$$

and thereby

$$\begin{aligned}|\,\mathrm{Pr}_{\mathbb{C}}\,u(t)|_e &\geq |T(t)\,\mathrm{Pr}_{\mathbb{C}}\,u(0)|_e - |\psi_t|_e \\ &\geq |T(t)\,\mathrm{Pr}_{\mathbb{C}}\,u(0)|_e - c_e||r^{\odot *}||\,||\,\mathrm{Pr}^{\odot}\,||\,Me^{\lambda t}\int_0^t |g(s)|\,ds.\end{aligned} \tag{16.2}$$

Now we are ready for the

Proof of Proposition 16.3 Set $\sigma_0 = \{\lambda_0, \lambda_1, \overline{\lambda_1}\}$ and

$$c = c_e Me^{\lambda}||r^{\odot *}||\,||\,\mathrm{Pr}^{\odot}\,||.$$

Let $\epsilon \in (0, \mu_1 - \mu^*)$ be given. Using $f(0) = 0$ and $T_0W = L \oplus P$ we obtain $\gamma > 0$ such that

$$c_{||\cdot||}^{-1}c|f(\xi) - f'(0)\xi| \leq \epsilon|\xi| \qquad \text{for all } \xi \in (-\gamma, \gamma) \tag{16.3}$$

and

$$||\Pr_Q \phi|| \le |\Pr_{L\oplus P} \phi|_e \qquad \text{for all } \phi \in W \text{ with } |\phi|_e < \gamma. \tag{16.4}$$

Let $\phi \in W$ with $|\phi|_e < \gamma$ be given. Choose $\delta^0 \in (0,1]$ with $|F(t,\phi)|_e < \gamma$ for all $t \in [0,\delta^0]$. The function $x = x^\phi$ satisfies the linear inhomogeneous equation

$$\dot{x}(t) = -\mu x(t) + f'(0)x(t-1) + [f(x(t-1)) - f'(0)x(t-1)].$$

Using the estimates (16.2), (16.3), (16.4) and the relations between $D_2F(t,0)$, $T(t)$ as well as between $\Pr_{L\oplus P}$ and the spectral projection $\Pr_{\mathbb{C}}$ onto the generalized eigenspace of σ_0 and A, we conclude that for every $t \in [0,\delta^0]$ we have

$$\begin{aligned} |F(t,\phi)|_e &= \max\{||\Pr_Q F(t,\phi)||, |\Pr_L F(t,\phi)|_L, ||\Pr_P F(t,\phi)||\} \\ &= |\Pr_{L\oplus P} F(t,\phi)|_e \ge |D_2F(t,0)\Pr_{L\oplus P}\phi|_e - \epsilon c_{||\cdot||}\int_0^t ||F(s,\phi)||\,ds \\ &\ge |D_2F(t,0)\Pr_{L\oplus P}\phi|_e - \epsilon\int_0^t |F(s,\phi)|_e\,ds. \end{aligned} \tag{16.5}$$

For every $t \ge 0$, set $y_t = D_2F(t,0)\Pr_{L\oplus P}\phi$. Then there exist reals a,b,c so that for all $t \ge 0$ and $s \in [-1,0]$,

$$y_t(s) = ae^{\lambda_0(t+s)} + be^{\mu_1(t+s)}\sin(\nu_1(t+s)+c),$$

and

$$\begin{aligned} &\Pr_P y_t(s) = ae^{\lambda_0(t+s)}, \quad ||\Pr_P y_t|| = |a|e^{\lambda_0 t}, \\ &\Pr_L y_t(s) = be^{\mu_1(t+s)}\sin(\nu_1(t+s)+c), \quad |\Pr_L y_t|_L = |b|e^{\mu_1 t}. \end{aligned}$$

Thus, $|y_t|_e = \max\{|a|e^{\lambda_0 t}, |b|e^{\mu_1 t}\}$ for every $t \ge 0$. It follows that

$$\lim_{\delta\to 0+}\frac{1}{\delta}[|y_\delta|_e - |y_0|_e] = \begin{cases} \lambda_0|a| \ge \mu_1|y_0|_e & \text{if } |a| \ge |b|, \\ \mu_1|b| = \mu_1|y_0|_e & \text{if } |a| < |b|. \end{cases} \tag{16.6}$$

Using (16.5) and $|\phi|_e = |\Pr_{L\oplus P}\phi|_e$, which follows from (16.4), together with (16.6) we infer

$$\begin{aligned} \frac{1}{\delta}(|x_\delta^\phi|_e - |\phi|_e) &= \frac{1}{\delta}(|x_\delta^\phi|_e - |\Pr_{L\oplus P}\phi|_e) \\ &\ge \frac{1}{\delta}(|D_2F(t,0)\Pr_{L\oplus P}\phi|_e - \epsilon\int_0^\delta |x_s^\phi|_e\,ds - |\Pr_{L\oplus P}\phi|_e) \\ &= \frac{1}{\delta}(|y_\delta|_e - |y_0|_e) - \frac{\epsilon}{\delta}\int_0^\delta |x_s^\phi|_e\,ds \qquad \text{for every } \delta \in (0,\delta^0]. \end{aligned}$$

Therefore,

$$\liminf_{\delta\to 0+}\frac{1}{\delta}(|x_\delta^\phi|_e - |\phi|_e) \ge \mu_1|y_0|_e - \epsilon|x_0^\phi|_e = \mu_1|\Pr_{L\oplus P}\phi|_e - \epsilon|\phi|_e = (\mu_1-\epsilon)|\phi|_e \ge \mu^*|\phi|_e.$$

□

Recall from Chapter 4 that N_L, N_P are open neighbourhoods of 0 in L, P, respectively, and that $w_{loc} : N_L + N_P \to Q$ is a C^1-map with $w_{loc}(0) = 0$, $Dw_{loc}(0) = 0$, and

$$W_{loc} = \{\chi + w_{loc}(\chi) : \chi \in N_L + N_P\}.$$

Recall also from Chapters 4, 9 and 10 that there exist $\eta > 0$, $\epsilon > 0$ and a C^1-map $\operatorname{Sep}_\eta : L_\eta \to Q \oplus P$ so that $\operatorname{Sep}_\eta(L_\eta) \subset Q + P_\epsilon$, $\operatorname{Sep}_\eta(0) = 0$, $D\operatorname{Sep}_\eta(0) = 0$, and

$$(W_{loc} \cap S) \cap (Q + L_\eta + P_\epsilon) = \{\chi + \operatorname{Sep}_\eta(\chi) : \chi \in L_\eta\}.$$

Proposition 16.4 *There exists $\delta > 0$ with the following properties.*

(i) For every $\phi \in W \setminus \{0\}$ there exists a unique $t_\phi \in \mathbb{R}$ so that $|F_W(t_\phi, \phi)|_e = \delta$ and $|F_W(t, \phi)|_e < \delta$ for all $t < t_\phi$, $|F_W(t, \phi)|_e > \delta$ for all $t > t_\phi$.

(ii) We have $c_- < c_P(F(t, \phi)) < c_+$ for all $t \in [0, 1]$ and for all $\phi \in C$ with $|\phi|_e \le \delta$.

(iii) $\{\chi \in L \oplus P : |\chi|_e \le \delta\} \subset N_L + N_P$ and $\{\chi \in L : |\chi|_e \le \delta\} \subset L_\eta$.

(iv) We have $|w_{loc}(\chi)|_e \le |\chi|_e$ for all $\chi \in L \oplus P$ with $|\chi|_e \le \delta$; and $|\operatorname{Sep}_\eta(\chi)|_e \le |\chi|_e$ for all $\chi \in L$ with $|\chi|_e \le \delta$.

(v) $\{\phi \in W : |\phi|_e = \delta\} = \{\chi + w_{loc}(\chi) : \chi \in L \oplus P,\ |\chi|_e = \delta\}$.

(vi) $\{\phi \in W \cap S : |\phi|_e = \delta\} = \{\chi + \operatorname{Sep}_\eta(\chi) : \chi \in L,\ |\chi|_e = \delta\}$.

Proof 1. Proof of (i). Choose $\mu^* \in (0, \mu_1)$ and $\gamma > 0$ according to Proposition 16.3. Select $\delta^0 > 0$ so that in case $\mu > 0$,

$$\delta^0 < \min\left\{\gamma, \min_{t \in [0,\omega]} |p_t|_e, |\xi_-|_e, |\xi_+|_e\right\},$$

and in case $\mu = 0$,

$$\delta^0 < \min\left\{\gamma, \min_{t \in [0,\omega]} |p_t|_e\right\}.$$

Let $\delta \in (0, \delta^0]$ and $\phi \in W \setminus \{0\}$. From the limiting behaviour of $F_W(t, \phi)$ as $t \to -\infty$ and $t \to \infty$, and from the choice of δ^0 we obtain that $|F_W(t, \phi)|_e < \delta$ for all sufficiently large negative t, and that $|F_W(t, \phi)|_e > \delta$ for all sufficiently large positive t. Let $t_\phi = \min\{t \in \mathbb{R} : |F_W(t, \phi)|_e = \delta\}$. Then $|F_W(t, \phi)|_e < \delta$ for all $t < t_\phi$. Using Proposition 16.3 it is easy to show that there is $t_1 > t_\phi$ with $|F_W(t, \phi)|_e > \delta$ for all $t \in (t_\phi, t_1)$. If $|F_W(t, \phi)|_e > \delta$ for all $t > t_\phi$ does not hold, then there is $t_2 \ge t_1 > t_\phi$ so that $|F_W(t_2, \phi)|_e = \delta$ and $|F_W(t, \phi)|_e > \delta$ for all $t \in (t_\phi, t_2)$. Then we find $t_3 \in (t_\phi, t_2)$ such that $|F_W(t_3, \phi)|_e < \gamma$ and $|F_W(t, \phi)|_e < |F_W(t_3, \phi)|_e$ for all $t \in (t_3, t_2)$. Hence

$$\liminf_{\delta \to 0+} \frac{1}{\delta}(|F_W(t_3 + \delta, \phi)|_e - |F_W(t_3, \phi)|_e) \le 0,$$

a contradiction to Proposition 16.3. This proves assertion (i).

2. From the continuous dependence of solutions of equation (1.1) on initial data, from the continuity of c_P, from $c_P(0) = 0$, and from $c_- < 0 < c_+$ we conclude that there exists $\delta^1 > 0$ so that assertion (ii) holds for all $\delta \in (0, \delta^1]$.

3. As N_L, L_η and N_P are open sets containing 0 in L and P, respectively, we find $\delta^2 > 0$ such that assertion (iii) is satisfied for all $\delta \in (0, \delta^2]$.

4. The properties $w_{loc}(0) = 0$, $Dw_{loc}(0) = 0$, $\mathrm{Sep}_\eta(0) = 0$ and $D\,\mathrm{Sep}_\eta(0) = 0$ of w_{loc} and Sep_η yield a real $\delta^3 > 0$ so that assertion (iv) holds for all $\delta \in (0, \delta^3]$.

5. The set W_{loc} is open in W containing 0, see Corollary 6.7. Then there exists $\delta^4 \in (0, \min\{\delta^2, \delta^3\})$ so that $\{\phi \in W : |\phi|_e \le \delta^4\} \subset W_{loc}$. Let $\delta \in (0, \delta^4]$ and $\phi \in W$ with $|\phi|_e = \delta$ be given. We have $\phi = \chi + w_{loc}(\chi)$ for some $\chi \in N_L + N_P$. From the definition of $|\cdot|_e$,

$$|\phi|_e = \max\{|\operatorname{Pr}_Q \phi|_e, |\operatorname{Pr}_{L\oplus P} \phi|_e\} = \max\{|w_{loc}(\chi)|_e, |\chi|_e\}.$$

Using the result of part 4 and $\delta \le \delta^4 \le \delta^3$, we obtain $|\phi|_e = |\chi|_e$. Therefore,

$$\{\phi \in W : |\phi|_e = \delta\} \subset \{\chi + w_{loc}(\chi) : \chi \in L \oplus P,\ |\chi|_e = \delta\}$$

for all $\delta \in (0, \delta^4]$. Let $\delta \in (0, \delta^4]$ and $\chi \in L \oplus P$ with $|\chi|_e = \delta$ be given. By part 3 and $\delta^4 \le \delta^2$, $\phi = \chi + w_{loc}(\chi) \in W_{loc} \subset W$. Part 4 and $\delta \le \delta^4 \le \delta^3$ yield $|\phi|_e = \max\{|w_{loc}(\chi)|_e, |\chi|_e\} = |\chi|_e = \delta$. Consequently, assertion (v) holds for all $\delta \in (0, \delta^4]$.

6. Choose $\delta^5 \in (0, \min\{\delta^0, \delta^1, \dots, \delta^4\})$ so small that for every $\phi \in C$ with $|\phi|_e \le \delta^5$, $\phi \in Q + L_\eta + P_\epsilon$. Let $\delta \in (0, \delta^5]$ and $\phi \in W \cap S$ with $|\phi|_e = \delta$ be given. Due to a remark in part 5 we get $\phi \in W_{loc} \cap S$. Using this and $\phi \in Q + L_\eta + P_\epsilon$ we find $\chi \in L_\eta$ with $\phi = \chi + \mathrm{Sep}_\eta(\chi)$. The equations

$$\begin{aligned}\delta = |\phi|_e &= \max\{||\operatorname{Pr}_Q \phi||, |\operatorname{Pr}_L \phi|_L, ||\operatorname{Pr}_P \phi||\}\\ &= \max\{||\operatorname{Pr}_Q \mathrm{Sep}_\eta(\chi)||, |\chi|_L, ||\operatorname{Pr}_P \mathrm{Sep}_\eta(\chi)||\} = \max\{|\mathrm{Sep}_\eta(\chi)|_e, |\chi|_L\} = |\chi|_e\end{aligned}$$

yield $\phi = \chi + \mathrm{Sep}_\eta(\chi)$ with $|\chi|_e = \delta$. Conversely, if $\chi \in L$ with $|\chi|_e = \delta$ is given, then $\chi \in L_\eta$, and therefore $\phi = \chi + \mathrm{Sep}_\eta(\chi)$ belongs to $W_{loc} \cap S \subset W \cap S$, with $|\phi|_e = |\chi|_e = \delta$ as above. This proves the statement in assertion (vi) for $\delta \in (0, \delta^5]$.

7. For $\delta = \min\{\delta^0, \delta^1, \dots, \delta^5\}$ all statements in assertions (i)–(vi) hold. □

Choose $\delta > 0$ according to Proposition 16.4. Set

$$\begin{aligned} S_2^\delta &= \{\phi \in W : |\phi|_e = \delta\},\\ S_1^\delta &= \{\phi \in W \cap S : |\phi|_e = \delta\}.\end{aligned}$$

Proposition 16.5 *S_1^δ is homeomorphic to S^1, S_2^δ is homeomorphic to S^2. $S_2^\delta \setminus S_1^\delta$ has exactly two connected components. In case $\mu > 0$ these components are given by*

$$\begin{aligned}(S_2^\delta \setminus S_1^\delta)^+ &= \{\phi \in S_2^\delta \setminus S_1^\delta : \omega(\phi) = \{\xi_+\}\},\\ (S_2^\delta \setminus S_1^\delta)^- &= \{\phi \in S_2^\delta \setminus S_1^\delta : \omega(\phi) = \{\xi_-\}\}.\end{aligned}$$

In case $\mu = 0$ these components are given by

$$\begin{aligned}(S_2^\delta \setminus S_1^\delta)^+ &= \{\phi \in S_2^\delta \setminus S_1^\delta : x^\phi(t) \to +\infty \text{ as } t \to +\infty\},\\ (S_2^\delta \setminus S_1^\delta)^- &= \{\phi \in S_2^\delta \setminus S_1^\delta : x^\phi(t) \to -\infty \text{ as } t \to +\infty\}.\end{aligned}$$

The sets $(S_2^\delta \setminus S_1^\delta)^+$ and $(S_2^\delta \setminus S_1^\delta)^-$ are homeomorphic to the open unit disk in $\mathbb{R}^2$, and the sets $(S_2^\delta \setminus S_1^\delta)^+ \cup S_1^\delta$ and $(S_2^\delta \setminus S_1^\delta)^- \cup S_1^\delta$ are homeomorphic to D^2.

Proof 1. By Proposition 16.4(v), S_2^δ is homeomorphic to $\{\chi \in L \oplus P : |\chi|_e = \delta\}$ which is homeomorphic to S^2. By Proposition 16.4(vi), S_1^δ is homeomorphic to $\{\chi \in L : |\chi|_e = \delta\}$ which is homeomorphic to S^1. Since the inclusion $i : S_1^\delta \to S_2^\delta$ is continuous and injective, we can apply Theorem VIII.1(ii) to conclude that $S_2^\delta \setminus S_1^\delta$ has two connected components, that each of these is homeomorphic to the open unit disk in $\mathbb{R}^2$, and that the union of each of these components with S_1^δ is homeomorphic to D^2.

2. Assume $\mu > 0$. Since $S_2^\delta \cap S = S_1^\delta$, Proposition 4.5 implies that for every $\phi \in S_2^\delta \setminus S_1^\delta$, we have either $\omega(\phi) = \{\xi_+\}$ or $\omega(\phi) = \{\xi_-\}$. We claim that in case $\phi \in S_2^\delta \setminus S_1^\delta$ and $\omega(\phi) = \{\xi_+\}$, there exists a neighbourhood $O(\phi)$ of ϕ in $S_2^\delta \setminus S_1^\delta$ such that $\omega(\psi) = \{\xi_+\}$ for all $\psi \in O(\phi)$. Let $\phi \in S_2^\delta \setminus S_1^\delta$ with $\omega(\phi) = \{\xi_+\}$ be given. Choose $\eta \in W$ according to Proposition 4.1(ii) so that $x_t^\eta \to \xi_+$ as $t \to \infty$, and $x_t^\eta \to 0$ as $t \to -\infty$. Then $\omega(\phi) = \{\xi_+\}$ implies that there exists $t^* > 0$ with $x_{t^*}^\phi \gg \eta$. Using the continuous dependence on initial data, we find a neighbourhood $O(\phi)$ of ϕ in $S_2^\delta \setminus S_1^\delta$ such that $x_{t^*}^\psi \gg \eta$ for all $\psi \in O(\phi)$. Thus, the monotonicity of the semiflow implies $x^\psi(t^* + t) \geq x^\eta(t)$ for all $\psi \in O(\phi)$ and for all $t \geq 0$. Using this and $x^\psi(t^* + t) \leq \xi^+$ for all $t \in \mathbb{R}$ we obtain $x^\psi(t) \to \xi^+$ as $t \to \infty$. An analogous result holds for $\phi \in S_2^\delta \setminus S_1^\delta$ with $\omega(\phi) = \{\xi_-\}$.

3. We claim that for every ϕ, ψ in $S_2^\delta \setminus S_1^\delta$ with $\omega(\phi) \neq \omega(\psi)$, ϕ and ψ belong to different components of $S_2^\delta \setminus S_1^\delta$. Suppose ϕ, ψ are in the same component of $S_2^\delta \setminus S_1^\delta$ and $\omega(\phi) \neq \omega(\psi)$. Then there is a continuous curve $\gamma : [0,1] \to S_2^\delta \setminus S_1^\delta$ such that $\gamma([0,1])$ lies entirely in this component and $\gamma(0) = \phi, \gamma(1) = \psi$. Let $\Gamma_1 = \{s \in [0,1] : \omega(\gamma(s)) = \{\xi_+\}\}$ and $\Gamma_2 = \{s \in [0,1] : \omega(\gamma(s)) = \{\xi_-\}\}$. Then Γ_1 and Γ_2 are nonempty, disjoint, and the result in part 2 shows that Γ_1 and Γ_2 are open in $[0,1]$. Therefore, $[0,1] \setminus (\Gamma_1 \cup \Gamma_2) \neq \emptyset$. On the other hand, all points of $|\gamma|$ belong to $W \setminus S$. This fact and Proposition 4.5 yield $\Gamma_1 \cup \Gamma_2 = [0,1]$, a contradiction.

4. Recall that for all $\phi \in S_1^\delta \subset W \cap S \setminus \{0\}$, $\omega(\phi) = \mathcal{O} \subset C \setminus \{\xi_-, \xi_+\}$. Using Proposition 16.4(i) and Proposition 4.1(i) and (ii) one finds ϕ, ψ in S_2^δ with $\omega(\phi) = \{\xi_+\}$ and $\omega(\psi) = \{\xi_-\}$. In case $\mu > 0$ it now becomes easy to identify the two components of $S_2^\delta \setminus S_1^\delta$ as $(S_2^\delta \setminus S_1^\delta)^+$ and $(S_2^\delta \setminus S_1^\delta)^-$.

5. The proof of the assertions in case $\mu = 0$ is analogous. □

Proposition 16.6

(i) There exists a continuous map $\tau_0 : (S_2^\delta \setminus S_1^\delta)^+ \to \mathbb{R}$ such that $F_W(\tau_0(\phi), \phi) \in c_P^{-1}(c_+)$ for all $\phi \in (S_2^\delta \setminus S_1^\delta)^+$; $\tau_0((S_2^\delta \setminus S_1^\delta)^+) \subset (1, \infty)$; and $F_W(t, \phi) \notin c_P^{-1}(c_+)$ for all $(t, \phi) \in \mathbb{R} \times (S_2^\delta \setminus S_1^\delta)^+$ with $t \neq \tau_0(\phi)$.

(ii) There exists a continuous map $\tau_0^- : (S_2^\delta \setminus S_1^\delta)^- \to \mathbb{R}$ such that $F_W(\tau_0^-(\phi), \phi) \in c_P^{-1}(c_-)$ for all $\phi \in (S_2^\delta \setminus S_1^\delta)^-$; $\tau_0^-((S_2^\delta \setminus S_1^\delta)^-) \subset (1, \infty)$; and $F_W(t, \phi) \notin c_P^{-1}(c_-)$ for all $(t, \phi) \in \mathbb{R} \times (S_2^\delta \setminus S_1^\delta)^-$ with $t \neq \tau_0^-(\phi)$.

Proof Let $\phi \in (S_2^\delta \setminus S_1^\delta)^+$ be given. Proposition 16.4(ii) yields $c_P(\phi) < c_+$. The limiting behaviour of x_t^ϕ as $t \to \infty$ and the choice of c_+ show that $c_P(x_t^\phi) > c_+$ for all sufficiently large t. Then by continuity there exists $\tau_0(\phi) \in \mathbb{R}$ with $c_P(F_W(\tau_0(\phi), \phi)) = c_+$. The uniqueness of $\tau_0(\phi)$ is a consequence of Proposition 16.2(iii). The inequality $\tau_0(\phi) > 1$ follows from Proposition 16.4(ii). To prove the continuity of the map $\tau_0 : (S_2^\delta \setminus S_1^\delta)^+ \ni \phi \mapsto \tau_0(\phi) \in (1, \infty)$, we note that for each $\phi \in (S_2^\delta \setminus S_1^\delta)^+$ there exists a C^1-map τ_ϕ from a neighbourhood U_ϕ of

ϕ in C such that $F(\tau_\phi(\eta),\eta) \in c_P^{-1}(c_+)$ for all $\eta \in U_\phi$ and $\tau_\phi(\phi) = \tau_0(\phi)$ since Proposition 16.2(iii) permits to apply the Implicit Function Theorem to the zeroset of the mapping $(1,\infty)\times C \ni (t,\phi) \mapsto c_P(F(t,\phi)) - c_+ \in \mathbb{R}$ and to its zero $(\tau_0(\phi),\phi)$. By the uniqueness of τ_0, we must have $\tau_\phi(\eta) = \tau_0(\eta)$ for all $\eta \in (S_2^\delta \setminus S_1^\delta)^+ \cap U_\phi$ from which the continuity of τ_0 follows.

The proof of assertion (ii) is analogous. □

The diameter diam(A) of a subset A of $\mathbb{R}^3$ is defined by

$$\operatorname{diam}(A) = \sup\{|a-b|_{\mathbb{R}^3} : a \in A,\ b \in B\}.$$

A subset A of $\mathbb{R}^3$ is called uniformly locally 1-connected if for every $\epsilon > 0$ there exists $\delta > 0$ such that every continuous map $a : \partial([0,1]\times[0,1]) \to A$ with $\operatorname{diam}(a(\partial([0,1]\times[0,1]))) < \delta$ can be extended to a continuous map $b : [0,1]\times[0,1] \to A$ with $\operatorname{diam}(b([0,1]\times[0,1])) < \epsilon$.

With the above preparation, we can now proceed to the main results of this chapter.

Theorem 16.7 *In case $\mu > 0$ there is a homeomorphism h from $\overline{W}$ onto D^3 with $h(\operatorname{bd} W) = S^2$.*

Proof It suffices to show that there is a homeomorphism from $\Pi_3\overline{W}$ onto D^3 such that the homeomorphism sends $\Pi_3(\operatorname{bd} W)$ onto S^2, since Π_3 defines a homeomorphism from $\overline{W}$ onto $\Pi_3\overline{W}$.

Theorem 15.3 and Proposition 6.2 imply that $\Pi_3(\operatorname{bd} W)$ is homeomorphic to S^2. Therefore the Jordan–Brouwer Separation Theorem (Theorem VIII.2) shows that the set $\mathbb{R}^3 \setminus \Pi_3(\operatorname{bd} W)$ has two connected components, one bounded and the other one unbounded. Denote the bounded component by $\operatorname{int}(\Pi_3(\operatorname{bd} W))$ and the unbounded component by $\operatorname{ext}(\Pi_3(\operatorname{bd} W))$.

We claim that we have $\operatorname{int}(\Pi_3(\operatorname{bd} W)) = \Pi_3 W$, from which we arrive at

$$\Pi_3\overline{W} = \Pi_3(\operatorname{bd} W) \cup \operatorname{int}(\Pi_3(\operatorname{bd} W)).$$

To justify the claim, we recall that $\Pi_3 W$ is open in $\mathbb{R}^3$ by Proposition 6.6. We also have

$$\Pi_3(\operatorname{bd} W) = \Pi_3(\overline{W} \setminus W) = \Pi_3(\overline{W}) \setminus \Pi_3 W = \overline{\Pi_3 W} \setminus \Pi_3 W$$

and thereby $\Pi_3 W \subset \operatorname{int}(\Pi_3(\operatorname{bd} W)) \cup \operatorname{ext}(\Pi_3(\operatorname{bd} W))$. As $\Pi_3 W$ is connected we obtain $\Pi_3 W \subset \operatorname{int}(\Pi_3(\operatorname{bd} W))$ or $\Pi_3 W \subset \operatorname{ext}(\Pi_3(\operatorname{bd} W))$. In either case, $\Pi_3 W$ is open in the relative topology of the component in question. $\Pi_3 W$ is also closed in the relative topology since for every sequence in $\Pi_3 W$ converging to some ξ in the component, i.e., $\xi \in \mathbb{R}^3 \setminus \Pi_3(\operatorname{bd} W)$, we have

$$\xi \in \overline{\Pi_3 W} \setminus \Pi_3(\operatorname{bd} W) = \Pi_3\overline{W} \setminus \Pi_3(\operatorname{bd} W) = \Pi_3(\overline{W} \setminus \operatorname{bd} W) = \Pi_3 W.$$

It follows that $\Pi_3 W$ coincides with one of the components. As $\Pi_3 W$ is bounded we get

$$\Pi_3 W = \operatorname{int}(\Pi_3(\operatorname{bd} W))$$

as claimed.

We are going to apply Theorem VIII.3 with $A = \Pi_3(\operatorname{bd} W)$. We want to show that $\operatorname{int}(\Pi_3(\operatorname{bd} W))$ is uniformly locally 1-connected. Suppose there exist $\epsilon > 0$ and a sequence $(\gamma^n)_0^\infty$ of continuous maps $\gamma^n : \partial([0,1] \times [0,1]) \to \Pi_3 W$ such that

$$\operatorname{diam}(\gamma^n(\partial([0,1] \times [0,1]))) \to 0 \qquad \text{as } n \to \infty$$

and for every $n \in \mathbb{N}$, γ^n can not be extended to a continuous map from $[0,1] \times [0,1]$ into a subset of $\Pi_3 W$ with diameter less than ϵ. For every $n \in \mathbb{N}$, choose a point x^n on the image of γ^n. Since $\overline{\Pi_3 W}$ is compact, we may assume without loss of generality that $x^n \to x \in \overline{\Pi_3 W} = \Pi_3(\operatorname{bd} W) \cup \Pi_3 W$ as $n \to \infty$.

We distinguish three cases.

Case 1: $x \in \Pi_3 W$. Since $\Pi_3 W$ is open, there exists $\beta \in (0, \frac{\epsilon}{2})$ with $x + \mathbb{R}^3_\beta \subset \Pi_3 W$. There exists $n \in \mathbb{N}$ with $\gamma^n(\partial([0,1] \times [0,1])) \subset x + \mathbb{R}^3_\beta$. Thus the map $\Gamma^n : [0,1] \times [0,1] \to x + \mathbb{R}^3_\beta$ given by

$$\Gamma^n\left(ts + (1-t)\left(\frac{1}{2}, \frac{1}{2}\right)^{tr}\right) = t\gamma^n(s) + (1-t)x, \qquad t \in [0,1],\ s \in \partial([0,1] \times [0,1])$$

is a continuous extension of γ^n to $[0,1] \times [0,1]$ with image in $x + \mathbb{R}^3_\beta$. Since the diameter of $x + \mathbb{R}^3_\beta$ is less than ϵ, we get a contradiction.

Case 2: $x \in \Pi_3(\operatorname{bd} W) \setminus \{\Pi_3\xi_+, \Pi_3\xi_-\}$. Let $\phi = \Pi_3^{-1}x \in \operatorname{bd} W \setminus \{\xi_+, \xi_-\}$. Then Theorem 14.1 implies that there exist a 2-dimensional subspace G_3^2 of G_3, $\xi \in G_3 \setminus G_3^2$, open neighbourhoods V_1 of 0 in $G_3^1 = \mathbb{R}\xi$ and V_2 of 0 in G_3^2, a C^1-function $d : V_2 \to V_1$ with $d(0) = 0$ and $Dd(0) = 0$, and $m \in \{-1, 1\}$ so that

$$(P_3\phi + V_1 + V_2 + E) \cap W = \{\chi + w(\chi) : \chi \in V^*\}$$

where

$$V^* = \{P_3\phi + \chi_1 + \chi_2 : \chi_1 \in V_1,\ \chi_2 \in V_2,\ \chi_1 - d(\chi_2) \in (0,\infty)m\xi\}.$$

Choose $\kappa > 0$ so that the set

$$V^*_\kappa = \{P_3\phi + \chi_1 + \chi_2 : \chi_2 \in V_2,\ ||\chi_2|| < \kappa,\ \chi_1 \in d(\chi_2) + (0,\kappa)m\xi\}$$

is contained in V^* and satisfies

$$\operatorname{diam}(\Pi_3(\{\chi + w(\chi) : \chi \in V^*_\kappa\})) < \epsilon.$$

Then

$$\{\chi + w(\chi) : \chi \in V^*_\kappa\} \subset W,$$

and for every sequence $(\phi_n)_0^\infty$ in W with $\phi_n \to \phi \in \operatorname{bd} W \setminus \{\xi_-, \xi_+\}$ as $n \to \infty$ there exists $n_\kappa \in \mathbb{N}$ with $\phi_n \in \{\chi + w(\chi) : \chi \in V^*_\kappa\}$ for all integers $n \geq n_\kappa$. It follows that there exists $n \in \mathbb{N}$ so that

$$\Pi_3^{-1}(\gamma^n(\partial([0,1] \times [0,1]))) \subset \{\chi + w(\chi) : \chi \in V^*_\kappa\}.$$

The set V^*_κ is homeomorphic to the convex set

$$V^{*c}_\kappa = \{P_3\phi + \chi_1 + \chi_2 : \chi_1 \in (0,\kappa)m\xi,\ \chi_2 \in V_2,\ ||\chi_2|| < \kappa\}.$$

A homeomorphism is given by

$$l : V_\kappa^* \ni \chi \mapsto \chi - d(\mathrm{Pr}_{G_3^2}(\chi - P_3\phi)) \in V_\kappa^{*c},$$

where $\mathrm{Pr}_{G_3^2}$ denotes the projection of G_3 onto G_3^2 along G_3^1. Then $l \circ P_3 \circ \Pi_3^{-1} \circ \gamma^n$ is a continuous mapping from $\partial([0,1] \times [0,1])$ into the convex set V_κ^{*c}. Let $q \in V_\kappa^{*c}$. By the convexity of V_κ^{*c}, the continuous map $\Gamma^n : [0,1] \times [0,1] \to G_3$ given by

$$\Gamma^n\left(ts + (1-t)\left(\frac{1}{2}, \frac{1}{2}\right)^{tr}\right) = tl(P_3(\Pi_3^{-1}(\gamma^n(s)))) + (1-t)q$$

for $t \in [0,1]$ and $s \in \partial([0,1] \times [0,1])$, has its range in V_κ^{*c}. The map

$$[0,1] \times [0,1] \ni (\tilde{t}, \tilde{s}) \mapsto \Pi_3\left(l^{-1}(\Gamma^n(\tilde{t}, \tilde{s})) + w(l^{-1}(\Gamma^n(\tilde{t}, \tilde{s})))\right) \in \mathbb{R}^3$$

is a continuous extension of γ^n with range in $\Pi_3(\{\chi + w(\chi) : \chi \in V_\kappa^*\}) \subset \Pi_3 W$. As $\mathrm{diam}(\Pi_3(\{\chi + w(\chi) : \chi \in V_\kappa^*\})) < \epsilon$, we get a contradiction.

Case 3: $x = \Pi_3\xi_+$. Choose an open neighbourhood U_+ of ξ_+ in C and reals $r > 0$, $k_+ \in (0, c_P(\xi_+))$ with $\{\phi \in \overline{W} : c_P(\phi) \geq k_+\} \subset \xi_+ + C_r \subset U_+$ so that in addition

$$\mathrm{diam}(\Pi_3(\xi_+ + C_r)) < \epsilon \quad \text{and} \quad (\xi_+ + \overline{C_r}) \cap ((\overline{W} \cap S) \cup \{\xi_-\}) = \emptyset$$

are satisfied. Then

$$\mathrm{diam}(\Pi_3(\{\phi \in \overline{W} : c_P(\phi) \geq c_+\})) < \epsilon.$$

From now on, we fix a positive integer n such that $\hat{\gamma} = \Pi_3^{-1} \circ \gamma^n$ satisfies

$$\hat{\gamma}(\partial([0,1] \times [0,1])) \subset \{\phi \in W : c_P(\phi) > c_+\} \setminus (\overline{W} \cap S).$$

We claim that for every $s \in \partial([0,1] \times [0,1])$ there exists a unique $\tau_1(s) \in \mathbb{R}$ such that $F_W(\tau_1(s), \hat{\gamma}(s)) \in c_P^{-1}(c_+)$, and that the map $\tau_1 : \partial([0,1] \times [0,1]) \ni s \mapsto \tau_1(s) \in \mathbb{R}$ is continuous with range in $(-\infty, 0)$. Existence and negativity follow from $c_P(\hat{\gamma}(s)) > c_+$, $\alpha(\hat{\gamma}(s)) = \{0\}$ and $c_P(0) = 0$. The uniqueness is a direct consequence of the transversality of F_W with $\overline{W} \cap c_P^{-1}(c_+)$ (Proposition 16.2(iii)). To prove the continuity of τ_1 let a sequence $(s^k)_0^\infty$ in $\partial([0,1] \times [0,1])$ and $s \in \partial([0,1] \times [0,1])$ with $s^k \to s$ as $k \to \infty$ be given. Let $\tau > 0$. The uniqueness of $\tau_1(s)$ yields

$$c_P(F_W(\tau_1(s) - \tau, \hat{\gamma}(s))) < c_+ < c_P(F_W(\tau_1(s) + \tau, \hat{\gamma}(s))).$$

From $\hat{\gamma}(s^k) \to \hat{\gamma}(s)$ as $k \to \infty$ and from the continuity of F_W we infer

$$c_P(F_W(\tau_1(s) - \tau, \hat{\gamma}(s^k))) < c_+ < c_P(F_W(\tau_1(s) + \tau, \hat{\gamma}(s^k)))$$

for all sufficiently large $k \in \mathbb{N}$. This inequality combined with the uniqueness of $\tau_1(s^k)$ gives $\tau_1(s) - \tau < \tau_1(s^k) < \tau_1(s) + \tau$ for all sufficiently large $k \in \mathbb{N}$. Hence $\tau_1(s^k) \to \tau_1(s)$ as $k \to \infty$ since $\tau > 0$ was arbitrary.

We claim that for every $s \in \partial([0,1] \times [0,1])$ there exists a unique $\tau_2(s) \in \mathbb{R}$ such that $|F_W(\tau_2(s), \hat{\gamma}(s))|_e = \delta$, and that the map $\tau_2 : \partial([0,1] \times [0,1]) \to \mathbb{R}$ is continuous with range in $(-\infty, 0)$. Existence and negativity follow from $c_P(\hat{\gamma}(s)) > c_+ > c_P(\phi)$ for all $\phi \in C$ with $|\phi|_e \leq \delta$ (Proposition 16.4(ii)), and $\alpha(\hat{\gamma}(s)) = \{0\}$. The uniqueness follows from Proposition 16.4(i). To prove the continuity of τ_2, consider a sequence $(s^k)_0^\infty$ in $\partial([0,1] \times [0,1])$ and $s \in \partial([0,1] \times [0,1])$ such that $s^k \to s$ as $k \to \infty$. Then Proposition 16.4(i) and the continuity of F_W combined show that for every $\tau > 0$ we have an integer N so that

$$|F_W(\tau_2(s) - \tau, \hat{\gamma}(s^k))|_e < \delta \qquad \text{for all integers } k \geq N,$$
$$|F_W(\tau_2(s) + \tau, \hat{\gamma}(s^k))|_e > \delta \qquad \text{for all integers } k \geq N.$$

It follows that $\tau_2(s) - \tau < \tau_2(s^k) < \tau_2(s) + \tau$ for all integers $k \geq N$. Since $\tau > 0$ was arbitrary, we conclude $\tau_2(s^k) \to \tau_2(s)$ as $k \to \infty$.

We show that for every $s \in \partial([0,1] \times [0,1])$,

$$F_W(\tau_2(s), \hat{\gamma}(s)) \in (S_2^\delta \setminus S_1^\delta)^+.$$

For such s, $\hat{\gamma}(s) \in W$ and $c_P(\hat{\gamma}(s)) > c_+$. The positive invariance of the set $\{\phi \in \overline{W} : c_P(\phi) > c_+\}$ (see Proposition 16.2(iv)) yields

$$F_W([0,\infty) \times \{\hat{\gamma}(s)\}) \subset \{\phi \in \overline{W} : c_P(\phi) > c_+\} \subset \xi_+ + C_r.$$

It follows that

$$\omega(\hat{\gamma}(s)) \subset \xi_+ + \overline{C_r} \subset C \setminus (\mathcal{O} \cup \{\xi_-, 0\}).$$

Consequently, $\omega(\hat{\gamma}(s)) = \{\xi_+\}$ which gives

$$\omega(F_W(\tau_2(s), \hat{\gamma}(s))) = \{\xi_+\}.$$

Using Proposition 16.5 and $S_1^\delta \subset \overline{W} \cap S$, $\omega(\psi) = \mathcal{O}$ for $\psi \in \overline{W} \cap S \setminus \{0\}$, $\omega(\psi) = \{\xi_-\}$ or $\omega(\psi) = \{\xi_+\}$ for $\psi \in \overline{W} \setminus S$, one gets

$$F_W(\tau_2(s), \hat{\gamma}(s)) \in (S_2^\delta \setminus S_1^\delta)^+.$$

Proposition 16.5 shows that there is a homeomorphism j from $(S_2^\delta \setminus S_1^\delta)^+$ onto the open unit disk in $\mathbb{R}^2$. Consider the extension $\hat{\Gamma} : [0,1] \times [0,1] \to C$ of $\hat{\gamma}$ given by

$$\hat{\Gamma}\left(ts + (1-t)\left(\frac{1}{2}, \frac{1}{2}\right)^{tr}\right) = F\Big(\tau_0\big(j^{-1}(2tj \circ F_W(\tau_2(s), \hat{\gamma}(s)))\big), j^{-1}(2tj \circ F_W(\tau_2(s), \hat{\gamma}(s)))\Big)$$

for $s \in \partial([0,1] \times [0,1])$ and $0 \leq t \leq \frac{1}{2}$, and

$$\hat{\Gamma}\left(ts + (1-t)\left(\frac{1}{2}, \frac{1}{2}\right)^{tr}\right) = F_W(2(1-t)\tau_1(s), \hat{\gamma}(s))$$

for $s \in \partial([0,1] \times [0,1])$ and $\frac{1}{2} < t \leq 1$. Note that for every $s \in \partial([0,1] \times [0,1])$ and for $t = \frac{1}{2}$ we have

$$\hat{\Gamma}\left(\frac{s}{2} + \frac{1}{2}\left(\frac{1}{2}, \frac{1}{2}\right)^{tr}\right) = F\Big(\tau_0\big(F_W(\tau_2(s), \hat{\gamma}(s))\big), F_W(\tau_2(s), \hat{\gamma}(s))\Big) = F_W(\tau_1(s), \hat{\gamma}(s)),$$

where the second equality follows from

$$c_P^{-1}(c_+) \ni F\Big(\tau_0\big(F_W(\tau_2(s),\hat{\gamma}(s))\big), F_W(\tau_2(s),\hat{\gamma}(s))\Big)$$
$$= F_W\Big(\tau_0\big(F_W(\tau_2(s),\hat{\gamma}(s))\big) + \tau_2(s), \hat{\gamma}(s)\Big)$$

and from the uniqueness of $\tau_1(s)$. This, together with the continuity of all functions involved, implies that $\hat{\Gamma}$ is continuous. Clearly, $\hat{\Gamma}(s) = \hat{\gamma}(s)$ for all $s \in \partial([0,1] \times [0,1])$.

Using $\hat{\gamma}(\partial([0,1] \times [0,1])) \subset \{\phi \in W : c_P(\phi) > c_+\}$, the positive invariance of $\{\phi \in W : c_P(\phi) > c_+\}$, and the definition of τ_1 we infer

$$\hat{\Gamma}\left(ts + (1-t)\left(\frac{1}{2},\frac{1}{2}\right)^{tr}\right) \subset \{\phi \in W : c_P(\phi) > c_+\}$$

for all $s \in \partial([0,1]\times[0,1])$ and $1/2 < t \le 1$. For $s \in \partial([0,1]\times[0,1])$ and $0 \le t \le 1/2$,

$$\hat{\Gamma}\left(ts + (1-t)\left(\frac{1}{2},\frac{1}{2}\right)^{tr}\right) \subset \{\phi \in W : c_P(\phi) = c_+\}.$$

Therefore, $\hat{\Gamma}$ is a continuous extension of $\hat{\gamma}$ to $[0,1] \times [0,1]$ with image in $\{\phi \in W : c_P(\phi) \ge c_+\}$. It follows that the map $\Gamma : [0,1] \times [0,1] \to \mathbb{R}^3$ given by

$$\Gamma\left(ts + (1-t)\left(\frac{1}{2},\frac{1}{2}\right)^{tr}\right) = \Pi_3\left(\hat{\Gamma}\left(ts + (1-t)\left(\frac{1}{2},\frac{1}{2}\right)^{tr}\right)\right)$$

for $s \in \partial([0,1]\times[0,1])$ and $0 \le t \le 1$, is a continuous extension of γ^n with image in $\Pi_3(\{\phi \in W : c_P(\phi) \ge c_+\}) \subset \Pi_3 W$. Using $\operatorname{diam}(\Pi_3(\{\phi \in \overline{W} : c_P(\phi) \ge c_+\})) < \epsilon$, we arrive at $\operatorname{diam}(\Gamma([0,1] \times [0,1])) < \epsilon$, again a contradiction.

Case 4: $x = \Pi_3\xi_-$. Proceed as in case 3. □

In case $\mu = 0$ set

$$Z_+ = \left\{\phi \in \overline{W} : c_P(\phi) \ge c_+\right\},$$
$$Z_0 = \left\{\phi \in \overline{W} : c_- \le c_P(\phi) \le c_+\right\},$$
$$Z_- = \left\{\phi \in \overline{W} : c_P(\phi) \le c_-\right\}.$$

Theorem 16.8 *In case $\mu = 0$ there is a homeomorphism h from $\overline{W}$ onto $\mathbb{R} \times D^2$ with $h(\operatorname{bd} W) = \mathbb{R} \times S^1$.*

Proof Details which have analogues in the proof of Theorem 16.7 are only sketched.

1. We prove that there exist homeomorphisms from $\overline{W} \cap c_P^{-1}(c_+)$ and from $\overline{W} \cap c_P^{-1}(c_-)$ onto D^2 which send $\operatorname{bd} W \cap c_P^{-1}(c_+)$ and $\operatorname{bd} W \cap c_P^{-1}(c_-)$ onto S^1.

Using Proposition 16.4(i) we obtain that the continuous map

$$(S_2^\delta \setminus S_1^\delta)^+ \ni \phi \mapsto F(\tau_0(\phi),\phi) \in W \cap c_P^{-1}(c_+)$$

given by Proposition 16.6(i) is injective. In order to show surjectivity let $\phi \in W \cap c_P^{-1}(c_+)$ be given. Recall $\alpha(\phi) = \{0\}$. By Proposition 16.4(ii), $|\phi|_e > \delta$. It follows that for some $t < 0$,

$$F_W(t, \phi) \in S_2^\delta.$$

Using Proposition 16.2(ii) and equation (1.1) we infer $x^\phi(t) \to \infty$ as $t \to \infty$. Now the definitions of S_2^δ and S_1^δ, the fact that the invariant set $W \cap S$ is bounded, and Proposition 16.5 combined yield $F_W(t, \phi) \in (S_2^\delta \setminus S_1^\delta)^+$. Since $(S_2^\delta \setminus S_1^\delta)^+$ is homeomorphic to the open unit disk, we conclude that $W \cap c_P^{-1}(c_+)$ is the image of the open unit disk under a continuous and injective map.

We next prove that $\operatorname{bd} W \cap c_P^{-1}(c_+)$ is homeomorphic to S^1.

Recall the function g from Proposition 12.1 and the choice of $\delta_2, \delta_3, \epsilon$ at the beginning of Chapter 15. Choose $s_0 \in (0, \epsilon)$ so that $s_1 = g(s_0)$ and $s_1^- = g(-s_0)$ belong to $(0, \epsilon)$ and $(-\epsilon, 0)$, respectively, and the curves

$$\gamma : [s_0, s_1] \ni s \mapsto p_0 + s\psi_u + w^u(s\psi_u) \in C,$$
$$\gamma^- : [s_1^-, -s_0] \ni s \mapsto p_0 + s\psi_u + w^u(s\psi_u) \in C$$

satisfy

(i) $c_P(F(t, \gamma(s))) < c_+$ for all $t \in [0, 1]$ and $s \in [s_0, s_1]$,
$c_- < c_P(F(t, \gamma^-(s)))$ for all $t \in [0, 1]$ and $s \in [s_1^-, -s_0]$.

According to the results in Chapter 15, in particular, Proposition 15.1, we have $\gamma^-([s_1^-, -s_0]) \cup \gamma([s_0, s_1]) \subset \operatorname{bd} W$ and

(ii) $\gamma(s_1) = F(\omega, \gamma(s_0))$, $\gamma^-(s_1^-) = F(\omega, \gamma^-(-s_0))$,

(iii) $x^{\gamma(s)}(t) \to \infty$ as $t \to \infty$ for all $s \in [s_0, s_1]$, and
$x^{\gamma^-(s)}(t) \to -\infty$ as $t \to \infty$ for all $s \in [s_1^-, -s_0]$,

(iv) $\gamma(s) \ll \gamma(t)$ for all s, t in $[s_0, s_1]$ with $s < t$, and
$\gamma^-(s) \ll \gamma^-(t)$ for all s, t in $[s_1^-, -s_0]$ with $s < t$.

Arguing as in the proof of Proposition 16.6 we obtain continuous maps

$$\tau_3 : [s_0, s_1] \to (1, \infty), \quad \tau_3^- : [s_1^-, -s_0] \to (1, \infty)$$

so that

$$F(\tau_3(s), \gamma(s)) \in c_P^{-1}(c_+) \qquad \text{for all } s \in [s_0, s_1],$$
$$F(\tau_3^-(s), \gamma^-(s)) \in c_P^{-1}(c_-) \qquad \text{for all } s \in [s_1^-, -s_0],$$

and

$$F_W(t, \gamma(s)) \notin c_P^{-1}(c_+) \qquad \text{for all } s \in [s_0, s_1] \text{ and } t \in \mathbb{R} \setminus \{\tau_3(s)\},$$
$$F_W(t, \gamma^-(s)) \notin c_P^{-1}(c_-) \qquad \text{for all } s \in [s_1^-, -s_0] \text{ and } t \in \mathbb{R} \setminus \{\tau_3^-(s)\}.$$

Using property (ii) above, we infer

$$\tau_3(s_0) = \tau_3(s_1) + \omega, \quad \tau_3^-(-s_0) = \tau_3^-(s_1^-) + \omega,$$

and conclude that the continuous curves

$$[s_0, s_1] \ni s \mapsto F(\tau_3(s), \gamma(s)) \in C, \quad [s_1^-, -s_0] \ni s \mapsto F(\tau_3^-(s), \gamma^-(s)) \in C$$

with trace in $c_P^{-1}(c_+)$ and $c_P^{-1}(c_-)$, respectively, are closed. We show that they are simple closed. Consider the curve with domain $[s_0, s_1]$. Suppose there exist s, t in $[s_0, s_1]$ with $s < t$, $t - s < s_1 - s_0$, and

$$F(\tau_3(s), \gamma(s)) = F(\tau_3(t), \gamma(t)).$$

Then $F_W(\tau_3(s) - \tau_3(t), \gamma(s)) = \gamma(t)$, and thereby

$$F_W(-n\omega, F(\tau_3(s) - \tau_3(t), \gamma(s))) = F_W(-n\omega, \gamma(t)) \qquad \text{for all } n \in \mathbb{Z}.$$

Since $\gamma(t), \gamma(s)$ belong to $W^u(p_0, F(\omega, \cdot))$, we have

$$F_W(-n\omega, \gamma(t)) \to p_0, \quad F_W(-n\omega, \gamma(s)) \to p_0$$

and

$$\begin{aligned} F_W(-n\omega, F_W(\tau_3(s) - \tau_3(t), \gamma(s))) &= F_W(\tau_3(s) - \tau_3(t), F_W(-n\omega, \gamma(s))) \\ &\to F_W(\tau_3(s) - \tau_3(t), p_0) \end{aligned}$$

as $n \to \infty$. Consequently,

$$F_W(\tau_3(s) - \tau_3(t), p_0) = p_0.$$

It follows that $\tau_3(s) - \tau_3(t) = n\omega$ for some integer n. The inequalities $\gamma(s_0) \le \gamma(s) \ll \gamma(t) \le \gamma(s_1)$, the monotonicity of F and c_P, and Proposition 16.2(iv) combined imply

$$\tau_3(s_1) + \omega = \tau_3(s_0) \ge \tau_3(s) > \tau_3(t) \ge \tau_3(s_1).$$

Note that in case $s_0 < s$ we have $\tau_3(s_0) > \tau_3(s)$ instead of $\tau_3(s_0) \ge \tau_3(s)$, and in case $t < s_1$ we have $\tau_3(t) > \tau_3(s_1)$ instead of $\tau_3(t) \ge \tau_3(s_1)$. The inequalities $s_0 \le s < t \le s_1$ and $t - s < s_1 - s_0$ combined imply either $s = s_0$ and $t < s_1$ or $s > s_0$. It follows that $0 < \tau_3(s) - \tau_3(t) < \omega$, a contradiction to $\tau_3(s) - \tau_3(t) \in \mathbb{Z}\omega$. Thus $[s_0, s_1] \ni s \mapsto F(\tau_3(s), \gamma(s)) \in C$ is a simple closed curve.

Recall the set $\operatorname{bd}^+ W = \{\phi \in \operatorname{bd} W : x^\phi(t) \to \infty \text{ as } t \to \infty\}$ from Chapter 15. We show the equality

$$\operatorname{bd} W \cap c_P^{-1}(c_+) = \operatorname{bd}^+ W \cap c_P^{-1}(c_+) = \{F(\tau_3(s), \gamma(s)) : s_0 \le s < s_1\}. \tag{16.7}$$

Proposition 15.1(ii) and the properties of τ_3 give

$$F(\tau_3(s), \gamma(s)) \in \operatorname{bd}^+ W \cap c_P^{-1}(c_+) \subset \operatorname{bd} W \cap c_P^{-1}(c_+) \qquad \text{for all } s \in [s_0, s_1).$$

Conversely, let $\phi \in \operatorname{bd} W \cap c_P^{-1}(c_+)$ be given. Using Proposition 16.2(ii) and equation (1.1) we get $x^\phi(t) \to \infty$ as $t \to \infty$, or $\phi \in \operatorname{bd}^+ W$. Proposition 15.1(i) and the properties of τ_3 yield $\phi = F(\tau_3(s), \gamma(s))$ for some $s \in [s_0, s_1)$.

It follows that the set $\operatorname{bd} W \cap c_P^{-1}(c_+)$ is homeomorphic to S^1, and Π_3 maps $\operatorname{bd} W \cap c_P^{-1}(c_+)$ onto the trace Γ of a simple closed curve in the plane

$$\Pi_3(c_P^{-1}(c_+)) = \{x \in \mathbb{R}^3 : x_3 = c_+\}.$$

One easily sees that

$$\Pi_3(W \cap c_P^{-1}(c_+)) = \Pi_3 W \cap \{x \in \mathbb{R}^3 : x_3 = c_+\}.$$

Recall that $\Pi_3 W$ is an open subset of $\mathbb{R}^3$. Hence $\Pi_3(W \cap c_P^{-1}(c_+))$ is an open subset of the plane $\{x \in \mathbb{R}^3 : x_3 = c_+\}$. It is also connected, being the image of the open unit disk under a continuous map. Using this fact and the equation

$$\Pi_3(\mathrm{bd}\, W \cap c_P^{-1}(c_+)) \cap \Pi_3(W \cap c_P^{-1}(c_+)) = \emptyset$$

we infer that the set $\Pi_3(W \cap c_P^{-1}(c_+))$ is either contained in the bounded component $\mathrm{int}_{c_+}(\Gamma)$ of $\{x \in \mathbb{R}^3 : x_3 = c_+\} \setminus \Gamma$ or contained in the unbounded component of $\{x \in \mathbb{R}^3 : x_3 = c_+\} \setminus \Gamma$. In either case, $\Pi_3(W \cap c_P^{-1}(c_+))$ is also closed with respect to the topology of the component it is contained in. In order to show this consider a sequence $(x_n)_0^\infty$ in $\Pi_3(W \cap c_P^{-1}(c_+))$ converging to a limit x in the component containing $\Pi_3(W \cap c_P^{-1}(c_+))$. Then $x \in \mathbb{R}^3 \setminus \Pi_3(\mathrm{bd}\, W \cap c_P^{-1}(c_+))$. The boundedness of $W \cap c_P^{-1}(c_+)$ (see Proposition 16.2(i)) combined with the Lipschitz continuity of Π_3^{-1} on bounded subsets of $\Pi_3 \overline{W}$ yields that the preimages $\phi_n = \Pi_3^{-1}(x_n)$, $n \in \mathbb{N}$, form a Cauchy sequence converging to some ϕ in the closed set $\overline{W} \cap c_P^{-1}(c_+)$. By continuity, $x = \Pi_3 \phi \in \Pi_3(\overline{W} \cap c_P^{-1}(c_+))$, and we find $x \in \Pi_3(\overline{W} \cap c_P^{-1}(c_+)) \setminus \Pi_3(\mathrm{bd}\, W \cap c_P^{-1}(c_+)) = \Pi_3(W \cap c_P^{-1}(c_+))$.

It follows that the bounded set $\Pi_3(W \cap c_P^{-1}(c_+))$ coincides with the bounded component $\mathrm{int}_{c_+}(\Gamma)$. Hence $\Pi_3(\overline{W} \cap c_P^{-1}(c_+)) = \Gamma \cup \mathrm{int}_{c_+}(\Gamma)$, and $\Pi_3(\mathrm{bd}\, W \cap c_P^{-1}(c_+)) = \Gamma$. Using Theorem VIII.1(i) we conclude that there is a homeomorphism h_1^+ from $\overline{W} \cap c_P^{-1}(c_+)$ onto D^2 which maps $\mathrm{bd}\, W \cap c_P^{-1}(c_+)$ onto S^1.

Analogously we obtain a homeomorphism h_1^- from $\overline{W} \cap c_P^{-1}(c_-)$ onto D^2 which maps $\mathrm{bd}\, W \cap c_P^{-1}(c_-)$ onto S^1.

2. The subset $\{\phi \in \mathrm{bd}\, W : c_- \le c_P(\phi) \le c_+\}$ of $\overline{W}$ is closed and bounded, due to Proposition 16.2. Proposition 4.2 guarantees that it is compact. We show that there exists a homeomorphism

$$h_2 : \{\phi \in \mathrm{bd}\, W : c_- \le c_P(\phi) \le c_+\} \to [-1,1] \times S^1$$

with

$$h_2(\mathrm{bd}\, W \cap c_P^{-1}(c_-)) = \{-1\} \times S^1, \quad h_2(\mathrm{bd}\, W \cap c_P^{-1}(c_+)) = \{1\} \times S^1.$$

Recall the homeomorphisms $h^0 : \mathrm{bd}^+ W \cup \mathcal{O} \to [0,\infty) \times S^1_{\mathbb{C}}$ and $h^{0-} : \mathrm{bd}^- W \cup \mathcal{O} \to [0,\infty) \times S^1_{\mathbb{C}}$ from the proof of Theorem 15.4. The definition of h^0 and the equalities (16.7) combined yield

$$\begin{aligned} h^0\left(\mathrm{bd}^+ W \cap c_P^{-1}(c_+)\right) &= G^0\left(\tau_3(s), i_{FG^0}(\gamma(s))\right) \\ &= \left(e^{\tau_3(s)}\left(\frac{s - s_0}{s_1 - s_0} e^{\omega} + \frac{s_1 - s}{s_1 - s_0}\right), e^{i\frac{2\pi}{\omega}\tau_3(s)}\right), s_0 \le s < s_1. \end{aligned}$$

Property (iv) of γ in part 1, the monotonicity of F and c_P, and the definition of τ_3 combined yield that the function $\tau_3 : [s_0, s_1] \to (1, \infty)$ is strictly decreasing. From this and from the equation $\tau_3(s_0) = \tau_3(s_1) + \omega$ we obtain $\tau_3([s_0, s_1]) =$

$[\tau_3(s_1), \tau_3(s_1)+\omega]$. Let $\tau_3^{-1} : [\tau_3(s_1), \tau_3(s_1)+\omega] \to \mathbb{R}$ be given by the inverse of the map $[s_0, s_1] \ni s \mapsto \tau_3(s) \in [\tau_3(s_1), \tau_3(s_1)+\omega]$. Define

$$a(x) = e^x \left(\frac{\tau_3^{-1}(x) - s_0}{s_1 - s_0} e^\omega + \frac{s_1 - \tau_3^{-1}(x)}{s_1 - s_0} \right) \qquad \text{for } x \in [\tau_3(s_1), \tau_3(s_1)+\omega],$$

and

$$A_- = \left\{ \left(t, e^{i\frac{2\pi}{\omega}x}\right) : 0 \le t < a(x),\ \tau_3(s_1) < x \le \tau_3(s_1)+\omega \right\},$$
$$A_+ = \left\{ \left(t, e^{i\frac{2\pi}{\omega}x}\right) : t > a(x),\ \tau_3(s_1) < x \le \tau_3(s_1)+\omega \right\}.$$

Then

$$h^0\left(\mathrm{bd}^+ W \cap c_P^{-1}(c_+)\right) = \left\{ \left(a(x), e^{i\frac{2\pi}{\omega}x}\right) : \tau_3(s_1) < x \le \tau_3(s_1)+\omega \right\}$$

and

$$[0,\infty) \times S^1_{\mathbb{C}} = A_- \cup h^0\left(\mathrm{bd}^+ W \cap c_P^{-1}(c_+)\right) \cup A_+.$$

Observing $a(\tau_3(s_1)) = a(\tau_3(s_1)+\omega)$ it is easy to see that $h^0(\mathrm{bd}^+ W \cap c_P^{-1}(c_+))$ is the image of a simple closed curve, and A_-, A_+ are the open connected components of $[0,\infty) \times S^1_{\mathbb{C}} \setminus h^0(\mathrm{bd}^+ W \cap c_P^{-1}(c_+))$.

Let $\phi \in \mathrm{bd}^+ W$. The relations $\alpha(\phi) = \mathcal{O}$ and $x^\phi(t) \to \infty$ as $t \to \infty$ combined with the choice of c_+ and Proposition 16.2(iii) and (iv) imply that there exists $s_\phi \in \mathbb{R}$ with $c_P(F_W(s_\phi, \phi)) = c_+$, $c_P(F_W(t,\phi)) < c_+$ for all $t < s_\phi$, $c_P(F_W(t,\phi)) > c_+$ for all $t > s_\phi$. It follows that the curve $\mathbb{R} \ni t \mapsto h^0(F_W(t,\phi)) \in [0,\infty) \times S^1_{\mathbb{C}}$ intersects $h^0(\mathrm{bd}^+ W \cap c_P^{-1}(c_+))$ only at $t = s_\phi$, and that each of the connected sets $h^0(\{F_W(t,\phi) : t < s_\phi\})$ and $h^0(\{F_W(t,\phi) : t > s_\phi\})$ is contained in A_- or A_+. Using the definition of h^0 and the relation $x^\phi(t) \to \infty$ as $t \to \infty$ it is not hard to show $h^0(F_W(t,\phi)) \in A_+$ for $t \in \mathbb{R}$ sufficiently large. Using $\alpha(\phi) = \mathcal{O}$ and $h^0(\mathcal{O}) = \{0\} \times S^1_{\mathbb{C}} \subset A_-$ we infer $h^0(F_W(t,\phi)) \in A_-$ for all sufficiently large negative t. It follows that

$$h^0\left(\{F_W(t,\phi) : t > s_\phi\}\right) \subset A_+, \quad h^0\left(\{F_W(t,\phi) : t < s_\phi\}\right) \subset A_-.$$

As for every $\phi \in \mathrm{bd}^+ W$ with $c_P(\phi) > c_+$ we have $s_\phi < 0$, due to Proposition 16.2(iv), we infer

$$h^0\left(\left\{\phi \in \mathrm{bd}^+ W : c_P(\phi) > c_+\right\}\right) \subset A_+.$$

Similarly we get

$$h^0\left(\left\{\phi \in \mathrm{bd}^+ W : c_P(\phi) < c_+\right\}\right) \subset A_-.$$

For every $x \in A_- \subset [0,\infty) \times S^1_{\mathbb{C}} = h^0(\mathrm{bd}^+ W \cup \mathcal{O})$ we have $x = h^0(\phi)$ with $\phi \in \mathrm{bd}^+ W$ or $\phi \in \mathcal{O}$; in case $\phi \in \mathrm{bd}^+ W$ the assumption $c_P(\phi) > c_+$ leads to $x = h^0(\phi) \in A_+$ which is a contradiction. Consequently,

$$A_- \subset h^0\left(\left\{\phi \in \mathrm{bd}^+ W : c_P(\phi) \le c_+\right\} \cup \mathcal{O}\right).$$

Using the last inclusion we easily obtain

$$h^0\left(\left\{\phi \in \mathrm{bd}^+ W : c_P(\phi) \le c_+\right\} \cup \mathcal{O}\right) = A_- \cup h^0\left(\mathrm{bd}^+ W \cap c_P^{-1}(c_+)\right)$$

where the set on the right-hand side equals

$$A_{\leq} = \left\{ \left(t, e^{i\frac{2\pi}{\omega}x}\right) : 0 \leq t \leq a(x),\ \tau_3(s_1) < x \leq \tau_3(s_1) + \omega \right\}.$$

Analogously we obtain a continuous map $a^- : [\tau_3^-(s_1^-), \tau_3^-(s_1^-) + \omega] \to (0, \infty)$ so that

$$\begin{aligned} &h^{0-}\left(\{\phi \in \mathrm{bd}^- W : c_P(\phi) \geq c_-\} \cup \mathcal{O}\right) \\ &\qquad = \left\{ \left(t, e^{i\frac{2\pi}{\omega}x}\right) : 0 \leq t \leq a^-(x),\ \tau_3^-(s_1^-) < x \leq \tau_3^-(s_1^-) + \omega \right\}. \end{aligned}$$

We have

$$h^0(\phi) = h^{0-}(\phi) \qquad \text{for all } \phi \in \mathcal{O}.$$

The maps

$$\begin{aligned} b : &\left\{ \left(t, e^{i\frac{2\pi}{\omega}x}\right) : 0 \leq t \leq a(x),\ \tau_3(s_1) < x \leq \tau_3(s_1) + \omega \right\} \\ &\qquad \ni \left(t, e^{i\frac{2\pi}{\omega}x}\right) \mapsto \left(\frac{t}{a(x)}, \begin{pmatrix} \cos x \\ \sin x \end{pmatrix} \right) \in [0,1] \times S^1 \end{aligned}$$

and

$$\begin{aligned} b^- : &\left\{ \left(t, e^{i\frac{2\pi}{\omega}x}\right) : 0 \leq t \leq a^-(x),\ \tau_3^-(s_1^-) < x \leq \tau_3^-(s_1^-) + \omega \right\} \\ &\qquad \ni \left(t, e^{i\frac{2\pi}{\omega}x}\right) \mapsto \left(-\frac{t}{a^-(x)}, \begin{pmatrix} \cos x \\ \sin x \end{pmatrix} \right) \in [-1,0] \times S^1 \end{aligned}$$

are homeomorphisms. Using Proposition 16.2(iii) and (iv) we obtain

$$\begin{aligned} &\{\phi \in \mathrm{bd}^- W : c_P(\phi) \geq c_-\} \cup \{\phi \in \mathrm{bd}^+ W : c_P(\phi) \leq c_+\} \\ &\qquad \subset \{\phi \in \mathrm{bd}\, W : c_- \leq c_P(\phi) \leq c_+\}. \end{aligned}$$

It follows that $\{\phi \in \mathrm{bd}\, W : c_- \leq c_P(\phi) \leq c_+\}$ is the union of the mutually disjoint sets $\mathcal{O}$, $\{\phi \in \mathrm{bd}^+ W : c_P(\phi) \leq c_+\}$, $\{\phi \in \mathrm{bd}^- W : c_P(\phi) \geq c_-\}$. Note that for every sequence $(\phi_n)_0^\infty$ in $\{\psi \in \mathrm{bd}\, W : c_- \leq c_P(\psi) \leq c_+\}$ which converges to some $\phi \in \mathrm{bd}^+ W$ there exist $t \in \mathbb{R}$ and $n \in \mathbb{N}$ with $F_W(t, \phi_k) \gg 0$ for all integers $k \geq n$; it follows that for all such k, $x^{\phi_k}(s) \to \infty$ as $s \to \infty$, and furthermore $\phi_k \in \{\psi \in \mathrm{bd}^+ W : c_P(\psi) \leq c_+\}$. An analogous statement holds for sequences in $\{\psi \in \mathrm{bd}\, W : c_- \leq c_P(\psi) \leq c_+\}$ converging to points $\phi \in \mathrm{bd}^- W$. Recall the equalities

$$\begin{aligned} &b\left(h^0\left(\mathrm{bd}^+ W \cap c_P^{-1}(c_+)\right)\right) \\ &\qquad = \left\{ \left(1, \begin{pmatrix} \cos x \\ \sin x \end{pmatrix}\right) : \tau_3(s_1) < x \leq \tau_3(s_1) + \omega \right\} = \{1\} \times S^1, \end{aligned}$$

$$\begin{aligned} &b^-\left(h^{0-}\left(\mathrm{bd}^- W \cap c_P^{-1}(c_-)\right)\right) \\ &\qquad = \left\{ \left(-1, \begin{pmatrix} \cos x \\ \sin x \end{pmatrix}\right) : \tau_3^-(s_1^-) < x \leq \tau_3^-(s_1^-) + \omega \right\} = \{-1\} \times S^1, \end{aligned}$$

$$h^0\big|_{\mathcal{O}} = h^{0-}\big|_{\mathcal{O}},$$
$$\operatorname{bd} W \cap c_P^{-1}(c_-) = \operatorname{bd}^- W \cap c_P^{-1}(c_-),$$
$$\operatorname{bd} W \cap c_P^{-1}(c_+) = \operatorname{bd}^+ W \cap c_P^{-1}(c_+),$$

and the fact that the maps $h^0 : \operatorname{bd}^+ W \cup \mathcal{O} \to [0,\infty) \times S^1_{\mathbb{C}}$, $h^{0-} : \operatorname{bd}^- W \cup \mathcal{O} \to [0,\infty) \times S^1_{\mathbb{C}}$ are homeomorphisms. Combining all this information we deduce that the map $h_2 : \{\phi \in \operatorname{bd} W : c_- \le c_P(\phi) \le c_+\} \to [-1,1] \times S^1$ given by

$$h_2(\phi) = b(h^0(\phi)) \qquad \text{for } \phi \in \{\psi \in \operatorname{bd}^+ W : c_P(\psi) \le c_+\} \cup \mathcal{O}$$

and

$$h_2(\phi) = b^-(h^{0-}(\phi)) \qquad \text{for } \phi \in \{\psi \in \operatorname{bd}^- W : c_P(\psi) \ge c_-\}$$

is bijective and continuous with

$$h_2(\operatorname{bd} W \cap c_P^{-1}(c_-)) = \{-1\} \times S^1, \quad h_2(\operatorname{bd} W \cap c_P^{-1}(c_+)) = \{1\} \times S^1.$$

The compactness of its domain implies that h_2 is a homeomorphism.

3. We claim that there exists a homeomorphism h_3 from the set

$$S_0 = \{\phi \in \operatorname{bd} W : c_- \le c_P(\phi) \le c_+\} \cup \left(\overline{W} \cap c_P^{-1}(c_-)\right) \cup \left(\overline{W} \cap c_P^{-1}(c_+)\right)$$

onto $\partial([-1,1] \times D^2)$.

The homeomorphisms

$$h_{12}^+ : S^1 \ni x \mapsto h_2\left((h_1^+)^{-1}(x)\right) \in \{1\} \times S^1$$

and

$$h_{12}^- : S^1 \ni x \mapsto h_2\left((h_1^-)^{-1}(x)\right) \in \{-1\} \times S^1$$

extend to homeomorphisms $\widetilde{h_{12}^+} : D^2 \to \{1\} \times D^2$, $\widetilde{h_{12}^-} : D^2 \to \{-1\} \times D^2$ given by

$$\widetilde{h_{12}^+}(ts) = \left(1, t \operatorname{Pr}_2 h_{12}^+(s)\right) \qquad \text{for } 0 \le t \le 1,\ s \in S^1,$$
$$\widetilde{h_{12}^-}(ts) = \left(-1, t \operatorname{Pr}_2 h_{12}^-(s)\right) \qquad \text{for } 0 \le t \le 1,\ s \in S^1,$$

respectively. Of course, $\operatorname{Pr}_2(u,x) = x$ for all $(u,x) \in \mathbb{R} \times S^1$. Let the map $h_3 : S_0 \to \partial([-1,1] \times D^2)$ be given by

$$h_3(\phi) = h_2(\phi) \qquad \text{for } \phi \in \{\psi \in \operatorname{bd} W : c_- < c_P(\psi) < c_+\},$$
$$h_3(\phi) = \widetilde{h_{12}^+}(h_1^+(\phi)) \qquad \text{for } \phi \in \overline{W} \cap c_P^{-1}(c_+),$$
$$h_3(\phi) = \widetilde{h_{12}^-}(h_1^-(\phi)) \qquad \text{for } \phi \in \overline{W} \cap c_P^{-1}(c_-).$$

Using the properties of h_2 derived in part 2 and the fact that h_1^+, $\widetilde{h_{12}^+}$, h_1^-, $\widetilde{h_{12}^-}$ are homeomorphisms we obtain the injectivity of h_3 and the equality

$$h_3(S_0) = \partial([-1,1] \times D^2).$$

The continuity of h_3 is a consequence of the following observations. Suppose a sequence $(\phi_n)_0^\infty$ in S_0 converges to some $\phi \in S_0$. If $c_- < c_P(\phi) < c_+$ then there exists $n \in \mathbb{N}$ with $c_- < c_P(\phi_k) < c_+$ for all integers $k \geq n$, and using $\phi \in S_0$ we get $\phi_k \in \text{bd}\, W$ for such k. If $\phi \in W \cap c_P^{-1}(c_+)$ then there exists $n \in \mathbb{N}$ with $\phi_k \in W = \overline{W} \setminus \text{bd}\, W$ and $c_P(\phi_k) > c_-$ for all integers $k \geq n$, and using $\phi_k \in S_0$ we get $c_P(\phi_k) = c_+$ for such k. Analogously we find in case $\phi \in W \cap c_P^{-1}(c_-)$ an integer $n \in \mathbb{N}$ with $\phi_k \in W \cap c_P^{-1}(c_-)$ for $n \leq k \in \mathbb{N}$. If $\phi \in \text{bd}\, W \cap c_P^{-1}(c_+)$ then

$$h_3(\phi) = \widetilde{h_{12}^+}(h_1^+(\phi)) = h_{12}^+(h_1^+(\phi)) = h_2((h_1^+)^{-1}(h_1^+(\phi))) = h_2(\phi),$$

and analogously

$$h_3(\phi) = h_2(\phi) \qquad \text{for all } \phi \in \text{bd}\, W \cap c_P^{-1}(c_-).$$

The continuous, bijective map h_3 is a homeomorphism since its domain is compact, due to Propositions 16.2 and 4.2.

4. Since Π_3 defines a homeomorphism from $\overline{W}$ onto $\Pi_3\overline{W}$ and $\partial([-1,1] \times D^2)$ is homeomorphic to S^2 we obtain from the result of part 3 that there is a homeomorphism from $\Pi_3 S_0$ onto S^2, and Theorem VIII.2 guarantees that $\mathbb{R}^3 \setminus \Pi_3\, S_0$ is the disjoint union of the bounded connected open set $\text{int}(\Pi_3 S_0)$ with the unbounded connected open set $\text{ext}(\Pi_3 S_0)$. Arguments similar to those used in the first part of the proof of Theorem 16.7 yield

$$\text{int}(\Pi_3 S_0) = \{\Pi_3\phi : \phi \in W,\ c_- < c_P(\phi) < c_+\}$$

which gives

$$\Pi_3 Z_0 = \Pi_3 S_0 \cup \text{int}(\Pi_3 S_0).$$

We need a homeomorphism h_4 from $\Pi_3 Z_0$ onto D^3 which maps $\Pi_3 S_0$ onto S^2. Theorem VIII.3 says that a sufficient condition for this is that the set $\text{int}(\Pi_3 S_0)$ is locally uniformly 1-connected. Suppose the contrary, i.e., there exist $\epsilon > 0$ and a sequence $(\gamma^n)_0^\infty$ of maps $\gamma^n : \partial([0,1] \times [0,1]) \to \text{int}(\Pi_3 S_0)$, $n \in \mathbb{N}$, with

$$\text{diam}\,(\gamma^n(\partial([0,1] \times [0,1]))) \to 0 \qquad \text{as } n \to \infty$$

so that each γ^n, $n \in \mathbb{N}$, can not be extended to a continuous map from $[0,1] \times [0,1]$ into a subset of $\text{int}(\Pi_3 S_0)$ with diameter less than ϵ. Choose a sequence $(x^n)_0^\infty$ in $\mathbb{R}^3$ with

$$x^n \in \gamma^n(\partial([0,1] \times [0,1])) \qquad \text{for every } n \in \mathbb{N}.$$

The boundedness of the component $\text{int}(\Pi_3 S_0)$ implies the existence of a subsequence which converges to some

$$x \in \overline{\text{int}(\Pi_3 S_0)} = \Pi_3 S_0 \cup \text{int}(\Pi_3 S_0).$$

It is sufficient to consider the case that the sequence $(x^n)_0^\infty$ itself is convergent.

In the subcases

$$x \in \{\Pi_3\phi : \phi \in W,\ c_- < c_P(\phi) < c_+\}$$

and

$$x \in \{\Pi_3\phi : \phi \in \text{bd}\, W,\ c_- < c_P(\phi) < c_+\}$$

we obtain contradictions as in the cases 1 and 2 from the proof of Theorem 16.7.

In the subcase

$$x \in \{\Pi_3\phi : \phi \in W,\ c_P(\phi) \in \{c_-, c_+\}\} = \Pi_3 W \cap \{y \in \mathbb{R}^3 : y_3 \in \{c_-, c_+\}\}$$

the set $\text{int}(\Pi_3 S_0) = \Pi_3 W \cap \{y \in \mathbb{R}^3 : c_- < y_3 < c_+\}$ contains an open half-ball with center x and diameter less than ϵ since $\Pi_3 W$ is open. For some $n \in \mathbb{N}$, $\gamma^n(\partial([0,1] \times [0,1]))$ is contained in this half-ball whose convexity permits to find a continuous extension of γ^n to a map from $[0,1] \times [0,1]$ with range in the half-ball, as in case 2 from the proof of Theorem 16.7. This is a contradiction to the assumption above.

In the subcase

$$x \in \{\Pi_3\phi : \phi \in \text{bd}\, W,\ c_P(\phi) = c_+\}$$

there exist $\hat{\tau} < 0$, $\beta_1 > 0$ and $\hat{x} \in \Pi_3(\{\phi \in \text{bd}\, W : c_- < c_P(\phi) < c_+\})$ with $F_W(\hat{\tau}, \Pi_3^{-1}(x)) = \Pi_3^{-1}(\hat{x})$ and

$$\text{diam}\left(\{\Pi_3 F_W(t, \Pi_3^{-1}(y)) : \hat{\tau} \le t \le 0,\ y \in \left(x + \mathbb{R}^3_{\beta_1}\right) \cap \Pi_3 \overline{W}\}\right) < \frac{\epsilon}{2}.$$

The maps

$$\hat{\gamma}^n : \partial([0,1] \times [0,1]) \ni s \mapsto \Pi_3 F_W(\hat{\tau}, \Pi_3^{-1}(\gamma^n(s))) \in \mathbb{R}^3, \qquad n \in \mathbb{N},$$

satisfy $\hat{\gamma}^n(\partial([0,1] \times [0,1])) \subset \Pi_3 W$ for all $n \in \mathbb{N}$. Using $x^n \in \gamma^n(\partial([0,1] \times [0,1]))$ for all $n \in \mathbb{N}$ and $x^n \to x$ as $n \to \infty$ we deduce

$$\text{dist}\,(\hat{\gamma}^n(\partial([0,1] \times [0,1])), \hat{x}) \to 0 \qquad \text{as } n \to \infty.$$

Select $\beta_2 \in (0, \frac{\epsilon}{2})$ so that

$$c_- < c_P\left(\Pi_3^{-1}(y)\right) < c_+ \qquad \text{for all } y \in \left(\hat{x} + \mathbb{R}^3_{\beta_2}\right) \cap \Pi_3 \overline{W}.$$

Choose $n_0 \in \mathbb{N}$ such that $\gamma^n(\partial([0,1] \times [0,1])) \subset x + \mathbb{R}^3_{\beta_1}$ and $\hat{\gamma}^n(\partial([0,1] \times [0,1])) \subset \hat{x} + \mathbb{R}^3_{\beta_2}$ for all integers $n \ge n_0$. Arguments similar to those used in case 2 from the proof of Theorem 16.7 permit to find an integer $n \ge n_0$ and a continuous map

$$\hat{\Gamma}^n : [0,1] \times [0,1] \to \{\Pi_3\phi : \phi \in W,\ c_- < c_P(\phi) < c_+\}$$

with range in $\hat{x} + \mathbb{R}^3_{\beta_2}$ which coincides with $\hat{\gamma}^n$ on $\partial([0,1] \times [0,1])$. Consider the map $\tilde{\Gamma}^n : [0,1] \times [0,1] \to \mathbb{R}^3$ given by

$$\tilde{\Gamma}^n\left(ts + (1-t)\left(\frac{1}{2}, \frac{1}{2}\right)^{tr}\right) = \hat{\Gamma}^n\left(2ts + (1-2t)\left(\frac{1}{2}, \frac{1}{2}\right)^{tr}\right)$$

for $0 \le t \le \frac{1}{2}$, $s \in \partial([0,1] \times [0,1])$, and

$$\tilde{\Gamma}^n\left(ts + (1-t)\left(\frac{1}{2}, \frac{1}{2}\right)^{tr}\right) = \Pi_3 F_W\left(2(1-t)\hat{\tau}, \Pi_3^{-1}(\gamma^n(s))\right)$$

for $\frac{1}{2} < t \le 1$, $s \in \partial([0,1] \times [0,1])$. Proposition 16.2(iii) and (iv) and $\hat{\tau} < 0$ combined yield $c_- < c_P(F_W(2(1-t)\hat{\tau}, \Pi_3^{-1}(\gamma^n(s)))) < c_+$ for all $t \in \left[\frac{1}{2}, 1\right]$ and all $s \in \partial([0,1] \times [0,1])$. This gives

$$\tilde{\Gamma}^n\left(ts + (1-t)\left(\frac{1}{2}, \frac{1}{2}\right)^{tr}\right) \in \{\Pi_3\phi : \phi \in W,\ c_- < c_P(\phi) < c_+\} = \mathrm{int}(\Pi_3 S_0)$$

for all $t \in \left[\frac{1}{2}, 1\right]$ and all $s \in \partial([0,1] \times [0,1])$. Obviously we also have

$$\tilde{\Gamma}^n\left(ts + (1-t)\left(\frac{1}{2}, \frac{1}{2}\right)^{tr}\right) \in \mathrm{int}(\Pi_3 S_0)$$

for all $t \in \left[0, \frac{1}{2}\right]$ and all $s \in \partial([0,1] \times [0,1])$. By the choices of β_1, β_2 and by $\hat{\Gamma}^n([0,1] \times [0,1]) \subset (\hat{x} + \mathbb{R}^3_{\beta_2})$, $\mathrm{diam}(\tilde{\Gamma}^n([0,1] \times [0,1])) < \epsilon$. The map $\Gamma^n : [0,1] \times [0,1] \to \mathrm{int}(\Pi_3 S_0)$ given by $\tilde{\Gamma}^n$ is a continuous extension of γ^n from $\partial([0,1] \times [0,1])$ to $[0,1] \times [0,1]$. This is a contradiction.

The proof in the remaining subcase $x \in \{\Pi_3\phi : \phi \in \mathrm{bd}\, W,\ c_P(\phi) = c_-\}$ proceeds analogously.

Consequently, $\mathrm{int}(\Pi_3 S_0)$ is locally uniformly 1-connected, and there exists a homeomorphism h_4 from $\Pi_3 Z_0$ onto D^3 with $h_4(\Pi_3 S_0) = S^2$.

5. Using $h_4(\Pi_3 S_0) = S^2$ and $h_3(S_0) = \partial([-1,1] \times D^2)$, we see that the map $h_5 : D^3 \to [-1,1] \times D^2$ given by

$$h_5(ts) = t h_3\left(\Pi_3^{-1}(h_4^{-1}(s))\right) \qquad \text{for } 0 \le t \le 1,\ s \in S^2$$

is a homeomorphism mapping S^2 onto $\partial([-1,1] \times D^2)$.

6. Homeomorphisms from Z_+ onto $[1,\infty) \times D^2$ and from Z_- onto $(-\infty,-1] \times D^2$.

There exist continuous maps $\tau_4 : Z_+ \to \mathbb{R}$ and $\tau_4^- : Z_- \to \mathbb{R}$ with $\tau_4(Z_+) \subset (-\infty, 0]$, $\tau_4^-(Z_-) \subset (-\infty, 0]$ so that for every $\phi \in Z_+$,

$$F_W(\tau_4(\phi), \phi) \in c_P^{-1}(c_+) \quad \text{and} \quad F_W(t,\phi) \notin c_P^{-1}(c_+) \qquad \text{for all } t \in \mathbb{R} \setminus \{\tau_4(\phi)\},$$

and for every $\phi \in Z_-$,

$$F_W(\tau_4^-(\phi), \phi) \in c_P^{-1}(c_-) \quad \text{and} \quad F_W(t,\phi) \notin c_P^{-1}(c_-) \qquad \text{for all } t \in \mathbb{R} \setminus \{\tau_4^-(\phi)\}.$$

The maps $\tilde{h}_6 : Z_+ \to \mathbb{R} \times \mathbb{R}^2$ and $\tilde{h}_6^- : Z_- \to \mathbb{R} \times \mathbb{R}^2$ given by

$$\tilde{h}_6(\phi) = (1 - \tau_4(\phi), \mathrm{Pr}_2\, h_3(F_W(\tau_4(\phi), \phi))) \qquad \text{for } \phi \in Z_+,$$

and

$$\tilde{h}_6^-(\phi) = \left(\tau_4^-(\phi) - 1, \mathrm{Pr}_2\, h_3(F_W(\tau_4(\phi), \phi))\right) \qquad \text{for } \phi \in Z_-,$$

with $\mathrm{Pr}_2 : \mathbb{R} \times \mathbb{R}^2 \ni (t,x) \mapsto x \in \mathbb{R}^2$, are continuous with $\tilde{h}_6(Z_+) \subset [1,\infty) \times D^2$, $\tilde{h}_6^-(Z_-) \subset (-\infty,-1] \times D^2$.

Recall that we have

$$h_3(\overline{W} \cap c_P^{-1}(c_+)) = \{1\} \times D^2 \quad \text{and} \quad h_3(\overline{W} \cap c_P^{-1}(c_-)) = \{-1\} \times D^2.$$

Using Proposition 16.2(iii) and (iv) we infer that the continuous maps

$$g : [1,\infty) \times D^2 \ni (t,x) \mapsto F(t-1, h_3^{-1}(1,x)) \in C$$

and

$$g^- : (-\infty,-1] \times D^2 \ni (t,x) \mapsto F(-t-1, h_3^{-1}(-1,x)) \in C$$

map their domains into Z_+ and into Z_-, respectively. For every $\phi \in Z_+$ we obtain

$$\begin{aligned} g(\tilde{h}_6(\phi)) &= F(-\tau_4(\phi), h_3^{-1}(1, \mathrm{Pr}_2\, h_3(F_W(\tau_4(\phi),\phi)))) \\ &= F(-\tau_4(\phi), h_3^{-1}(h_3(F_W(\tau_4(\phi),\phi)))) \\ &= \phi. \end{aligned}$$

For every $(t,x) \in [1,\infty)\times D^2$ we have $\tilde{h}_6(g(t,x)) = \tilde{h}_6(F(t-1, h_3^{-1}(1,x)))$. Observe that $h_3^{-1}(1,x) \in \overline{W} \cap c_P^{-1}(c_+)$ yields $\tau_4(F(t-1,h_3^{-1}(1,x))) = 1-t$. Using this we get

$$\tilde{h}_6(g(t,x)) = \big(1-(1-t), \mathrm{Pr}_2\, h_3(h_3^{-1}(1,x))\big) = (t,x).$$

It follows that $\tilde{h}_6$ maps Z_+ injectively onto $[1,\infty)\times D^2$, and that the map $h_6 : Z_+ \to [1,\infty)\times D^2$ given by $\tilde{h}_6$ has a continuous inverse. Analogously one sees that $\tilde{h}_6^-$ defines a homeomorphism h_6^- onto $(-\infty,-1]\times D^2$.

7. Finally, consider the map $\tilde{h} : \overline{W} \to \mathbb{R}\times D^2$ given by

$$\begin{aligned} \tilde{h}(\phi) &= h_6^-(\phi) \qquad \text{for } \phi \in \{\psi \in Z_- : c_P(\psi) < c_-\}, \\ \tilde{h}(\phi) &= h_6(\phi) \qquad \text{for } \phi \in \{\psi \in Z_+ : c_P(\psi) > c_+\}, \end{aligned}$$

and

$$\tilde{h}(\phi) = h_5(h_4(\Pi_3\phi)) \qquad \text{for } \phi \in Z_0.$$

The continuity of $\tilde{h}$ in $\overline{W} \setminus c_P^{-1}(\{c_-,c_+\})$ follows from the continuity of the maps $h_4, h_5, h_6, h_6^-, \Pi_3$. At points $\phi \in \overline{W}\cap c_P^{-1}(c_+)$ we have $\tau_4(\phi)=0$, and thereby

$$h_6(\phi) = (1-\tau_4(\phi), \mathrm{Pr}_2\, h_3(F_W(\tau_4(\phi),\phi))) = h_3(\phi).$$

Using $\overline{W}\cap c_P^{-1}(c_+) \subset S_0$ and $h_4(\Pi_3 S_0) = S^2$ we infer $h_4(\Pi_3\phi) \in S^2$. The definition of h_5 yields

$$h_5(s) = h_3\left(\Pi_3^{-1}(h_4^{-1}(s))\right) \qquad \text{for } s \in S^2.$$

Thus

$$\tilde{h}(\phi) = h_5(h_4(\Pi_3\phi)) = h_3\left(\Pi_3^{-1}(h_4^{-1}(h_4(\Pi_3\phi)))\right) = h_3(\phi).$$

Consequently $\tilde{h}(\phi) = h_6(\phi)$. Analogously we get $\tilde{h}(\phi) = h_6^-(\phi)$ for all $\phi \in \overline{W}\cap c_P^{-1}(c_-)$. Now it is easy to show that $\tilde{h}$ is continuous also at the points in $\overline{W}\cap c_P^{-1}(\{c_-,c_+\})$. Using parts 5 and 6 and the equalities

$$\tilde{h}\big|_{Z_+} = h_6, \quad \tilde{h}\big|_{Z_-} = h_6^-, \quad \tilde{h}\big|_{Z_0} = h_5 \circ h_4 \circ \left(\Pi_3\big|_{Z_0}\right)$$

we infer that $\tilde{h}$ is surjective.

We have

$$h_6\left(\overline{W}\cap c_P^{-1}(c_+)\right) = \{1\}\times D^2 \quad \text{and} \quad h_6^-\left(\overline{W}\cap c_P^{-1}(c_-)\right) = \{-1\}\times D^2,$$

which implies that the sets

$$\tilde{h}\left(\{\phi\in Z_+ : c_P(\phi) > c_+\}\right), \quad \tilde{h}\left(Z_0\right), \quad \tilde{h}\left(\{\phi\in Z_- : c_P(\phi) < c_+\}\right)$$

are mutually disjoint. Using this and the injectivity of all maps involved in the definition of $\tilde{h}$ we obtain that $\tilde{h}$ is injective, with

$$\tilde{h}^{-1}(t,x) = \left\{\begin{matrix} g(t,x) \\ \Pi_3^{-1}\left(h_4^{-1}(h_5^{-1}(t,x))\right) \\ g^-(t,x)\end{matrix}\right\} \qquad \text{for all } (t,x)\in \left\{\begin{matrix}(1,\infty)\times D^2 \\ [-1,1]\times D^2 \\ (-\infty,-1)\times D^2\end{matrix}\right\}.$$

Obviously $\tilde{h}^{-1}$ is continuous at all points $(t,x)\in\mathbb{R}\times D^2$ with $|t|\neq 1$. At points $(1,x)\in\{1\}\times D^2$ we deduce from

$$\tilde{h}(\phi) = h_6(\phi) \qquad \text{for all } \phi\in\overline{W}\cap c_P^{-1}(c_+) = g(\{1\}\times D^2)$$

that we have

$$\tilde{h}(g(1,x)) = h_6(g(1,x)) = (1,x),$$

or

$$\tilde{h}^{-1}(1,x) = g(1,x).$$

Analogously we find

$$\tilde{h}^{-1}(-1,x) = g^-(-1,x) \qquad \text{for all } x\in D^2,$$

and it becomes easy to show that $\tilde{h}^{-1}$ is continuous also at the points $(1,x)$, $(-1,x)$ with $x\in D^2$. So, $\tilde{h}$ is a homeomorphism.

The definition of $\tilde{h}$ implies $\tilde{h}(\mathrm{bd}\,W) = \mathbb{R}\times S^1$. Therefore the map $h:\overline{W}\to\mathbb{R}\times D^2$ given by $\tilde{h}$ is a homeomorphism as desired. □

CHAPTER 17

Resumé

We studied the semiflow $F : \mathbb{R}^+ \times C \to C$ of the equation

$$\dot{x}(t) = -\mu x(t) + f(x(t-1)) \tag{1.1}$$

for $\mu \geq 0$ and a C^1-function $f : \mathbb{R} \to \mathbb{R}$ with the properties

(H1) $f(0) = 0$, $f'(\xi) > 0$ for all $\xi \in \mathbb{R}$;

(H2) $\mu = 0$ and $f'(0) > 3\pi/2$, or
$\mu > 0$ and $f'(0) > \mu/\cos\theta_\mu$ for $\theta_\mu \in (3\pi/2, 2\pi)$ with $\theta_\mu = -\mu\tan\theta_\mu$;

(H3) $\mu = 0$ and $-\infty < \inf f$ or $\sup f < \infty$, or
$\mu > 0$ and $f(\xi)/\xi < \mu$ outside a bounded neighbourhood of 0;

(H4) $\mu = 0$, or
$\mu > 0$ and $f'(\xi^-) < \mu$ for the largest negative zero ξ^- of $f - \mu\,\mathrm{id}$, $f'(\xi^+) < \mu$ for the smallest positive zero ξ^+ of $f - \mu\,\mathrm{id}$.

The semiflow F is monotone with respect to the cone K of nonnegative functions $\phi \in C$. The set S of initial data of solutions with unbounded zeroset separates the data for eventually positive solutions from those for eventually negative ones.

The leading eigenvalue λ_0 of the generator of the semigroup $(D_2F(t,0))_{t\geq 0}$, the next pair of eigenvalues $\lambda_1, \overline{\lambda}_1$, and the rest of the spectrum of the generator define realified generalized eigenspaces P, L, Q, respectively; the separatrix S is given by a Lipschitz map Sep: $Q \oplus L \to P$, and there is a 3-dimensional locally positively invariant C^1-submanifold W_{loc} of a local unstable manifold at the stationary point 0, with $T_0 W_{loc} = L \oplus P$. In case $\mu > 0$ the constant functions ξ_- and ξ_+ in C given by the zeros ξ^-, ξ^+ of $f - \mu\,\mathrm{id}$ are stationary points of F.

The next theorem collects our main results about the shape and smoothness of the closure of the forward extension

$$W = F(\mathbb{R}^+ \times W_{loc})$$

of W_{loc} and of the part $\overline{W} \cap S$ of the separatrix S in $\overline{W}$.

Theorem 17.1 *There exist a 3-dimensional space $G_3 \subset C$, a closed complementary space E of G_3 in C, a 2-dimensional subspace $G_2 \subset G_3$ and a 1-dimensional complementary subspace G_1 of G_2 in G_3, a closed neighbourhood D_W of 0 in G_3, a compact neighbourhood D_S of 0 in G_2, and maps*

$$w : D_W \to E, \quad w_S : D_S \to G_1 \oplus E$$

with the following properties.

(i) $\overline{W} = \{\chi + w(\chi) : \chi \in D_W\}$, $W = \{\chi + w(\chi) : \chi \in \overset{\circ}{D}_W\}$.

(ii) *In case* $\mu > 0$ *there exists a homeomorphism from* D_W *onto* D^3 *which maps* ∂D_W *onto* S^2*. The points* ξ_-, ξ_+ *belong to* $\overline{W} \setminus S$*, and*

$$\xi_- = \chi_- + w(\chi_-),\ \xi_+ = \chi_+ + w(\chi_+) \quad \text{with } \chi_-, \chi_+ \text{ in } D_W.$$

The set

$$D_W \setminus \{\chi_-, \chi_+\}$$

is a 3-dimensional C^1*-submanifold of* G *with smooth boundary* $\partial D_W \setminus \{\chi_-, \chi_+\}$*. The map* w *is Lipschitz continuous, and its restriction to* $D_W \setminus \{\chi_-, \chi_+\}$ *is* C^1*-smooth.*

(iii) *In case* $\mu = 0$ *there is a homeomorphism from* D_W *onto the solid cylinder* $\{x \in \mathbb{R}^3 : x_1^2 + x_2^2 \leq 1\}$ *which maps* ∂D_W *onto the cylinder* $\{x \in \mathbb{R}^3 : (x_1, x_2)^{tr} \in S^1\}$*.* D_W *is a 3-dimensional* C^1*-submanifold of* G *with smooth boundary* ∂D_W*, and* w *is* C^1*-smooth.*

(iv) $\overline{W} \cap S = \{\chi + w_s(\chi) : \chi \in D_S\}$, $W \cap S = \{\chi + w_s(\chi) : \chi \in \overset{\circ}{D}_S\}$.

(v) ∂D_S *is the trace of a simple closed* C^1*-curve* ζ*, and* $\overset{\circ}{D}_S = \operatorname{int}(\zeta)$*. There is a homeomorphism from* D_S *onto* D^2 *which maps* ∂D_S *onto* S^1*, and* w_S *is* C^1*-smooth.*

Before we proceed to the invariant stratification of $\overline{W}$ and to the dynamics on $\overline{W}$, let us supplement a simple fact concerning connectedness. In case $\mu > 0$, set $W^s(\xi_+) = \{\phi \in C : F(t,\phi) \to \xi_+ \text{ as } t \to +\infty\}$. Define $W^s(\xi_-)$ analogously. In case $\mu = 0$, set $W^s(\infty) = \{\phi \in C : x^\phi(t) \to \infty \text{ as } t \to \infty\}$. Define $W^s(-\infty)$ analogously.

Proposition 17.2

(i) *For* $\mu > 0$*, the sets* $W^s(\xi_+) \cap \overline{W}$ *and* $W^s(\xi_-) \cap \overline{W}$ *are arcwise connected.*

(ii) *For* $\mu = 0$*, the sets* $W^s(\infty) \cap \overline{W}$ *and* $W^s(-\infty) \cap \overline{W}$ *are arcwise connected.*

Proof Let $\mu > 0$. ξ_+ is above S, according to Proposition 4.5(ii). Using the homeomorphism from $\overline{W}$ onto the closed unit ball in $\mathbb{R}^3$ we obtain a neighbourhood N of ξ^+ in the open subset $\overline{W} \setminus S$ of $\overline{W}$ which is arcwise connected and above S. Proposition 4.5(i) yields $N \subset W^s(\xi_+) \cap \overline{W}$. For every $\phi \in W^s(\xi_+) \cap \overline{W}$ there exists $t \geq 0$ with $F(t,\phi) \in N$. The curve $[0,t] \ni s \mapsto F(s,\phi) \in C$ connects ϕ in $W^s(\xi_+) \cap \overline{W}$ to a point in N. Now it is obvious how to complete the argument. The proof for $W^s(\xi_-) \cap \overline{W}$ is analogous. The proof of assertion (ii) is similar. □

Theorem 17.3 *The solutions* $x : \mathbb{R} \to \mathbb{R}$ *of equation (1.1) with segments* $x_0 \in \overline{W}$ *define a continuous flow*

$$F_{\overline{W}} : \mathbb{R} \times \overline{W} \to \overline{W}.$$

The subsets W, $\operatorname{bd} W = \overline{W} \setminus W$, $\overline{W} \cap S$, $W \cap S$, $\operatorname{bd}(\overline{W} \cap S) = (\overline{W} \cap S) \setminus (W \cap S)$ *are invariant. The restricted flow*

$$\mathbb{R} \times W \ni (t,\phi) \mapsto F_{\overline{W}}(t,\phi) \in W$$

is C^1*-smooth. The set* $\overline{W} \cap S$ *contains the stationary point* 0 *and no other stationary point.* $\overline{W}$ *contains exactly one periodic orbit, namely*

$$\mathcal{O} = \operatorname{bd}(\overline{W} \cap S).$$

(i) In case $\mu > 0$, $\overline{W} \setminus S$ *contains no stationary points except for* ξ_-, ξ_+*. For every* $\phi \in W$, $F_W(t, \phi) \to 0$ *as* $t \to -\infty$*. For every* $\phi \in \operatorname{bd} W \setminus \{\xi_-, \xi_+\}$,

$$F_W(t, \phi) \to \mathcal{O} \qquad \text{as } t \to -\infty.$$

For every $\phi \in \overline{W} \cap S \setminus \{0\}$, $F_W(t, \phi) \to \mathcal{O}$ *as* $t \to \infty$*. The set* $\overline{W} \setminus S$ *is the union of the nonempty, connected and disjoint domains of attraction* $W^s(\xi_+) \cap \overline{W} = \{\phi \in \overline{W} : \phi \text{ above } S\}$ *towards* ξ_+ *and* $W^s(\xi_-) \cap \overline{W} = \{\phi \in \overline{W} : \phi \text{ below } S\}$ *towards* ξ_-*, which are both open in* $\overline{W}$.

(ii) In case $\mu = 0$ *there are no stationary points in* $\overline{W} \setminus S$*. For every* $\phi \in W$, $F_W(t, \phi) \to 0$ *as* $t \to -\infty$*. For every* $\phi \in \operatorname{bd} W$, $F_W(t, \phi) \to \mathcal{O}$ *as* $t \to -\infty$*. For every* $\phi \in \overline{W} \cap S \setminus \{0\}$, $F_W(t, \phi) \to \mathcal{O}$ *as* $t \to \infty$*. The set* $\overline{W} \setminus S$ *is the union of the nonempty, connected and disjoint domains of attraction* $W^s(\infty) \cap \overline{W} = \{\phi \in \overline{W} : \phi \text{ above } S\}$ *towards* ∞ *and* $W^s(-\infty) \cap \overline{W} = \{\phi \in \overline{W} : \phi \text{ below } S\}$ *towards* $-\infty$*, which are both open in* $\overline{W}$.

Recall also the result that local strong unstable manifolds of period maps at points of the periodic orbit $\mathcal{O}$ are one-dimensional, and that in case $\mu > 0$ the set $\operatorname{bd} W \setminus \{\xi_-, \xi_+\}$ coincides with the forward extension $W^u(\mathcal{O})$ of such a local invariant manifold, and $\operatorname{bd} W = W^u(\mathcal{O})$ for $\mu = 0$.

What remains to be done? Directly related to Theorems 17.1 and 17.3 are the following four problems and proposals.

- Can $\overline{W} \cap S$ also be represented by a smooth map from a subset of L into $Q \oplus P$?
- Are there diffeomorphisms from D_W onto the solid cylinder in case $\mu = 0$, and from $D_W \setminus \{\xi_-, \xi_+\}$ onto $\{x \in \mathbb{R}^3 : \sum_{j=1}^{3} x_j^2 \le 1, |x_3| < 1\}$ in case $\mu > 0$?
- In case $\mu > 0$ the shape of $\overline{W}$ close to the singularities ξ_-, ξ_+ should be described. At ξ_+, $T_{\xi_+}\overline{W} = \{0\}$ is expected, and for every C^1-curve $c : (-1, 0] \to C$ with range in $\overline{W}$ and $c(0) = \xi_+, c'(0)$ should be a multiple of the eigenvector $[-1, 0] \ni t \mapsto e^{\lambda_0^+ t} \in \mathbb{R}$ of the leading eigenvalue λ_0^+ of the generator of the semigroup $(D_2 F(t, \xi_+))_{t \ge 0}$.
- Remove the hypothesis (H4).

In the sequel we consider only the case $\mu > 0$. One can show that in this case F has a global attractor in the sense explained in Chapter 1. This is particularly simple if the function f in equation (1.1) is bounded, which is the case in most applications. Let us sketch the proof for the semiflow $F_g : \mathbb{R}^+ \times C \to C$ defined by the segments of the solutions $x : [-1, \infty) \to \mathbb{R}$ of the equation

$$\dot{x}(t) = -\mu x(t) + g(x(t-1)) \tag{17.1}$$

with $\mu > 0$ and with a continuous, injective and bounded function $g : \mathbb{R} \to \mathbb{R}$. F_g is continuous, all maps $F_g(t, \cdot), t \ge 0$, are injective, and $F_g(1, \cdot)$ maps bounded sets into sets with compact closure.

Proposition 17.4 *Let* $r > 1 + (\sup_{\xi \in \mathbb{R}} |g(\xi)|)/\mu$*. Then* $F_g(t, C_r) \subset C_r$ *for all* $t \ge 0$*, and for every* $\phi \in C$ *there exists* $t \ge 0$ *with* $F_g(t, \phi) \in C_r$*.*

Proof 1. We show that for every solution $x:[-1,\infty)\to\mathbb{R}$ of equation (17.1) and for every $t\geq 0$ with $|x(t)|<r$,

$$|x(s)|<r \qquad \text{for all } s>t.$$

Suppose $r\leq x(s)$ for some $s>t$. Then there exists $u\in(t,s]$ with $r\leq x(u)$ and $0\leq \dot{x}(u)$. Observe $\dot{x}(u)\leq -\mu r+\sup_{\xi\in\mathbb{R}}|g(\xi)|\leq -\mu<0$, which yields a contradiction. The argument in case $x(s)\leq -r$ for some $s>t$ is analogous.

2. Let $\phi\in C$. Suppose $F_g(t,\phi)\in C\setminus C_r$ for all $t\geq 0$. Part 1 yields $r\leq \min_{t\geq 0}|x(t)|$ for the solution $x:[-1,\infty)\to\mathbb{R}$ of equation (17.1) with $x_0=\phi$. It follows that $r\leq x(t)$ for all $t\geq 0$, or $x(t)\leq -r$ for all $t\geq 0$. In the first case, $\dot{x}(t)\leq -\mu r+\sup_{\xi\in\mathbb{R}}|g(\xi)|\leq -\mu$ for all $t>0$, and we obtain a contradiction. The argument in the second case is analogous. □

The existence of a global attractor A for the semiflow F_g is a consequence of Proposition 17.4 and of the next proposition which is a special case of abstract results from **[28]** about continuous semiflows $\Phi:\mathbb{R}^+\times M\to M$ on complete metric spaces M. Suppose such a semiflow is given, each map $\Phi(t,\cdot), t\geq 0$, is injective, $\Phi(1,\cdot)$ maps bounded sets into sets with compact closure, and there is a positively invariant bounded open set $B\subset M$ so that for every $y\in M$ there exists $t\geq 0$ with $\Phi(t,y)\in B$.

Proposition 17.5 *The set $A=\bigcap_{t\geq 0}\overline{\Phi(t,B)}$ is compact. For every $t\geq 0$, $\Phi(t,A)=A$. For every bounded set $X\subset M$ and for every open set $U\subset M$ with $A\subset U$ there exists $t\geq 0$ with*

$$\Phi(s,X)\subset U \qquad \textit{for all } s\geq t.$$

Proof 1. A is closed. For $s\geq t\geq 0$, $\Phi(s,B)=\Phi(t,\Phi(s-t,B))\subset\Phi(t,B)$, and thereby $\overline{\Phi(s,B)}\subset\overline{\Phi(t,B)}$. Hence

$$A=\bigcap_{t\geq s}\overline{\Phi(t,B)} \qquad \text{for all } s\geq 0,$$

and A as well as all $\overline{\Phi(t,B)}$, $t\geq 1$, are contained in $\overline{\Phi(1,B)}$, and are compact.

Observe that for all $t\geq 0$ and all $N\subset M$, $\Phi(t,\overline{N})\subset\overline{\Phi(t,N)}$. For $t\geq 0$ and $s\geq 1$ we have that

$$\Phi(t,\overline{\Phi(s,B)})\supset\Phi(t,\Phi(s,B))$$

is compact, and it follows that

$$\Phi(t,\overline{\Phi(s,B)})\supset\overline{\Phi(t,\Phi(s,B))}, \quad \Phi(t,\overline{\Phi(s,B)})=\overline{\Phi(t,\Phi(s,B))}.$$

For every open subset $U\supset A$ of M there is $t\geq 0$ with $\overline{\Phi(t,B)}\subset U$ since otherwise there would be an open set $V\supset A$ so that the nonempty compact sets $(M\setminus V)\cap\overline{\Phi(t,B)}$, $t\geq 1$, have a nonempty intersection equal to $(M\setminus V)\cap A$, in contradiction to $A\subset V$.

2. Let $t \geq 0$. Then

$$\Phi(t,A) = \Phi(t, \bigcap_{s\geq 1} \overline{\Phi(s,B)}) \subset \bigcap_{s\geq 1} \Phi(t, \overline{\Phi(s,B)})$$
$$= \bigcap_{s\geq 1} \overline{\Phi(t,\Phi(s,B))} = \bigcap_{s\geq 1} \overline{\Phi(t+s,B)} = \bigcap_{u\geq t+1} \overline{\Phi(u,B)} = A.$$

Proof of $A \subset \Phi(t,A)$: For every $t' \geq 0$,

$$A = \bigcap_{s\geq 1} \Phi(t', \overline{\Phi(s,B)}) \subset \Phi(t', \overline{\Phi(1,B)}) \subset \Phi(t', \overline{B}).$$

Let $y \in A$. For every $t' \geq 0$ there exists $y_{t'} \in \overline{B}$ with $y = \Phi(t', y_{t'})$. We show $y_t \in A$: Let $u \geq 0$. Set $t' = u + t$. We have $\Phi(t, y_t) = y = \Phi(t, \Phi(t'-t, y_{t'}))$. By injectivity,

$$y_t = \Phi(t'-t, y_t) \in \Phi(u, \overline{B}) \subset \overline{\Phi(u,B)}.$$

It follows that $y_t \in \bigcap_{u\geq 0} \overline{\Phi(u,B)} = A$.

3. Let $X \subset M$ be bounded, and $t \geq 0$. We show $\Phi(s,X) \subset \Phi(t,B)$ for all $s \geq 0$ outside a bounded set. The absorption property of B and openness imply that for each y in the compact set $\overline{\Phi(1,X)}$ there are an open neighbourhood U_y of y and $t_y \geq 0$ with $\Phi([t_y,\infty) \times U_y) \subset B$. There is a covering of $\overline{\Phi(1,x)}$ by sets U_y with y in a finite set $Y \subset \overline{\Phi(1,x)}$. For $u \geq \max_{y\in Y} t_y$, $\Phi(u, \Phi(1,X)) \subset B$. Consequently, $s \geq t + \max_{y\in Y} t_y + 1$ implies

$$\Phi(s,X) = \Phi(t, \Phi(s-t-1, \Phi(1,X))) \subset \Phi(t,B).$$

□

Recall from the introduction that $\overline{W} \subset \overline{W^u}$ is a subset of the global attractor A of the semiflow F of equation (1.1), and that we expect equality

$$A = \overline{W}$$

for bounded functions f which satisfy $\mathrm{Re}\lambda_2 < 0$, $f'(\xi) < f'(\eta)$ for $0 < \eta < \xi$, and $f'(\eta) < f'(\xi)$ for $\eta < \xi < 0$, in addition to (H1)–(H4). A proof would involve a new uniqueness result for periodic solutions of delay differential equations since the global attractor contains all ω-limit sets, in particular, all periodic orbits, and in $\overline{W}$ there is exactly one periodic orbit.

How may the global attractor be organized in case $\overline{W}$ is a proper subset of A? Let us first consider situations where A is a subset of the order interval $[\xi_-, \xi_+] = \{\phi \in C : \xi_- \leq \phi \leq \xi_+\}$. The segments of the solutions $x : \mathbb{R} \to \mathbb{R}$ with $V(x_t) = 2$ for all $t \in \mathbb{R}$ may form, together with the stationary point 0, a smooth disk-like submanifold A_2 in A which extends $\overline{W} \cap S$ beyond $\mathcal{O}$ and contains at least one additional periodic orbit, forming the manifold boundary. In this case, the unstable sets of the periodic orbits in A_2 should contain analogues of $\mathrm{bd}\, W \setminus \{\xi_-, \xi_+\}$, namely 2-dimensional invariant submanifolds given by heteroclinic connections from the periodic orbit to ξ_- and ξ_+. These submanifolds should subdivide the 3-dimensional subset $A_{\leq 2}$ of A formed by 0 and the segments of all solutions $x : \mathbb{R} \to \mathbb{R}$ with

$\xi^- \leq x(t) \leq \xi^+$ and $V(x_t) \leq 2$ for all $t \in \mathbb{R}$ into invariant layers; a section of $A_{\leq 2}$ containing ξ_- and ξ_+ might have the structure shown by a sliced onion. It may be that $A = A_{\leq 2}$. If A is strictly larger than $A_{\leq 2}$ then we have to expect a finite number of analogues of the smooth disk A_2, given by higher even values of V, and a more complicated variety of heteroclinic connections, between periodic orbits in the same disk, between periodic orbits in different disks, from periodic orbits and 0 to ξ_- and ξ_+, from 0 to periodic orbits and to ξ_- and ξ_+. Also connections from periodic orbits to 0 become possible. A Morse decomposition of A similar to the one constructed by Mallet-Paret **[42]** should be useful to describe a part of these heteroclinic connections.

Suppose now that A is not confined to the order interval $[\xi_-, \xi_+]$. Then there may exist zeros ξ^* of $f - \mu\,\mathrm{id}$ below ξ^- and above ξ^+ with $f'(\xi^*) > \mu$; one out of many possibilities is that the part of A in a certain neighbourhood of a stationary point $\xi_* \in C$ given by such a value ξ^* looks just as we began to sketch it for the case $A \subset [\xi_-, \xi_+]$.

Let us emphasize that the main results obtained in the preceding chapters and discussed here rely on the hypothesis that the feedback function f in equation (1.1) is monotone. In the class of equations of the same form, with $f(0) = 0$ and $f'(0) > 0$, but f not necessarily increasing everywhere, other phenomena are known to exist – for example, chaotic motion **[63, 65]**. There is recent evidence that non-monotone feedback occurs in single neurons of living beings and small assemblies of neurons **[13]**. See also **[24, 47, 58]** for problems with non-monotone feedback.

A large number of papers (see, e.g., **[15, 31, 38]** and references therein) considered nonautonomous generalizations of equation (1.1), and proved asymptotic behaviours like stability properties of steady states. It would be interesting to describe asymptotic behaviours of solutions for nonautonomous equations analogous to the above results.

APPENDIX I

Equivalent norms, invariant manifolds, Poincaré maps and asymptotic phases

We shall need an equivalent norm of a given Banach space in terms of the spectral decomposition of a given bounded linear operator.

Theorem I.1 *Consider a Banach space E and a bounded linear operator $L : E \to E$. Assume that there are closed subspaces E_1, E_2 of E with $E = E_1 \oplus E_2$, $LE_1 \subset E_1$, $LE_2 \subset E_2$, and that the spectra σ_i of the induced maps $L_i : E_i \ni x \mapsto Lx \in E_i$, $i \in \{1,2\}$, satisfy*

$$\sup_{\lambda\in\sigma_1} |\lambda| < \inf_{\lambda\in\sigma_2} |\lambda|.$$

Let $b = \sup_{\lambda\in\sigma_1} |\lambda|$ and $a = \inf_{\lambda\in\sigma_2} |\lambda|$. Then for every $\epsilon > 0$ there exists a norm $|\cdot|$, equivalent to the originally given one, such that

$$|\operatorname{Pr}_{E_1}| = |\operatorname{Pr}_{E_2}| = 1,$$

and for all $x \in E$,

$$\begin{aligned} |x| &= |\operatorname{Pr}_{E_1} x| + |\operatorname{Pr}_{E_2} x|, \\ |L\operatorname{Pr}_{E_1} x| &\le (b+\epsilon)|\operatorname{Pr}_{E_1} x|, \\ |L\operatorname{Pr}_{E_2} x| &\ge (a-\epsilon)|\operatorname{Pr}_{E_2} x|. \end{aligned}$$

We refer to Iooss **[37]** for a proof of this result.

We now consider an open subset U of a Banach space E with norm $||\cdot||$ and a C^1-map $f : U \to E$ with a fixed point p. Let σ denote the spectrum of $L = Df(p)$ and assume that E has the decomposition $E = E_s \oplus E_c \oplus E_u$ such that each of E_s, E_c, E_u is a closed subspace of E and $L(E_s) \subset E_s$, $L(E_c) \subset E_c$, $L(E_u) \subset E_u$. We assume that the spectra σ_s, σ_c, σ_u of the maps $L_s : E_s \ni x \mapsto Lx \in E_s$, $L_c : E_c \ni x \mapsto Lx \in E_c$, $L_u : E_u \ni x \mapsto Lx \in E_u$ are contained in $\mathbb{C}_1$, $S^1_{\mathbb{C}}$, $\{\mu \in \mathbb{C} : 1 < |\mu|\}$, respectively. Let $E_{sc} = E_s \oplus E_c$, $E_{cu} = E_c \oplus E_u$. Then the maps $L_{sc} : E_{sc} \ni x \mapsto Lx \in E_{sc}$, $L_{cu} : E_{cu} \ni x \mapsto Lx \in E_{cu}$ have spectra $\sigma_s \cup \sigma_c$ and $\sigma_c \cup \sigma_u$, respectively. Choose $\lambda \in (0,1)$ so that

$$\lambda > \max\left\{ \sup_{\mu\in\sigma_s} |\mu|, \sup_{\mu\in\sigma_u} \frac{1}{|\mu|} \right\}.$$

By Theorem I.1, there exist a norm $|\cdot|_s$ on E_s and a norm $|\cdot|_{cu}$ on E_{cu}, which are equivalent to the restrictions of the given norm $||\cdot||$ on E, so that

$$|L_s| = \sup_{|x|_s \le 1} |L_s x|_s < \lambda,$$

$$|L_{cu}^{-1}| = \sup_{|x|_{cu} \le 1} |L_{cu}^{-1} x|_{cu} < \frac{1}{\lambda},$$

and the norm on E defined by $||x||_s = \max\{|\operatorname{Pr}_{E_s} x|_s, |\operatorname{Pr}_{E_c \oplus E_u} x|_{cu}\}$ is equivalent to $||\cdot||$. Analogously, there exist a norm $|\cdot|_u$ on E_u and a norm $|\cdot|_{sc}$ on E_{sc}, which are equivalent to the restrictions of the given norm $||\cdot||$ on E, so that

$$|L_u^{-1}| = \sup_{|x|_u \le 1} |L_u^{-1} x|_s < \lambda,$$

$$|L_{sc}| = \sup_{|x|_{sc} \le 1} |L_{sc} x|_{sc} < \frac{1}{\lambda},$$

and the norm on E defined by $||x||_u = \max\{|\operatorname{Pr}_{E_u} x|_u, |\operatorname{Pr}_{E_s \oplus E_c} x|_{sc}\}$ is equivalent to $||\cdot||$.

Theorem I.2 *There exist $\alpha_s \in (0, \lambda)$, an open neighbourhood U^s of p in E, convex open neighbourhoods N_s of 0 in E_s and N_{cu} of 0 in E_{cu}, and a C^1-map $w_s : N_s \to E_{cu}$ such that the following properties hold.*

(i) $N^s = N_s + N_{cu}$ is a subset of $U - p$, $w_s(N_s) \subset N_{cu}$, $w_s(0) = 0$, $Dw_s(0) = 0$.

(ii) The set $W^s = W^s(p, f, N^s) = \{p + x + w_s(x) : x \in N_s\}$ is equal to the set $\{x \in p + N^s :$ There is a trajectory $(x_n)_{n=0}^{\infty}$ of f such that $x_0 = x$, $\lambda^{-n}(x_n - p) \in N^s$ for all $n \in \mathbb{N}$, and $\lambda^{-n}(x_n - p) \to 0$ as $n \to \infty\}$.

(iii) $f(W^s) \subset W^s$.

(iv) $||f(x) - f(y)||_s \le \alpha_s ||x - y||_s$ for all x, y in W^s, in particular, $\lambda^{-n}||f^n(x) - p||_s \to 0$ as $n \to \infty$ for all $x \in W^s$.

(v) $Df^n(x)v \to 0$ as $n \to \infty$ for all $x \in W^s$ and $v \in T_x W^s$.

(vi) For every trajectory $(x_n)_{n=0}^{\infty}$ of f in U^s such that $\lambda^{-n}(x_n - p) \in N^s$ for all $n \in \mathbb{N}$, $x_0 \in W^s$.

In case the fixed point p of f is hyperbolic, i.e.,

$$\sigma \cap S^1_{\mathbb{C}} = \emptyset,$$

Theorem I.2 holds with $E_c = \{0\}$ and with the statements

(ii') The set $W^s = W^s(p, f, N^s) = \{p + x + w_s(x) : x \in N_s\}$ is equal to the set $\{x \in p + N^s :$ There is a trajectory $(x_n)_0^{\infty}$ of f in $p + N^s$ with $x_0 = x$ and $x_n \to p$ as $n \to \infty\}$

and

(vi') For every trajectory $(x_n)_0^{\infty}$ of f in $p + N^s$, $x_0 \in W^s$

instead of (ii) and (vi), respectively.

Theorem I.3 *There exist $\alpha_u \in (0, \lambda)$, an open neighbourhood U^u of p in E, convex open neighbourhoods N_u of 0 in E_u and N_{sc} of 0 in E_{sc}, and a C^1-map $w_u : N_u \to E_{sc}$ such that the following properties hold.*

(i) $N^u = N_{sc} + N_u$ is a subset of $U - p$, $w_u(N_u) \subset N_{sc}$, $w_u(0) = 0$, $Dw_u(0) = 0$.

(ii) The set $W^u = W^u(p, f, N^u) = \{p + x + w_u(x) : x \in N_u\}$ is equal to the set $\{x \in p + N^u :$ There is a trajectory $(x_n)_{-\infty}^{0}$ of f such that $x_0 = x$, $\lambda^n(x_n - p) \in N^u$ for all $n \in -\mathbb{N}$, and $\lambda^n(x_n - p) \to 0$ as $n \to -\infty\}$.

(iii) *There is an open neighbourhood* $\widetilde{N^u} \subset N^u$ *of* 0 *in* E *such that* $f|_{W^u \cap (p+\widetilde{N^u})}$ *defines a* C^1*-diffeomorphism* f_u *onto* W^u *whose inverse satisfies*

$$||f_u^{-1}(x) - f_u^{-1}(y)||_u \leq \alpha_u ||x-y||_u \qquad \text{for all } x, y \text{ in } W^u.$$

(iv) *For every trajectory* $(x_n)_{-\infty}^0$ *of* f *in* U^u *with* $\lambda^n(x_n - p) \in N^u$ *for all* $n \in -\mathbb{N}$, $x_0 \in W^u$.

In case p is hyperbolic Theorem I.3 holds with $E_c = \{0\}$ and with the statements

(ii') The set $W^u = W^u(p, f, N^u) = \{p + x + w_u(x) : x \in N_u\}$ is equal to the set $\{x \in p + N^u :$ There is a trajectory $(x_n)_{-\infty}^0$ of f in $p + N^u$ with $x_0 = x$ and $x_n \to p$ as $n \to -\infty\}$

and

(iv') For every trajectory $(x_n)_{-\infty}^0$ of f in $p + N^u$, $x_0 \in W^u$

instead of (ii) and (iv), respectively.

The C^1-submanifolds W^s and W^u are called strong stable manifold of f at p and strong unstable manifold of f at p, respectively. For proofs of versions of Theorems I.2 and I.3 for the hyperbolic case $\sigma_c = \emptyset$ and $E_c = \{0\}$, see Section 3 in **[29]** and the very detailed exposition in **[48]**. In order to derive, say, Theorem I.2, replace f by the map $x \mapsto f(x+p) - p$ and follow the proof of Theorem 2.7 in **[48]**; work in the sequence space $E(\lambda) = \{(x_n)_0^\infty \in E^{\mathbb{N}} : \lambda^{-n} x_n \to 0 \text{ as } n \to \infty\}$ with norm given by $||(x_n)_0^\infty|| = \sup_{n \in \mathbb{N}} ||\lambda^{-n} x_n||_s$, instead of the sequence space used in **[48]**. The neighbourhoods N_s and N_{cu} are obtained as open balls with respect to the norm $||\cdot||_s$.

The aforementioned proof can also be modified to obtain the leading unstable manifolds which are used in the main body of our work. Consider f and p as before. Assume that there exist real numbers $a > 1$ and $b > a$ and closed subspaces $E_{<a}$ and $E_{b<}$ of E such that $E = E_{<a} \oplus E_{b<}$, $LE_{<a} \subset E_{<a}$, $LE_{b<} \subset E_{b<}$, and the spectra $\sigma_{<a}$ and $\sigma_{b<}$ of the induced maps $L_{<a} : E_{<a} \ni x \mapsto Lx \in E_{<a}$, $L_{b<} : E_{b<} \ni x \mapsto Lx \in E_{b<}$ satisfy $\sigma_{<a} \subset \mathbb{C}_a$, $\sigma_{b<} \subset \{z \in \mathbb{C} : b < |z|\}$, respectively. Fix $\beta \in (a, b)$. Then by Theorem I.1 there exists a norm $|\cdot|$ equivalent to the originally given one in E such that

$$|L_{b<}^{-1}| < \beta^{-1}, \quad |L_{<a}| < \beta.$$

For such β and any fixed open neighbourhood V of 0, define the β-unstable set of the fixed point p as

$$\begin{aligned} W_\beta(V) = \{x \in V + p :\ & \text{There is a trajectory } (x_n)_{-\infty}^0 \text{ of } f \text{ with } x_0 = x, \\ & \beta^{-n}(x_n - p) \in V \text{ for all } n \in -\mathbb{N}, \text{ and } \beta^{-n}(x_n - p) \to 0 \text{ as } n \to -\infty\}. \end{aligned}$$

Theorem I.4 *There exist open convex neighbourhoods* $N_{b<}$ *of* 0 *in* $E_{b<}$, $N_{<a}$ *of* 0 *in* $E_{<a}$ *and a* C^1*-map* $w : N_{b<} \to E_{<a}$ *with the following properties.*

(i) $N_{<a} + N_{b<} \subset U - p$, $w(N_{b<}) \subset N_{<a}$, $w(0) = 0$, $Dw(0) = 0$.

(ii) $W_\beta(N_{b<} + N_{<a}) = \{p + x + w(x) : x \in N_{b<}\}$.

(iii) *For every trajectory* $(x_n)_{-\infty}^0$ *of* f *in* U *so that* $\beta^{-n}(x_n - p) \in N_{b<} + N_{<a}$ *for all* $n \in -\mathbb{N}$, $\beta^{-n}(x_n - p) \to 0$ *as* $n \to -\infty$.

Let $S : \mathbb{R}^+ \times E \to E$ be a continuous semiflow, and let $y : \mathbb{R} \to E$ be a phase curve (i.e., $y(t+s) = S(t, y(s))$ for all $t \geq 0$ and $s \in \mathbb{R}$) which is not constant and periodic. Let $\omega > 0$ be a period of y. Assume S is continuously differentiable on $(t_0, \infty) \times E$ for some $t_0 \in (0, \omega)$. We recall the construction and basic properties of Poincaré return maps associated with y and ω. Let a closed linear subspace $H \subset E$ of codimension 1 be given so that $y'(\omega) = D_1S(\omega, y(0))1 \in E \setminus H$. Choose a continuous linear functional $e^* : E \to \mathbb{R}$ with

$$(e^*)^{-1}(0) = H.$$

The relation $e^*y'(\omega) \neq 0$ permits to apply the Implicit Function Theorem to the equation

$$e^*(S(t,x) - y(0)) = 0 \qquad \text{for } t \in \mathbb{R}^+,\ x \in E$$

in a neighbourhood of its solution $(\omega, y(0))$. It follows that there exist a convex bounded open neighbourhood N of $y(0)$ in E, $\epsilon \in (0, \omega - t_0)$, and a C^1-map $\tau : N \to \mathbb{R}$ with $\tau(y(0)) = \omega$ and $\tau(N) \subset (\omega - \epsilon, \omega + \epsilon)$ so that for every $(t, x) \in (\omega - \epsilon, \omega + \epsilon) \times N$, $S(t, x) \in y(0) + H$ if and only if $t = \tau(x)$. Moreover one can achieve that $D_1S(\tau(x), x)1 \in E \setminus H$ for all $x \in N$, and that $D\tau$ is bounded. The map

$$P_\tau : N \ni x \mapsto S(\tau(x), x) \in E$$

and the Poincaré map

$$P : N \cap (y(0) + H) \ni x \mapsto S(\tau(x), x) \in y(0) + H$$

are C^1-smooth, and $y(0)$ is a fixed point of P and P_τ. For every $x \in N$ the projection $\mathrm{Pr}_x : E \to E$ along $\mathbb{R}D_1S(\tau(x), x)1$ onto H satisfies

$$\mathrm{Pr}_x\,\eta = \eta - \frac{e^*(\eta)}{e^*(D_1S(\tau(x), x)1)} D_1S(\tau(x), x)1 \qquad \text{for all } \eta \in E.$$

Using the equation $e^*(S(\tau(x), x) - y(0)) = 0$ one computes

$$DP(x)\eta = \mathrm{Pr}_x\, D_2S(\tau(x), x)\eta$$

for all $x \in N \cap (y(0) + H)$ and $\eta \in T_x(N \cap (y(0) + H)) = H$, and

$$DP_\tau(x)\eta = \mathrm{Pr}_x\, D_2S(\tau(x), x)\eta \qquad \text{for all } x \in N \text{ and } \eta \in E.$$

In particular,

$$DP(y(0))\eta = \mathrm{Pr}_{y(0)}\, D_2S(\omega, y(0))\eta = D_2S(\omega, y(0))\eta - \frac{e^*(D_2S(\omega, y(0))\eta)}{e^*(y'(\omega))} y'(\omega)$$

for every $\eta \in H$, and

$$DP_\tau(y(0))\eta = D_2S(\omega, y(0))\eta - \frac{e^*(D_2S(\omega, y(0))\eta)}{e^*(y'(\omega))} y'(\omega) \quad \text{for all } \eta \in E.$$

It is also convenient to have a formula for derivatives of iterates of P. Let j be a positive integer. For x in the domain of P^j, define

$$\tau_j(x) = \sum_{k=0}^{j-1} \tau(P^k(x)).$$

Using the equation $e^*(S(\tau_j(x),x)-y(0)) = e^*(P^j(x)-y(0)) = 0$ one finds

$$DP^j(x)\eta = \mathrm{Pr}_{P^{j-1}(x)} D_2S(\tau_j(x),x)\eta \qquad \text{for all } \eta \in H.$$

Suppose the linearized period map, or monodromy operator, $M = D_2S(\omega, y(0))$ is compact. The nonzero points in the spectrum σ^M of M are called Floquet multipliers. The results of Section XIV.4 in **[22]** apply to σ^M and to the spectrum σ^P of the linearization $DP(y(0)) : H \to H$ of the Poincaré map. We have $1 \in \sigma^M$ and $My'(\omega) = y'(\omega)$, and deduce from Theorem XIV.4.5 the following result.

Proposition I.5

(i) $\sigma^P \setminus \{0,1\} = \sigma^M \setminus \{0,1\}$, *and for every* $\lambda \in \sigma^M \setminus \{0,1\}$ *the map* $\mathrm{Pr}_{y(0)}$ *defines an isomorphism from the realified generalized eigenspace of* λ *and* M *onto the realified generalized eigenspace of* λ *and* $DP(y(0))$.

(ii) If the generalized eigenspace $G(1,M)$ *associated with* $1 \in \sigma^M$ *and* M *is one-dimensional then* $1 \notin \sigma^P$.

(iii) If $1 < \dim G(1,M)$ *then* $1 \in \sigma^P$, *and the realified generalized eigenspaces* $G_{\mathbb{R}}(1,M)$ *and* $G_{\mathbb{R}}(1,P)$ *associated with* 1 *and* M *and with* 1 *and* $DP(y(0))$, *respectively, satisfy*

$$G_{\mathbb{R}}(1,P) = H \cap G_{\mathbb{R}}(1,M),$$
$$G_{\mathbb{R}}(1,M) = \mathbb{R}y'(\omega) \oplus G_{\mathbb{R}}(1,P).$$

For the next result we assume in addition that y is hyperbolic, i.e., $\sigma^M \cap S^1_{\mathbb{C}} = \{1\}$, and the generalized eigenspace associated with the Floquet multiplier 1 and M has dimension 1. We can apply Theorem I.2 in the case where $p = y(0)$ and f is the restriction of $S(\omega,\cdot)$ to a neighbourhood U of $y(0)$ in N so that f is C^1, or in the case where $p = y(0)$ and $f = P$, to obtain strong stable manifolds

$$W^s = W^s(y(0), S(\omega,\cdot), N^s), \quad W^s_P = W^s(y(0), P, N^s_H) \subset y(0) + H$$

with convex open neighbourhoods N^s of 0 in E and N^s_H of 0 in H. The following asymptotic phase theorem is a reformulation of Proposition 3.3 in **[41]**.

Theorem I.6 *Suppose, in addition, that* y *is Lipschitz continuous and that there exists a constant* $c > 0$ *such that*

$$||S(t,x) - S(t,\tilde{x})|| \le c||x-\tilde{x}|| \qquad \text{for all } x, \tilde{x} \text{ in } N \text{ and } t \in [0, \omega+\epsilon).$$

Then there exist a continuous map $\alpha : W^s_P \to \mathbb{R}$ *and an open neighbourhood* V *of* $y(0)$ *in* W^s_P *such that*

(i) $\alpha(y(0)) = 0$;

(ii) for every $x \in W^s_P$, $S(t,x) - y(t+\alpha(x)) \to 0$ *as* $t \to \infty$;

(iii) for every $x \in V$, $\alpha(x) < \omega$ *and* $S(\omega - \alpha(x), x) \in W^s$.

Suppose ω is the minimal period of y, i.e., $y(t) \ne y(s)$ for $0 \le t < s < \omega$, and let

$$\mathcal{O} = \{y(t) : 0 \le t \le \omega\}.$$

Proposition I.7 *Let* $x \in E$ *be given so that* $S(t,x) \to \mathcal{O}$ *as* $t \to \infty$. *Then for every neighbourhood* U *of* $y(0)$ *in* E *there exist* $t \ge 0$ *and a trajectory* $(x_n)_0^\infty$ *of* P *in* $(y(0)+H) \cap N \cap U$ *so that* $x_0 = S(t,x)$ *and* $x_n \to y(0)$ *as* $n \to \infty$.

Proof Choose $\tilde{\epsilon} \in \left(0, \frac{\omega}{2}\right)$ so small that for $0 < |s| < \tilde{\epsilon}$, $y(s) = y(s+\omega) \notin y(0) + H$. By the minimality of ω, $y(0) \notin \{y(t) : \tilde{\epsilon} \le t \le \omega - \tilde{\epsilon}\}$. Therefore, there are disjoint neighbourhoods U_0 of $y(0)$ in $N \cap U$ and $V_{\tilde{\epsilon}}$ of the compact set $\{y(t) : \tilde{\epsilon} \le t \le \omega - \tilde{\epsilon}\}$. Choose a neighbourhood V_0 of $y(0)$ in U_0 so that $\overline{V_0} \subset U_0$ and $S(\tau(z), z) \in U_0$ for all $z \in V_0$. For every $t \in (-\tilde{\epsilon}, \tilde{\epsilon}) \setminus \{0\}$ there exists a neighbourhood V^t of $y(t)$ in $E \setminus (y(0) + H)$. The set $V = V_0 \cup V_{\tilde{\epsilon}} \cup (\cup_{0<|t|<\tilde{\epsilon}} V^t)$ is a neighbourhood of $\mathcal{O}$, and there exists $t_V \ge 0$ so that $S(t,x) \in V$ for all $t \ge t_V$.

The sequence $(S(n,x))_0^\infty$ has a subsequence $(S(n_j,x))_0^\infty$ which converges to some $y(s)$, $0 \le s < \omega$. Then $S(n_j + \omega - s, x) \to y(\omega) = y(0)$ as $j \to \infty$, and hence for some $j \in \mathbb{N}$ with $n_j + \omega - s \ge t_V$, $x_* = S(n_j + \omega - s, x)$ belongs to V_0. Consequently, $x_0 = S(\tau(x_*), x_*) \in U_0 \cap (y(0) + H)$ and therefore, $x_0 \notin V_{\tilde{\epsilon}}$. Clearly, $x_0 \notin V^t$ for $0 < |t| < \tilde{\epsilon}$. Using $x_0 \in V$ we obtain $x_0 \in V_0$. Induction yields a sequence $(x_n)_0^\infty$ in $V_0 \subset U_0 \cap (y(0) + H) \subset (y(0) + H) \cap N \cap U$ with $x_{n+1} = P(x_n)$ for all $n \in \mathbb{N}$.

Suppose the sequence does not converge to $y(0)$. Then there are a subsequence $(x_{n_k})_0^\infty$ and $\delta > 0$ with $\delta \le ||y(0) - x_{n_k}||$ for all $k \in \mathbb{N}$. Convergence of $S(\cdot, x)$ to the compact set $\mathcal{O}$ implies that there is a subsequence $(x_{n_{k_m}})_0^\infty$ which converges to some $y(t)$ with $0 < t < \omega$. Necessarily, $y(t) \in (y(0) + H) \cap \overline{V_0}$. It follows that $\tilde{\epsilon} \le t \le \omega - \tilde{\epsilon}$ and $y(t) \notin V_{\tilde{\epsilon}}$, a contradiction. □

The last result of this appendix is not related to invariant manifolds and Poincaré maps. It is useful in the main body of our work to abbreviate proofs of smoothness of submanifolds.

Lemma I.8 *Let g be a C^1-map from a finite-dimensional C^1-manifold M into a C^1-manifold X modelled over a Banach space. Let $p \in M$. If $Dg(p)$ is injective then there exists a neighbourhood N of p so that for all open sets $U \subset N$, $g(U)$ is a C^1-submanifold of X.*

Proof Choose a manifold chart k at $g(p)$, from an open neighbourhood V of $g(p)$ onto an open set in a Banach space E, with $k(g(p)) = 0$. The derivative of the C^1-map $g^{-1}(V) \ni q \mapsto k(g(q)) \in E$ at p maps T_pM isomorphically onto a finite-dimensional subspace F of E with a closed complementary space G of F in E. Also the derivative of $h : g^{-1}(V) \ni q \mapsto \mathrm{Pr}_F(k(g(q))) \in E$ at p is injective. It follows that there is a neighbourhood N of p in M so that h maps open subsets U of N injectively onto open subsets of F, with a C^1-inverse $h_U^{-1} : h(U) \to M$. Let $U \subset N$ be open. The restriction of Pr_F to $k(g(U))$ is injective. This means that there is a map $r : h(U) \to G$ with $k(g(U)) = \{x + r(x) : x \in h(U)\}$. We have

$$r(x) = (\mathrm{id} - \mathrm{Pr}_F)k(g(h_U^{-1}(x))) \qquad \text{for all } x \in h(U),$$

which implies that r is C^1-smooth. So $k(g(U))$ is a C^1-submanifold of E, and it follows that $g(U)$ is a C^1-submanifold of X. □

APPENDIX II

Smooth center-stable manifolds for C^1-maps

Let $f : U \to E$ be a C^1-map on an open subset U of a Banach space E over $\mathbb{R}$, with a fixed point p. Let $L = Df(p)$ and assume that E has the following decomposition

$$E = E_s \oplus E_c \oplus E_u,$$

where $E_s \neq \{0\}$ is a closed subspace, $E_c \neq \{0\}$ and $E_u \neq \{0\}$ are finite-dimensional, $L(E_s) \subset E_s$, $L(E_c) \subset E_c$ and $L(E_u) \subset E_u$. We further assume that the spectrum σ_s of the induced map $L_s : E_s \ni x \mapsto Lx \in E_s$ is contained in a compact subset of $\{z \in \mathbb{C} : |z| < 1\}$, and that the spectra σ_c and σ_u of the induced maps $L_c : E_c \ni x \mapsto Lx \in E_c$ and $L_u : E_u \ni x \mapsto Lx \in E_u$ are contained in $S^1_{\mathbb{C}}$ and $\{z \in \mathbb{C} : |z| > 1\}$, respectively. Let $E_{sc} = E_s \oplus E_c$, $E_{cu} = E_c \oplus E_u$.

In the sequel, we give a detailed proof of the following result on a center-stable manifold for f at p.

Theorem II.1 *There exist convex open bounded neighbourhoods N_{sc} of 0 in E_{sc}, N_u of 0 in E_u, N of p in U, and a C^1-map*

$$w : N_{sc} \to E_u$$

with $w(0) = 0$, $Dw(0) = 0$ and $w(N_{sc}) \subset N_u$ so that the shifted graph

$$W = p + \{z + w(z) : z \in N_{sc}\}$$

satisfies

$$f(W \cap N) \subset W$$

and

$$\bigcap_{n=0}^{\infty} f^{-n}(p + N_{sc} + N_u) \subset W.$$

In order to prove Theorem II.1, we follow the approach used to obtain center manifolds for flows of vector fields and semiflows of functional differential equations in Vanderbauwhede and van Gils **[59]** and Diekmann, van Gils, Verduyn Lunel and Walther **[22]** as closely as possible. For another recent result, which requires additional conditions, see **[17]**.

Set

$$a = \sup_{\lambda \in \sigma_s} |\lambda|, \quad b = \inf_{\lambda \in \sigma_u} |\lambda|.$$

Then $a < 1 < b$. Fix $\epsilon > 0$ with

$$a + \epsilon < 1, \quad 1 + \epsilon < b - \epsilon.$$

Whenever convenient, we shall use abbreviations like

$$x_s = \operatorname{Pr}_{E_s} x,\ x_c = \operatorname{Pr}_{E_c} x,\ x_u = \operatorname{Pr}_{E_u} x,\ x_{sc} = \operatorname{Pr}_{E_{sc}} x,\ x_{cu} = \operatorname{Pr}_{E_{cu}} x.$$

By Theorem I.1, there exists a norm $|\cdot|$ on E which is equivalent to the originally given one and satisfies

$$\begin{aligned}
&|x| = |x_s| + |x_c| + |x_u|,\\
&|L \operatorname{Pr}_{E_s} x| \le (a+\epsilon)|\operatorname{Pr}_{E_s} x|,\\
&|L \operatorname{Pr}_{E_c} x| \le (1+\epsilon)|\operatorname{Pr}_{E_c} x|,\\
&|L \operatorname{Pr}_{E_u} x| \ge (b-\epsilon)|\operatorname{Pr}_{E_u} x|
\end{aligned}$$

for all $x \in E$. Then

$$|\operatorname{Pr}_{E_s}| = \sup_{|x|\le 1} |\operatorname{Pr}_{E_s} x| = 1.$$

Similarly, $|\operatorname{Pr}_{E_c}| = |\operatorname{Pr}_{E_u}| = 1$.

Set $V = U - p$. Consider the transformed map

$$g^* : V \ni x \mapsto f(x+p) - p \in E$$

with fixed point 0 and $Dg^*(0) = L$. Define the reminder map $r^* : V \to E$ by

$$r^*(x) = g^*(x) - Lx.$$

We extend r^* to a map $r : E \to E$ by $r(x) = 0$ for all $x \in E \setminus V$, and define

$$g = L + r.$$

Next, we construct modifications of g which are Lipschitz continuous and smooth on strips containing the center-unstable space E_{cu}. Fix a norm $|\cdot|_{cu}$ on E_{cu} which is C^1-smooth on $E_{cu} \setminus \{0\}$. This is possible since E_{cu} is finite-dimensional. The norm $||\cdot|| : E \ni x \mapsto \max\{|x_s|, |x_{cu}|_{cu}\} \in \mathbb{R}$ is equivalent to $|\cdot|$. For $\delta > 0$, set $E(\delta) = \{x \in E : ||x|| < \delta\}$. Choose a C^∞-function $\rho : \mathbb{R} \to \mathbb{R}$ with

$$\begin{aligned}
&\rho([0,\infty)) \subset [0,1],\\
&\rho(t) = 1 \qquad \text{for } 0 \le t \le 1,\\
&\rho(t) = 0 \qquad \text{for } t \ge 2.
\end{aligned}$$

For every $\delta > 0$, define

$$r_\delta : E \to E$$

by

$$r_\delta(x) = \rho\left(\frac{|x_{cu}|_{cu}}{\delta}\right) \rho\left(\frac{|x_s|}{\delta}\right) r(x),$$

and set

$$g_\delta = L + r_\delta.$$

Fix $\delta_0 > 0$ so that $\overline{E(2\delta_0)} \subset V$. Observe that for every $\delta \in (0, \delta_0)$, the restriction $r_\delta|_{\{x\in E:|x_s|<\delta\}}$ is given by

$$\rho\left(\frac{|x_{cu}|_{cu}}{\delta}\right) r(x);$$

it follows that $r_\delta|_{\{x\in E:|x_s|<\delta\}}$ is C^1-smooth.

Proposition II.2 *There exist $\delta_1 \in (0, \delta_0)$ and a nondecreasing function $\lambda : [0, \delta_1] \to [0, 1]$ with $\lim_{\delta \to 0+} \lambda(\delta) = 0 = \lambda(0)$ so that for each $\delta \in (0, \delta_1]$ and for all x, y in E,*

$$|r_\delta(x)| \leq \delta\lambda(\delta)$$

and

$$|r_\delta(x) - r_\delta(y)| \leq \lambda(\delta)|x - y|.$$

Proof 1. Recall that $r(0) = 0$, and that $r|_V$ is continuously differentiable with $Dr(0) = 0$. It follows that there exist $\alpha_0 \in (0, 2\delta_0)$ and a nondecreasing function $\lambda_0 : [0, \alpha_0] \to [0, 1]$ with $\lambda_0(0) = 0 = \lim_{\delta \to 0+} \lambda_0(\delta)$ so that

$$\{x \in E : ||x|| < \alpha_0\} \subset V,$$

and

$$||r(x) - r(y)|| \leq \lambda_0(\alpha)||x - y||$$

for all $\alpha \in (0, \alpha_0]$ and x, y in E with $||x|| < \alpha$, $||y|| < \alpha$.

Set $\delta_1 = \frac{\alpha_0}{4}$. Let $\delta \in (0, \delta_1]$ and $x \in E$. In case $||x|| \geq 2\delta$, either $|x_{cu}|_{cu} \geq 2\delta$ or $|x_s| \geq 2\delta$. In both subcases, $r_\delta(x) = 0$. In case $||x|| < 2\delta$, we have $||x|| < \alpha_0$ and thus $x \in V$, and it follows that

$$||r_\delta(x)|| \leq ||r(x)|| \leq \lambda_0(2\delta)||x|| \leq \lambda_0(2\delta)2\delta.$$

2. Let $x \in E$, $y \in E$. Observe that

$$\begin{aligned}||r_\delta(x) - r_\delta(y)|| \leq& ||r(x) - r(y)||\rho\left(\frac{|x_{cu}|_{cu}}{\delta}\right)\rho\left(\frac{|x_s|}{\delta}\right) \\ &+ ||r(y)||\left|\rho\left(\frac{|x_{cu}|_{cu}}{\delta}\right)\rho\left(\frac{|x_s|}{\delta}\right) - \rho\left(\frac{|y_{cu}|_{cu}}{\delta}\right)\rho\left(\frac{|y_s|}{\delta}\right)\right|.\end{aligned}$$

In case $||x|| \geq 2\delta$ and $||y|| \geq 2\delta$, we argue as in part 1 and get

$$||r_\delta(x) - r_\delta(y)|| = 0.$$

In case $||x|| \geq 2\delta$ and $||y|| < 2\delta$, it follows that

$$\begin{aligned}||r_\delta(x) - r_\delta(y)|| \leq& \lambda_0(2\delta)2\delta\left\{\left|\rho\left(\frac{|x_s|}{\delta}\right) - \rho\left(\frac{|y_s|}{\delta}\right)\right|\rho\left(\frac{|x_{cu}|_{cu}}{\delta}\right)\right. \\ &\left.+ \rho\left(\frac{|y_s|}{\delta}\right)\left|\rho\left(\frac{|x_{cu}|_{cu}}{\delta}\right) - \rho\left(\frac{|y_{cu}|_{cu}}{\delta}\right)\right|\right\} \\ \leq& \lambda_0(2\delta)2\delta\left\{\sup_{t\geq 0}|\rho'(t)|\frac{||x_s| - |y_s||}{\delta} + \sup_{t\geq 0}|\rho'(t)|\frac{||x_{cu}|_{cu} - |y_{cu}|_{cu}|}{\delta}\right\} \\ \leq& 4\lambda_0(2\delta)\sup_{t\geq 0}|\rho'(t)|\,||x - y||.\end{aligned}$$

The case $||x|| < 2\delta$ and $||y|| < 2\delta$ is similar; one finds the estimate

$$||r_\delta(x) - r_\delta(y)|| \leq \lambda_0(2\delta)||x - y|| + 4\lambda_0(2\delta)\sup_{t\geq 0}|\rho'(t)|\,||x - y||.$$

3. The equivalence of $||\cdot||$ and $|\cdot|$ and the estimates in parts 1 and 2 show how to define λ. □

It will be convenient to introduce the vector space $E^{\mathbb{N}}$ and to consider the substitution operator

$$R_\delta : E^{\mathbb{N}} \to E^{\mathbb{N}} \text{ given by } R_\delta(\xi) = (r_\delta(x_n))_0^\infty \text{ for } \xi = (x_n)_0^\infty \in E^{\mathbb{N}}.$$

In the next steps, we construct, for δ sufficiently small and for some $\eta \in (1+\epsilon, b-\epsilon)$, Lipschitz continuous maps $w_{\delta\eta} : E_{sc} \to E_u$ so that the set $W_{\delta\eta} = \{z + w_{\delta\eta}(z) : z \in E_{sc}\}$ is positively invariant under g_δ and consists of all points x whose trajectories $(g_\delta^n(x))_0^\infty$ do not grow faster than $(\eta^n)_0^\infty$ as $n \to \infty$. It will be important to observe that the projections into E_s of such trajectories are small provided the initial point $x = x_s + x_c + w_{\delta\eta}(x_s + x_c)$ has a small component $x_s \in E_s$.

For $\eta > 0$, let E_η denote the Banach space of all sequences $\xi = (x_n)_0^\infty \in E^{\mathbb{N}}$ with

$$\sup_{j\in\mathbb{N}} |x_j|\eta^{-j} < \infty$$

and norm

$$|\xi|_\eta = \sup_{j\in\mathbb{N}} |x_j|\eta^{-j}.$$

Following the Lyapunov-Perron approach, we shall formulate and solve a fixed point equation for trajectories of g_δ in E_η whose initial point has a given projection in E_{sc}. The map $w_{\delta\eta}$ will then easily be found from the fixed point.

It is convenient to begin with the existence and uniqueness of solutions $\xi \in E_\eta$ of the initial value problem

$$\begin{cases} x_{n+1} = Lx_n + f_n & \text{for } n \in \mathbb{N}, \\ \mathrm{Pr}_{E_{sc}} x_0 = z \end{cases} \tag{II.1}$$

for given $z \in E_{sc}$, $\phi = (f_n)_0^\infty \in E_\eta$ and $\eta \in (1+\epsilon, b-\epsilon)$.

The map L_u is a topological isomorphism. Observe that for every $\eta \in (1+\epsilon, b-\epsilon)$, for every $\phi = (f_j)_0^\infty \in E_\eta$ and for every $n \in \mathbb{N}$, $\sum_{j=n}^\infty L_u^{n-j-1} \mathrm{Pr}_{E_u} f_j$ is convergent in E with

$$\begin{aligned} \left| \sum_{j=n}^\infty L_u^{n-j-1} \mathrm{Pr}_{E_u} f_j \right| &\le \sum_{j=n}^\infty (b-\epsilon)^{n-j-1} |f_j| \\ &\le \sum_{j=n}^\infty (b-\epsilon)^{n-j-1} \eta^j |\phi|_\eta \\ &= \frac{\eta^n}{b-\epsilon} \frac{1}{1-\frac{\eta}{b-\epsilon}} |\phi|_\eta \\ &= \eta^n \frac{1}{b-\epsilon-\eta} |\phi|_\eta. \end{aligned}$$

Let L_{sc} denote the map $E_{sc} \ni x \mapsto Lx \in E_{sc}$.

Proposition II.3 *Let $z \in E_{sc}$, $1+\epsilon < \eta < b-\epsilon$ and $\phi \in E_\eta$, and suppose $\xi \in E_\eta$ satisfies (II.1). Then*

$$x_n = \sum_{j=0}^{n-1} L_{sc}^{n-j-1} \mathrm{Pr}_{E_{sc}} f_j - \sum_{j=n}^\infty L_u^{n-j-1} \mathrm{Pr}_{E_u} f_j + L_{sc}^n z \qquad \textit{for } 1 \le n \in \mathbb{N}$$

and

$$x_0 = z - \sum_{j=0}^{\infty} L_u^{-j-1} \operatorname{Pr}_{E_u} f_j.$$

Proof An application of $\operatorname{Pr}_{E_{sc}}$ to the difference equation yields

$$\operatorname{Pr}_{E_{sc}} x_n = \sum_{j=0}^{n-1} L_{sc}^{n-j-1} \operatorname{Pr}_{E_{sc}} f_j + L_{sc}^n z \qquad \text{for } 1 \le n \in \mathbb{N}.$$

An application of Pr_{E_u} and L_u^{-1} yields

$$\operatorname{Pr}_{E_u} x_n = L_u^{-1}(\operatorname{Pr}_{E_u} x_{n+1} - \operatorname{Pr}_{E_u} f_n) \qquad \text{for each } n \in \mathbb{N},$$

and iterating one finds

$$\operatorname{Pr}_{E_u} x_n = L_u^{-j} \operatorname{Pr}_{E_u} x_{n+j} - \sum_{i=n}^{n+j-1} L_u^{n-i-1} \operatorname{Pr}_{E_u} f_i \text{ for all } n \in \mathbb{N} \text{ and all integers } j \ge 1.$$

The estimate

$$|L_u^{-j} \operatorname{Pr}_{E_u} x_{n+j}| \le \left(\frac{1}{b-\epsilon}\right)^j |x_{n+j}| \le \left(\frac{1}{b-\epsilon}\right)^j \eta^{n+j} |\xi|_\eta$$

with $\frac{\eta}{b-\epsilon} < 1$ and the considerations before Proposition II.3 show that the term on the right-hand side in the equation for $\operatorname{Pr}_{E_u} x_n$ converges to $-\sum_{i=n}^{\infty} L_u^{n-i-1} \operatorname{Pr}_{E_u} f_i$ as $j \to \infty$. □

In particular, given $z \in E_{sc}$ and $\phi \in E_\eta$ there is at most one solution of (II.1) in E_η. In order to show existence, we consider the map

$$K : \{\xi \in E^{\mathbb{N}} : \xi \in E_\eta \text{ for some } \eta \in (1+\epsilon, b-\epsilon)\} \to E^{\mathbb{N}}$$

given by

$$(K\phi)_n = \sum_{j=0}^{n-1} L_{sc}^{n-j-1} \operatorname{Pr}_{E_{sc}} f_j - \sum_{j=n}^{\infty} L_u^{n-j-1} \operatorname{Pr}_{E_u} f_j \qquad \text{for } 1 \le n \in \mathbb{N}$$

and

$$(K\phi)_0 = -\sum_{j=0}^{\infty} L_u^{-j-1} \operatorname{Pr}_{E_u} f_j.$$

For $\eta \in (1+\epsilon, b-\epsilon)$, set

$$c(\eta) = \frac{1}{\eta - 1 - \epsilon} + \frac{1}{b - \epsilon - \eta}.$$

An estimate similar to the one preceding Proposition II.3 shows that for every $\eta \in (1+\epsilon, b-\epsilon)$, for every $\phi \in E_\eta$, and for all $j \in \mathbb{N}$,

$$|(K\phi)_j| \eta^{-j} \le c(\eta) |\phi|_\eta.$$

It follows that the linear map

$$K_\eta : E_\eta \to E_\eta$$

given by

$$K_\eta \phi = K\phi$$

is continuous with

$$|K_\eta| \leq c(\eta).$$

Observe that for every $z \in E_{sc}$ the sequence $\zeta = (L_{sc}^n z)_0^\infty$ satisfies

$$|L_{sc}^n z| \leq (1+\epsilon)^n |z| \qquad \text{for all } n \in \mathbb{N}.$$

So $\zeta \in E_\eta$ for each $\eta \in (1+\epsilon, b-\epsilon)$ with $|\zeta|_\eta \leq |z|$.

Proposition II.4 *For every $\eta \in (1+\epsilon, b-\epsilon)$, for every $z \in E_{sc}$ and for every $\phi \in E_\eta$, the sequence*

$$\xi = K_\eta \phi + (L_{sc}^n z)_0^\infty \in E_\eta$$

solves (II.1).

Proof For $1 \leq n \in \mathbb{N}$,

$$\begin{aligned} Lx_n + f_n &= L((K_\eta\phi)_n + L_{sc}^n z) + f_n \\ &= \sum_{j=0}^{n-1} L_{sc}^{n-j} \operatorname{Pr}_{E_{sc}} f_j - \sum_{j=n}^{\infty} L_u^{n-j} \operatorname{Pr}_{E_u} f_j + L_{sc}^{n+1} z + f_n \\ &= \sum_{j=0}^{n} L_{sc}^{n-j} \operatorname{Pr}_{E_{sc}} f_j - \sum_{j=n+1}^{\infty} L_u^{n-j} \operatorname{Pr}_{E_u} f_j + L_{sc}^{n+1} z \\ &= (K_\eta\phi)_{n+1} + L_{sc}^{n+1} z = x_{n+1}, \end{aligned}$$

and for $n = 0$,

$$\begin{aligned} Lx_0 + f_0 &= L((K_\eta\phi)_0 + L_{sc}^0 z) + f_0 \\ &= -\sum_{j=0}^{\infty} L_u^{-j} \operatorname{Pr}_{E_u} f_j + L_{sc} z + f_0 \\ &= -\sum_{j=1}^{\infty} L_u^{-j} \operatorname{Pr}_{E_u} f_j + L_{sc} z + \operatorname{Pr}_{E_{sc}} f_0 \\ &= (K_\eta\phi)_1 + L_{sc} z = x_1. \end{aligned}$$

□

For every $\eta \in (1+\epsilon, b-\epsilon)$, choose $\delta_\eta \in (0, \delta_1]$ with

$$\lambda(\delta_\eta) c(\eta) < 1.$$

Let $\eta \in (1+\epsilon, b-\epsilon)$ and $\delta \in (0, \delta_\eta)$. Observe that for each $\xi \in E_\eta$ and $\xi^* = (x_j^*)_0^\infty \in E_\eta$ and for all $j \in \mathbb{N}$,

$$\eta^{-j} |r_\delta(x_j) - r_\delta(x_j^*)| \leq \eta^{-j} \lambda(\delta) |x_j - x_j^*| \leq \lambda(\delta) |\xi - \xi^*|_\eta.$$

Therefore, $R_\delta(E_\eta) \subset E_\eta$, and the induced map

$$r_{\delta\eta} : E_\eta \ni \xi \mapsto R_\delta(\xi) \in E_\eta$$

is Lipschitz continuous with Lipschitz constant $\lambda(\delta)$.

Now we are ready to look for trajectories of g_δ in E_η, for given projections $z \in E_{sc}$ of the initial point.

Using Propositions II.3 and II.4, we conclude that for every $z \in E_{sc}$ and $\xi \in E_\eta$ the properties

$$x_{n+1} = g_\delta(x_n) \quad \text{for all } n \in \mathbb{N}, \quad \mathrm{Pr}_{E_{sc}}\, x_0 = z$$

are equivalent to the fixed point equation

$$\xi = K_\eta(r_{\delta\eta}(\xi)) + (L^j_{sc} z)_0^\infty .$$

The map

$$T_{\delta\eta} : E_\eta \times E_{sc} \to E_\eta$$

given by

$$T_{\delta\eta}(\xi, z) = K_\eta(r_{\delta\eta}(\xi)) + (L^j_{sc} z)_0^\infty$$

satisfies

$$|T_{\delta\eta}(\xi, z) - T_{\delta\eta}(\xi^*, z)|_\eta \le c(\eta)\lambda(\delta)|\xi - \xi^*|_\eta$$

for all ξ, ξ^* in E_η and for all $z \in E_{sc}$. So for every $z \in E_{sc}$ there is exactly one fixed point $\xi_{\delta\eta}(z) \in E_\eta$ of the contraction

$$T_{\delta\eta}(\cdot, z) : E_\eta \to E_\eta .$$

We have

$$\mathrm{Pr}_{E_{sc}}(\xi_{\delta\eta}(z))_0 = [K_\eta(r_{\delta\eta}(\xi_{\delta\eta}(z)))]_0 + z = z$$

since $[K(\phi)]_0 \in E_u$ for all ϕ in the domain of K. It follows that $\xi \in E_\eta$ is a trajectory of g_δ with $\mathrm{Pr}_{E_{sc}}\, x_0 = z$ if and only if $\xi = \xi_{\delta\eta}(z)$.

The map

$$\xi_{\delta\eta} : E_{sc} \ni z \mapsto \xi_{\delta\eta}(z) \in E_\eta$$

is Lipschitz continuous since the fixed point equations for $T_{\delta\eta}(\cdot, z)$, $T_{\delta\eta}(\cdot, z^*)$ yield

$$\begin{aligned}
&|\xi_{\delta\eta}(z) - \xi_{\delta\eta}(z^*)|_\eta \\
&\qquad = \sup_{j\ge 0} \eta^{-j}\{|[K_\eta(r_{\delta\eta}(\xi_{\delta\eta}(z)))]_j - [K_\eta(r_{\delta\eta}(\xi_{\delta\eta}(z^*)))]_j| + |L^j_{sc}(z - z^*)|\} \\
&\qquad \le |K_\eta(r_{\delta\eta}(\xi_{\delta\eta}(z))) - K_\eta(r_{\delta\eta}(\xi_{\delta\eta}(z^*)))|_\eta + \sup_{j\ge 0}\left(\frac{1+\epsilon}{\eta}\right)^j |z - z^*| \\
&\qquad \le c(\eta)\lambda(\delta)|\xi_{\delta\eta}(z) - \xi_{\delta\eta}(z^*)|_\eta + |z - z^*|,
\end{aligned}$$

which implies

$$|\xi_{\delta\eta}(z) - \xi_{\delta\eta}(z^*)|_\eta \le \frac{1}{1 - c(\eta)\lambda(\delta)}|z - z^*|$$

for z, z^* in E_{sc}. Consequently,

$$w_{\delta\eta} : E_{sc} \ni z \mapsto \mathrm{Pr}_{E_u}(\xi_{\delta\eta}(z))_0 \in E_u$$

is Lipschitz continuous. Moreover, from $T_{\delta\eta}(0,0) = 0$, we get $\xi_{\delta\eta}(0) = 0$ and thereby

$$w_{\delta\eta}(0) = 0.$$

The graph

$$W_{\delta\eta} = \{z + w_{\delta\eta}(z) : z \in E_{sc}\}$$

coincides with the set

$$\{x \in E : (g^j_\delta(x))_0^\infty \in E_\eta\}$$

which is positively invariant with respect to g_δ. Recall $E(\delta) \subset V$, and $g_\delta(x) = g^*(x)$ for all $x \in E(\delta)$. It follows that

$$g^*(W_{\delta\eta} \cap E(\delta)) \subset W_{\delta\eta}.$$

Initial points of trajectories $(x_j)_0^\infty$ of g with $x_j \in E(\delta)$ for all $j \in \mathbb{N}$ define trajectories of g_δ which belong to E_η. Consequently,

$$\bigcap_{j=0}^{\infty} g^{-j}(E(\delta)) \subset W_{\delta\eta}.$$

The next result is crucial for the proof of differentiability of $w_{\delta\eta}$.

Proposition II.5 *If $0 < \delta < \delta_\eta$ and $\lambda(\delta) < \frac{(1-a-\epsilon)^2}{2}$, then for every $z \in E_{sc}$ with $|\operatorname{Pr}_{E_s} z| < \frac{\delta}{2}$ and for all $j \in \mathbb{N}$,*

$$|\operatorname{Pr}_{E_s}(\xi_{\delta\eta}(z))_j| < \frac{\delta}{2}.$$

Proof The fixed point equation for $\xi_{\delta\eta}(z)$ yields

$$(\xi_{\delta\eta}(z))_n = \sum_{j=0}^{n-1} L_{sc}^{n-j-1} \operatorname{Pr}_{E_{sc}}[r_{\delta\eta}(\xi_{\delta\eta}(z))]_j - \sum_{j=n}^{\infty} L_u^{n-j-1} \operatorname{Pr}_{E_u}[r_{\delta\eta}(\xi_{\delta\eta}(z))]_j + L_{sc}^n z$$

for $1 \leq n \in \mathbb{N}$. It follows that

$$\begin{aligned} |\operatorname{Pr}_{E_s}(\xi_{\delta\eta}(z))_n| &\leq \sum_{j=0}^{n-1} (a+\epsilon)^{n-j-1} |r_\delta((\xi_{\delta\eta}(z))_j)| + (a+\epsilon)^n |\operatorname{Pr}_{E_s} z| \\ &\leq \frac{1}{1-a-\epsilon}\delta\lambda(\delta) + (a+\epsilon)\frac{\delta}{2} < \frac{\delta}{2}. \end{aligned}$$

For $n = 0$ we have

$$|\operatorname{Pr}_{E_s}(\xi_{\delta\eta}(z))_0| = |\operatorname{Pr}_{E_s} \operatorname{Pr}_{E_{sc}}(\xi_{\delta\eta}(z))_0| = |\operatorname{Pr}_{E_s} z| < \frac{\delta}{2}.$$

□

We now interrupt the investigation of the maps $\xi_{\delta\eta}$ and $w_{\delta\eta}$ and introduce three abstract results which will later be employed to show that $w_{\delta\eta}$ is C^1-smooth for certain η and δ. The first of these results deals with smoothness of substitution operators between E_η and $E_{\tilde{\eta}}$ with $\eta < \tilde{\eta}$.

Lemma II.6 *Let a bounded map $q : \mathcal{O} \to E$, $\mathcal{O} \subset E$, be given. Consider the substitution operator*

$$\hat{q} : \mathcal{O}^{\mathbb{N}} \to E^{\mathbb{N}},\ \hat{q}(\xi) = (q(x_n))_0^\infty \qquad \textit{for } \xi = (x_n)_0^\infty \in \mathcal{O}^{\mathbb{N}}.$$

(i) If $0 < \eta$ and $1 < \tilde{\eta}$, then $\hat{q}(\mathcal{O}^{\mathbb{N}} \cap E_\eta) \subset E_{\tilde{\eta}}$.

(ii) If $\mathcal{O}$ is open, if q is C^1-smooth with Dq bounded, and if $0 < \eta < \tilde{\eta}$, then for every $\xi \in \mathcal{O}^{\mathbb{N}}$, the linear map

$$A_q(\xi) : E^{\mathbb{N}} \ni \hat{\xi} \mapsto (Dq(x_j)\hat{x}_j)_0^\infty \in E^{\mathbb{N}}$$

satisfies

$$A_q(\xi)E_\eta \subset E_{\tilde{\eta}}, \qquad \sup_{|\hat{\xi}|_\eta \leq 1} |A_q(\xi)\hat{\xi}|_{\tilde{\eta}} \leq \sup_{x \in \mathcal{O}} |Dq(x)|,$$

the induced linear maps

$$A(\xi)_{q\eta\tilde{\eta}} : E_\eta \to E_{\tilde{\eta}}$$

are continuous, and the map

$$A_{q\eta\tilde{\eta}} : \mathcal{O}^{\mathbb{N}} \cap E_\eta \ni \xi \mapsto A(\xi)_{q\eta\tilde{\eta}} \in L(E_\eta, E_{\tilde{\eta}})$$

is continuous.

(iii) If, in addition, $\mathcal{O}$ is convex and $1 < \tilde{\eta}$ then for every $\tilde{\epsilon} > 0$ and $\xi \in \mathcal{O}^{\mathbb{N}} \cap E_\eta$ there exists $\tilde{\delta} > 0$ so that for every $\tilde{\xi} \in \mathcal{O}^{\mathbb{N}} \cap E_\eta$ with $|\xi - \tilde{\xi}|_\eta \leq \tilde{\delta}$,

$$|\hat{q}(\tilde{\xi}) - \hat{q}(\xi) - A_{q\eta\tilde{\eta}}(\xi)[\tilde{\xi} - \xi]|_{\tilde{\eta}} \leq \tilde{\epsilon}|\tilde{\xi} - \xi|_\eta.$$

We remark that the assertion in (iii) differs from the usual notion of differentiability of $\hat{q}$ since $\mathcal{O}^{\mathbb{N}} \cap E_\eta$ is not required to be an open subset of E_η.

Proof 1. Proof of (i). The inclusion follows from

$$|q(x_j)|\tilde{\eta}^{-j} \leq \sup_{x \in \mathcal{O}} |q(x)| \qquad \text{for all } j \in \mathbb{N}.$$

2. Proof of (ii). For $\xi \in \mathcal{O}^{\mathbb{N}}$ and $\hat{\xi} \in E_\eta$,

$$|Dq(x_j)\hat{x}_j|\tilde{\eta}^{-j} \leq \sup_{x \in \mathcal{O}} |Dq(x)||\hat{x}_j|\eta^{-j} \left(\frac{\tilde{\eta}}{\eta}\right)^{-j}.$$

It follows that $A_q(\xi)E_\eta \subset E_{\tilde{\eta}}$, and

$$\sup_{|\hat{\xi}|_\eta \leq 1} |A_q(\xi)\hat{\xi}|_{\tilde{\eta}} \leq \sup_{x \in \mathcal{O}} |Dq(x)|.$$

In order to show the continuity of $A_{q\eta\tilde{\eta}}$, let $\xi \in \mathcal{O}^{\mathbb{N}} \cap E_\eta$ be given and let $\tilde{\epsilon} > 0$. Choose an integer $j_0 \geq 0$ such that for all integers $j \geq j_0$,

$$2 \sup_{x \in \mathcal{O}} |Dq(x)| \left(\frac{\tilde{\eta}}{\eta}\right)^{-j} < \frac{\tilde{\epsilon}}{2}.$$

Next choose $\tilde{\delta} > 0$ such that for all $j \in \{0, 1, \ldots, j_0\}$ and for all $y \in \mathcal{O}$ with $|y - x_j| < \tilde{\delta}\eta^j$,

$$|Dq(y) - Dq(x_j)| < \frac{\tilde{\epsilon}}{2} \left(\frac{\tilde{\eta}}{\eta}\right)^j.$$

Now consider $\hat{\xi} \in \mathcal{O}^{\mathbb{N}} \cap E_\eta$ with $|\hat{\xi} - \xi|_\eta < \tilde{\delta}$ and consider $\tilde{\xi} \in E_\eta$ with $|\tilde{\xi}|_\eta \leq 1$. For all integers $j \geq j_0$,

$$|(Dq(\hat{x}_j) - Dq(x_j))\tilde{x}_j|\tilde{\eta}^{-j} \leq 2 \sup_{x \in \mathcal{O}} |Dq(x)||\tilde{x}_j|\eta^{-j} \left(\frac{\tilde{\eta}}{\eta}\right)^{-j} < \frac{\tilde{\epsilon}}{2}.$$

For all $j \in \{0, \ldots, j_0\}$,

$$|(Dq(\hat{x}_j) - Dq(x_j))\tilde{x}_j|\tilde{\eta}^{-j} \leq |Dq(\hat{x}_j) - Dq(x_j)||\tilde{x}_j|\eta^{-j} \left(\frac{\tilde{\eta}}{\eta}\right)^{-j} < \frac{\tilde{\epsilon}}{2}.$$

It follows that

$$\begin{aligned}
&|A_{q\eta\tilde\eta}(\hat\xi) - A_{q\eta\tilde\eta}(\xi)|_{L(E_\eta,E_{\tilde\eta})}\\
&\qquad = \sup_{\tilde\xi\in E_\eta, |\tilde\xi|_\eta\le 1} |(A_{q\eta\tilde\eta}(\hat\xi) - A_{q\eta\tilde\eta}(\xi))\tilde\xi|_{\tilde\eta}\\
&\qquad = \sup_{\tilde\xi\in E_\eta, |\tilde\xi|_\eta\le 1} \sup_{j\in\mathbb{N}} |Dq(\hat x_j)\tilde x_j - Dq(x_j)\tilde x_j|\tilde\eta^{-j} \le \frac{\tilde\epsilon}{2} < \tilde\epsilon.
\end{aligned}$$

3. Proof of (iii). For all $\xi, \tilde\xi$ in $\mathcal{O}^{\mathbb{N}} \cap E_\eta$ and for all integers $j \ge 0$ we have

$$\begin{aligned}
&|q(\tilde x_j) - q(x_j) - Dq(x_j)(\tilde x_j - x_j)|\tilde\eta^{-j}\\
&\qquad = \left|\int_0^1 [Dq(x_j + t(\tilde x_j - x_j)) - Dq(x_j)](\tilde x_j - x_j)\,dt\right|\tilde\eta^{-j}\\
&\qquad \le \max_{t\in[0,1]} |Dq(x_j + t(\tilde x_j - x_j)) - Dq(x_j)||\tilde x_j - x_j|\eta^{-j}\left(\frac{\tilde\eta}{\eta}\right)^{-j}.
\end{aligned}$$

Let $\tilde\epsilon > 0$ and $\xi \in \mathcal{O}^{\mathbb{N}} \cap E_\eta$ be given. Choose $j_0 \in \mathbb{N}$ so that for all integers $j \ge j_0$,

$$2\sup_{x\in\mathcal{O}} |Dq(x)|\left(\frac{\tilde\eta}{\eta}\right)^{-j} \le \tilde\epsilon.$$

Next choose $\tilde\delta > 0$ so that for all $j \in \{0, \dots, j_0\}$, the ball $B_j = \{y \in E : |y - x_j| < \tilde\delta\eta^j\}$ is contained in $\mathcal{O}$, and for all $y \in B_j$,

$$|Dq(y) - Dq(x_j)|\left(\frac{\tilde\eta}{\eta}\right)^{-j} \le \tilde\epsilon.$$

Let $\tilde\xi \in \mathcal{O}^{\mathbb{N}} \cap E_\eta$ be given with $|\tilde\xi - \xi|_\eta < \tilde\delta$. For all integers $j \ge j_0$, the estimate above yields

$$|q(\tilde x_j) - q(x_j) - Dq(x_j)(\tilde x_j - x_j)|\tilde\eta^{-j} \le 2\sup_{x\in\mathcal{O}} |Dq(x)||\tilde\xi - \xi|_\eta\left(\frac{\tilde\eta}{\eta}\right)^{-j} \le \tilde\epsilon|\tilde\xi - \xi|_\eta.$$

For $j \in \{0, \dots, j_0\}$, we get $|\tilde x_j - x_j| < \tilde\delta\eta^j$ and therefore

$$|q(\tilde x_j) - q(x_j) - Dq(x_j)(\tilde x_j - x_j)|\tilde\eta^{-j} \le \tilde\epsilon\left(\frac{\tilde\eta}{\eta}\right)^{j}|\tilde x_j - x_j|\eta^{-j}\left(\frac{\tilde\eta}{\eta}\right)^{-j} \le \tilde\epsilon|\tilde\xi - \xi|_\eta.$$

It follows that

$$|\hat q(\tilde\xi) - \hat q(\xi) - A_{q\eta\tilde\eta}(\xi)(\tilde\xi - \xi)|_{\tilde\eta} \le \tilde\epsilon|\tilde\xi - \xi|_\eta.$$

□

Observe that for a convex set $\mathcal{O} \subset E$ also the set $\mathcal{O}^{\mathbb{N}} \cap E_\eta$, $0 < \eta$, is convex. Next we consider fixed points of contractions depending on a parameter.

Lemma II.7 *Let Y, Λ be Banach spaces over $\mathbb{R}$ and let an open set $P \subset \Lambda$, a map $h : Y \times P \to Y$ and a constant $\kappa \in [0, 1)$ be given with*

$$|h(y,p) - h(\tilde{y},p)| \le \kappa|y - \tilde{y}|$$

for all $y, \tilde{y}$ in Y and all $p \in P$. Consider a subset $M \subset Y$ and a map $\Phi : P \to M$ so that for every $p \in P$, $\Phi(p)$ is the unique fixed point of $h(\cdot, p) : Y \to Y$. Suppose the restriction $h_0 = h|_{M\times P}$ has a partial derivative

$$D_2h_0 : M \times P \to L(\Lambda, Y)$$

so that for each $y \in M$, the map $D_2h_0(y, \cdot) : P \to L(\Lambda, Y)$ is continuous. Then Φ is locally Lipschitz continuous.

Proof Let $p \in P$. Choose $c > |D_2h_0(\Phi(p), p)|$. The continuity of $D_2h_0(\Phi(p), \cdot)$ permits to find $\delta > 0$ so that each $\tilde{p} \in \Lambda$ with $|\tilde{p} - p| < \delta$ is in P, and

$$c > \max_{t\in[0,1]} |D_2h_0(\Phi(p), p + t(\tilde{p} - p))|.$$

It follows that

$$\begin{aligned}|\Phi(\tilde{p}) - \Phi(p)| &\le |h(\Phi(\tilde{p}), \tilde{p}) - h(\Phi(p), \tilde{p})| + |h(\Phi(p), \tilde{p}) - h(\Phi(p), p)| \\ &\le \kappa|\Phi(\tilde{p}) - \Phi(p)| + \max_{t\in[0,1]} |D_2h_0(\Phi(p), p + t(\tilde{p} - p))||\tilde{p} - p|,\end{aligned}$$

or

$$|\Phi(\tilde{p}) - \Phi(p)| \le \frac{c}{1-\kappa}|\tilde{p} - p|.$$

□

The third abstract result yields C^1-dependence of fixed points on the parameter.

Lemma II.8 *Suppose the hypotheses of Lemma II.7 are fulfilled, and the following hold.*

(i) M is convex.

(ii) There are a Banach space Y_1 over $\mathbb{R}$ and a continuous injective linear map

$$j : Y \to Y_1$$

so that the map $k = j \circ h_0$ is continuously differentiable with respect to Y in the sense that there is a continuous map

$$A : M \times P \to L(Y, Y_1)$$

so that for every $(y, p) \in M \times P$ and every $\epsilon^* > 0$ there exists $\tilde{\delta} > 0$ with

$$|k(\tilde{y}, p) - k(y, p) - A(y, p)(\tilde{y} - y)| \le \epsilon^*|\tilde{y} - y|$$

for all $\tilde{y} \in M$ with $|\tilde{y} - y| \le \tilde{\delta}$.

(iii) There exist maps

$$\begin{aligned}h^{(1)} &: M \times P \to L(Y, Y),\\ h_1^{(1)} &: M \times P \to L(Y_1, Y_1)\end{aligned}$$

such that

$$A(y,p)\hat{y} = jh^{(1)}(y,p)\hat{y} = h_1^{(1)}(y,p)j\hat{y} \qquad \text{on } M \times P \times Y$$

and

$$|h^{(1)}(y,p)| \le \kappa, \ |h_1^{(1)}(y,p)| \le \kappa \qquad \text{on } M \times P.$$

(iv) The map $M \times P \ni (y,p) \mapsto j \circ h^{(1)}(y,p) \in L(Y, Y_1)$ *is continuous.*
Then the map $j \circ \Phi : P \to Y_1$ *is differentiable, with*

$$D(j \circ \Phi)(p) = h_1^{(1)}(\Phi(p), p) \circ D(j \circ \Phi)(p) + j \circ D_2 h_0(\Phi(p), p) \qquad \text{for all } p \in P.$$

If, in addition, $D_2 h_0$ *is continuous then* $j \circ \Phi$ *is* C^1*-smooth.*

Proof 1. Suppose the hypotheses of Lemma II.7 are fulfilled, and (i)-(iv) hold. For every $p \in P$, the map

$$L(\Lambda, Y) \ni B \mapsto h^{(1)}(\Phi(p), p) \circ B + D_2 h_0(\Phi(p), p) \in L(\Lambda, Y)$$

is a contraction. Denote its unique fixed point by $B(p)$.

2. The definition of k implies that k has a partial derivative

$$\begin{aligned} &D_2 k : M \times P \to L(\Lambda, Y_1), \\ &D_2 k(y,p) = j D_2 h_0(y,p) \qquad \text{on } M \times P, \end{aligned}$$

so that for each $y \in M$ the map

$$D_2 k(y, \cdot) : P \to L(\Lambda, Y_1)$$

is continuous.

3. Let $p \in P$. We show that $j \circ B(p) \in L(\Lambda, Y_1)$ is the derivative of $j \circ \Phi$ at p. Choose a convex open neighbourhood N of p in P. For every $\tilde{p} \in N$, the fixed point equation for $B(p)$ and the result of part 2 yield

$$\begin{aligned} &j\Phi(\tilde{p}) - j\Phi(p) - jB(p)(\tilde{p} - p) \\ &\quad = jh(\Phi(\tilde{p}), \tilde{p}) - jh(\Phi(p), p) - jh^{(1)}(\Phi(p), p)B(p)(\tilde{p} - p) \\ &\qquad - jD_2 h_0(\Phi(p), p)(\tilde{p} - p) \\ &\quad = k(\Phi(\tilde{p}), \tilde{p}) - k(\Phi(p), p) - jh^{(1)}(\Phi(p), p)B(p)(\tilde{p} - p) - D_2 k(\Phi(p), p)(\tilde{p} - p) \\ &\quad = k(\Phi(\tilde{p}), \tilde{p}) - k(\Phi(p), \tilde{p}) - jh^{(1)}(\Phi(p), p)B(p)(\tilde{p} - p) \\ &\qquad + k(\Phi(p), \tilde{p}) - k(\Phi(p), p) - D_2 k(\Phi(p), p)(\tilde{p} - p). \end{aligned}$$

Denote the sum after the last equality sign by $T(\tilde{p}, p)$, and set

$$\begin{aligned} R(\tilde{p}, p) = &\int_0^1 \{jh^{(1)}(\Phi(p) + t(\Phi(\tilde{p}) - \Phi(p)), \tilde{p}) - jh^{(1)}(\Phi(p), p)\}(\Phi(\tilde{p}) - \Phi(p))\, dt \\ &+ \int_0^1 \{D_2 k(\Phi(p), p + t(\tilde{p} - p)) - D_2 k(\Phi(p), p)\}(\tilde{p} - p)\, dt. \end{aligned}$$

Observe that the smoothness property of k implies that the curve

$$\gamma : [0,1] \ni t \mapsto k(\Phi(p) + t(\Phi(\tilde{p}) - \Phi(p)), \tilde{p}) \in Y_1$$

connecting $\gamma(0) = k(\Phi(p), \tilde{p})$ to $\gamma(1) = k(\Phi(\tilde{p}), \tilde{p})$ is continuously differentiable, with

$$\begin{aligned} \gamma'(t) &= A(\Phi(p) + t(\Phi(\tilde{p}) - \Phi(p)), \tilde{p})(\Phi(\tilde{p}) - \Phi(p)) \\ &= jh^{(1)}(\Phi(p) + t(\Phi(\tilde{p}) - \Phi(p)), \tilde{p})(\Phi(\tilde{p}) - \Phi(p)) \end{aligned}$$

for all $t \in [0,1]$. It follows that

$$T(\tilde{p},p) = jh^{(1)}(\Phi(p),p)[\Phi(\tilde{p}) - \Phi(p) - B(p)(p-\tilde{p})] + R(\tilde{p},p),$$

and the equation in (iii) yields

$$T(\tilde{p},p) = h_1^{(1)}(\Phi(p),p)(j\Phi(\tilde{p}) - j\Phi(p) - jB(p)(\tilde{p}-p)) + R(\tilde{p},p).$$

Lemma II.7 guarantees that there is an open neighbourhood N^* of p in N so that $\Phi|_{N^*}$ is Lipschitz continuous with a Lipschitz constant c. Let $\tilde{\epsilon} > 0$. Set $\epsilon^* = \frac{1-\kappa}{1+c}\tilde{\epsilon}$. The continuity of Φ and property (iv) combined imply that there exists $\delta^* > 0$ such that $\tilde{p} \in N^*$ and

$$\sup_{t\in[0,1]} |jh^{(1)}(\Phi(p) + t(\Phi(\tilde{p}) - \Phi(p)),\tilde{p}) - jh^{(1)}(\Phi(p),p)| < \epsilon^*$$

for all $\tilde{p} \in \Lambda$ with $|\tilde{p} - p| \le \delta^*$. The continuity of $D_2k(\Phi(p),\cdot)$ permits to find $\tilde{\delta} \in (0,\delta^*)$ with

$$\sup_{t\in[0,1]} |D_2k(\Phi(p),p+t(\tilde{p}-p)) - D_2k(\Phi(p),p)| < \epsilon^*$$

for all $\tilde{p} \in \Lambda$ with $|\tilde{p}-p| \le \tilde{\delta}$. Using $|h_1^{(1)}(\Phi(p),p)| \le \kappa$ and employing all estimates obtained, we find

$$|j\Phi(\tilde{p}) - j\Phi(p) - jB(p)(\tilde{p}-p)| \le \frac{1+c}{1-\kappa}\epsilon^*|\tilde{p}-p| = \tilde{\epsilon}|\tilde{p}-p|$$

for all $\tilde{p} \in \Lambda$ with $|\tilde{p}-p| \le \tilde{\delta}$.

4. Suppose, in addition, that D_2h_0 is continuous. The argument in part 2 shows that D_2k is continuous. In order to prove the continuity of $D(j\circ\Phi) : P \to L(\Lambda, Y_1)$, it remains to deduce that the map

$$P \ni p \mapsto j\circ B(p) \in L(\Lambda, Y_1)$$

is continuous. For all $\tilde{p}$ and p in P, the fixed point equations for $B(\tilde{p})$ and $B(p)$ yield

$$\begin{aligned}
j\circ B(\tilde{p}) - j\circ B(p) &= j\circ h^{(1)}(\Phi(\tilde{p}),\tilde{p})\circ B(\tilde{p}) + j\circ D_2h_0(\Phi(\tilde{p}),\tilde{p}) \\
&\quad - j\circ h^{(1)}(\Phi(p),p)\circ B(p) - j\circ D_2h_0(\Phi(p),p) \\
&= j\circ h^{(1)}(\Phi(\tilde{p}),\tilde{p})\circ B(\tilde{p}) + D_2k(\Phi(\tilde{p}),\tilde{p}) \\
&\quad - j\circ h^{(1)}(\Phi(p),p)\circ B(p) - D_2k(\Phi(p),p) \qquad \text{(see part 2)} \\
&= h_1^{(1)}(\Phi(\tilde{p}),\tilde{p})\circ [j\circ B(\tilde{p}) - j\circ B(p)] \\
&\quad + [j\circ h^{(1)}(\Phi(\tilde{p}),\tilde{p}) - j\circ h^{(1)}(\Phi(\tilde{p}),p)]\circ B(p) \\
&\quad + [j\circ h^{(1)}(\Phi(\tilde{p}),p) - j\circ h^{(1)}(\Phi(p),p)]\circ B(p) \\
&\quad + D_2k(\Phi(\tilde{p}),\tilde{p}) - D_2k(\Phi(p),p).
\end{aligned}$$

It follows that

$$\begin{aligned}
&|j\circ B(\tilde{p}) - j\circ B(p)| \\
&\le \frac{1}{1-\kappa}\{[|j\circ h^{(1)}(\Phi(\tilde{p}),\tilde{p}) - j\circ h^{(1)}(\Phi(\tilde{p}),p)| \\
&\quad + |j\circ h^{(1)}(\Phi(\tilde{p}),p) - j\circ h^{(1)}(\Phi(p),p)|]|B(p)| + |D_2k(\Phi(\tilde{p}),\tilde{p}) - D_2k(\Phi(p),p)|\}.
\end{aligned}$$

The last estimate, the continuity of Φ and of D_2k, and property (iv) combined imply that $p \mapsto j \circ B(p)$ is continuous. $\square$

We now return to the maps $\xi_{\delta\eta} : E_{sc} \to E_\eta$ and $w_{\delta\eta} : E_{sc} \to E_u$ with $1+\epsilon < \eta < b-\epsilon$ and

$$0 < \delta < \delta_\eta, \quad \lambda(\delta) < \frac{(1-a-\epsilon)^2}{2}.$$

Fix $\eta \in (1+\epsilon, b-\epsilon)$, $\tilde{\eta} \in (\eta, b-\epsilon)$ and $\delta > 0$ so that

$$\delta < \min\{\delta_{\tilde{\eta}}, \delta_\eta\}, \quad \lambda(\delta) < \frac{(1-a-\epsilon)^2}{2}.$$

In the next steps we employ Lemma II.8 to prove that the restriction of $\xi_{\delta\tilde{\eta}}$ to the open subset

$$P = \left\{x \in E_{sc} : |x_s| < \frac{\delta}{2}\right\}$$

in the Banach space $\Lambda = E_{sc}$ is C^1-smooth.

Proposition II.2 and the fact that $r_\delta|_{\{x \in E : |x_s| < \delta\}}$ is C^1-smooth yield

$$\sup\{|Dr_\delta(x)| : |x_s| < \delta\} \le \lambda(\delta).$$

An application of Lemma II.6 to the open convex set

$$\mathcal{O} = \left\{x \in E : |x_s| < \frac{\delta}{2}\right\}$$

and to the bounded C^1-map $q = r_\delta|_{\mathcal{O}}$ shows that the linear maps

$$A_{r_\delta}(\xi) : E^{\mathbb{N}} \ni \hat{\xi} \mapsto (Dr_\delta(x_j)\hat{x}_j)_0^\infty \in E^{\mathbb{N}}, \qquad \xi \in \mathcal{O}^{\mathbb{N}},$$

induce a continuous map $A_{r_\delta\eta\tilde{\eta}}$ from the convex set

$$M = \{\xi \in E_\eta : |\operatorname{Pr}_{E_s} x_j| < \frac{\delta}{2} \text{ for all } j \in \mathbb{N}\} \subset E_\eta$$

into $L(E_\eta, E_{\tilde{\eta}})$ so that for every $\xi \in M$ and for every $\tilde{\epsilon} > 0$ there exists $\tilde{\delta} > 0$ such that

$$R_\delta(\xi) \in E_{\tilde{\eta}},\ R_\delta(\tilde{\xi}) \in E_{\tilde{\eta}},\ |R_\delta(\tilde{\xi}) - R_\delta(\xi) - A_{r_\delta\eta\tilde{\eta}}(\xi)(\tilde{\xi}-\xi)|_{\tilde{\eta}} \le \tilde{\epsilon}|\tilde{\xi}-\xi|_\eta \quad \text{(II.2)}$$

for all $\tilde{\xi} \in M$ with $|\tilde{\xi}-\xi|_\eta \le \tilde{\delta}$.

We verify the hypotheses of Lemma II.7 with $Y = E_\eta$, $h = T_{\delta\eta}|_{Y\times P}$ and

$$\kappa = \lambda(\delta)\max\{c(\eta), c(\tilde{\eta})\}.$$

Proposition II.5 yields $\xi_{\delta\eta}(P) \subset M$.

Define $\Phi : P \to M$ by $\Phi(z) = \xi_{\delta\eta}(z)$. Then

$$h(\Phi(p), p) = \Phi(p) \qquad \text{for all } p \in M.$$

The map $h_0 = h|_{M\times P}$ satisfies

$$h_0(\xi, z) = T_{\delta\eta}(\xi, z) = K(R_\delta(\xi)) + (L_{sc}^j z)_0^\infty$$

which shows that for every $(\xi, z) \in M \times P$ the derivative $D_2h_0(\xi, z)$ exists and is given by

$$D_2h_0(\xi, z)\tilde{z} = (L_{sc}^j \tilde{z})_0^\infty \in E_\eta.$$

This derivative is constant on $M \times P$, and therefore continuous.

We proceed to other hypotheses of Lemma II.8. We have already noted that M is convex. Set $Y_1 = E_{\tilde{\eta}}$. Recall $E_\eta \subset E_{\tilde{\eta}}$ and $|\xi|_{\tilde{\eta}} \leq |\xi|_\eta$ on E_η. Define $j : Y \to Y_1$ by $j(\xi) = \xi$. Then j is continuous and injective, and the map $k = j \circ h_0$ is given by

$$k(\xi, z) = T_{\delta\eta}(\xi, z) = K(R_\delta(\xi)) + (L_{sc}^j z)_0^\infty,$$

and $A : M \times P \ni (\xi, z) \mapsto K_{\tilde{\eta}} \circ A_{r_\delta \eta \tilde{\eta}}(\xi) \in L(Y, Y_1)$ is continuous.

Let $\epsilon^* > 0$ and $(\xi, z) \in M \times P$ be given. Set

$$\tilde{\epsilon} = \frac{\epsilon^*}{1 + |K_{\tilde{\eta}}|}.$$

Choose $\tilde{\delta} > 0$ so that property (II.2) holds for all $\tilde{\xi} \in M$ with $|\tilde{\xi} - \xi|_\eta \leq \tilde{\delta}$. For all such $\tilde{\xi} \in M$ we get

$$\begin{aligned}|k(\tilde{\xi}, z) - k(\xi, z) - A(\xi, z)(\tilde{\xi} - \xi)| &= |K(R_\delta(\tilde{\xi})) - K(R_\delta(\xi)) - K(A_{r_\delta \eta \tilde{\eta}}(\xi)(\tilde{\xi} - \xi))|_{\tilde{\eta}} \\ &= |K_{\tilde{\eta}}[R_\delta(\tilde{\xi}) - R_\delta(\xi) - A_{r_\delta \eta \tilde{\eta}}(\xi)(\tilde{\xi} - \xi)]|_{\tilde{\eta}} \\ &\leq |K_{\tilde{\eta}}|\tilde{\epsilon}|\tilde{\xi} - \xi|_\eta \leq \epsilon^* |\tilde{\xi} - \xi|_\eta.\end{aligned}$$

So, hypothesis (ii) of Lemma II.8 is satisfied.

In order to define maps $h^{(1)}$ and $h_1^{(1)}$ as in hypothesis (iii) of Lemma II.8, note first that for all $\xi \in M$, $\hat{\xi} \in Y = E_\eta$, and $j \in \mathbb{N}$ we have

$$|(A_{r_\delta}(\xi)\hat{\xi})_j|\eta^{-j} = |Dr_\delta(x_j)\hat{x}_j|\eta^{-j} \leq \lambda(\delta)|\hat{x}_j|\eta^{-j}.$$

Therefore, each $A_{r_\delta}(\xi)$, $\xi \in M$, defines an element

$$A_{r_\delta \eta\eta}(\xi) \in L(Y, Y) \qquad \text{with } |A_{r_\delta \eta\eta}(\xi)| \leq \lambda(\delta).$$

Analogously, we obtain for each $\xi \in M$ a map

$$A_{r_\delta \tilde{\eta}\tilde{\eta}}(\xi) \in L(Y_1, Y_1) \qquad \text{with } |A_{r_\delta \tilde{\eta}\tilde{\eta}}(\xi)| \leq \lambda(\delta).$$

Define

$$h^{(1)} : M \times P \to L(Y, Y)$$

by $h^{(1)}(\xi, z) = K_\eta \circ A_{r_\delta \eta\eta}(\xi)$ and

$$h_1^{(1)} : M \times P \to L(Y_1, Y_1)$$

by $h_1^{(1)}(\xi, z) = K_{\tilde{\eta}} \circ A_{r_\delta \tilde{\eta}\tilde{\eta}}(\xi)$. For all $(\xi, z, \hat{\xi}) \in M \times P \times Y$, we obtain

$$A(\xi, z)\hat{\xi} = K(A_{r_\delta}(\xi)\hat{\xi}) = \begin{cases} jh^{(1)}(\xi, z)\hat{\xi}, \\ h_1^{(1)}(\xi, z)j\hat{\xi}, \end{cases}$$

and for all $(\xi, z) \in M \times P$,

$$\max\left\{|h^{(1)}(\xi, z)|, |h_1^{(1)}(\xi, z)|\right\} \leq \max\{c(\eta), c(\tilde{\eta})\}\lambda(\delta) = \kappa.$$

Hypothesis (iv) of Lemma II.8 is fulfilled because we have

$$(j \circ h^{(1)}(\xi, z))\hat{\xi} = (j \circ K_\eta \circ A_{r_\delta \eta\eta}(\xi))\hat{\xi} = K(A_{r_\delta}(\xi)\hat{\xi}) = A(\xi, z)\hat{\xi}$$

on $M \times P \times Y$, and $A : M \times P \to L(Y, Y_1)$ is continuous.

We saw earlier that $D_2 h_0$ is continuous. So Lemma II.8 yields that $j \circ \Phi = j \circ (\xi_{\delta\eta}|_P)$ is C^1-smooth. We have $j \circ \Phi = \xi_{\delta\tilde{\eta}}|_P$, since the equations

$$\begin{aligned} T_{\delta\tilde{\eta}}(j\Phi(z), z) &= K(R_\delta(\xi_{\delta\eta}(z))) + (L^j_{sc} z)_0^\infty \\ &= T_{\delta\eta}(\xi_{\delta\eta}(z), z) \\ &= \xi_{\delta\eta}(z) = j\Phi(z) \qquad \text{for every } z \in P \end{aligned}$$

show that each $j\Phi(z)$, $z \in P$, is the unique fixed point of $T_{\delta\tilde{\eta}}(\cdot, z)$ in $E_{\tilde{\eta}} = Y_1$. It follows that $\xi_{\delta\tilde{\eta}}|_P$ is C^1-smooth.

Define $ev \in L(E_{\tilde{\eta}}, E)$ by $ev(\xi) = x_0$. Thus,

$$w_{\delta\tilde{\eta}}|_P = \mathrm{Pr}_{E_u} \circ ev \circ \xi_{\delta\tilde{\eta}}|_P$$

is C^1-smooth.

We now prove $Dw_{\delta\tilde{\eta}}(0) = 0$. Recall from Lemma II.8 that $B = D\xi_{\delta\tilde{\eta}}(0) = D(j \circ \Phi)(0)$ satisfies the equation

$$B = h_1^{(1)}(\Phi(0), 0) \circ B + j \circ D_2 h_0(\Phi(0), 0).$$

Using $D_2 h_0(\Phi(0), 0) z = (L^j_{sc} z)_0^\infty$ for all $z \in E_{sc}$ and the definition of $h_1^{(1)}$, we get

$$Bz = K(A_{r_\delta}(0) B z) + (L^j_{sc} z)_0^\infty \qquad \text{for all } z \in E_{sc}.$$

By $Dr(0) = 0$, $A_{r_\delta}(0) = 0$. Hence,

$$Dw_{\delta\tilde{\eta}}(0) z = (\mathrm{Pr}_{E_u} \circ ev \circ B)(z) = \mathrm{Pr}_{E_u} \circ ev(0 + (L^j_{sc} z)_0^\infty) = 0.$$

To complete the proof of Theorem II.1, we choose convex open neighbourhoods N_{sc} of 0 in E_{sc} and N_u of 0 in E_u so that

$$\begin{aligned} N_{sc} &\subset \left\{ z \in E_{sc} : |\mathrm{Pr}_{E_s} z| < \frac{\delta}{2} \right\}, \\ N_u &\subset \{ y \in E_u : |y|_u < \delta \}, \\ N_{sc} + N_u &\subset E(\delta) (\subset V). \end{aligned}$$

Recall $g^*(x) = g_\delta(x)$ on $E(\delta) \supset N_{sc} + N_u$. Define $w : N_{sc} \to E_u$ by $w(z) = w_{\delta\tilde{\eta}}(z)$. Then w is C^1-smooth with $w(0) = 0$ and $Dw(0) = 0$. Let

$$W = p + \{ z + w(z) : z \in N_{sc} \}.$$

Choose an open neighbourhood N of p in U so small that

$$N - p \subset N_{sc} + N_u$$

and

$$\mathrm{Pr}_{E_{sc}} g^*(N - p) \subset N_{sc}.$$

For every $x \in W \cap N$, we then get $x - p \in W_{\delta\tilde{\eta}} \cap E(\delta)$, and

$$f(x) - p = g^*(x - p) = g_\delta(x - p) \in W_{\delta\tilde{\eta}} \cap (N_{sc} + E_u) = W - p,$$

hence

$$f(W \cap N) \subset W.$$

Finally, if $x \in \cap_{n=0}^\infty f^{-n}(p + N_{sc} + N_u)$ then the trajectory $(f^n(x) - p)_0^\infty$ of g^* in $N_{sc} + N_u \subset E(\delta)$ is also a trajectory of g_δ in $E_{\tilde{\eta}}$. It follows that $x - p \in W_{\delta\tilde{\eta}}$. Using also $\mathrm{Pr}_{E_{sc}}(x - p) \in N_{sc}$ we find $x - p = z + w(z)$ for some $z \in N_{sc}$, or $x \in W$. This completes the proof of Theorem II.1.

APPENDIX III

Smooth generalized center-unstable manifolds for C^1-maps

Let $f : U \to E$ be a C^1-map on an open subset U of a Banach space E over $\mathbb{R}$, with a fixed point p. Let $L = Df(p)$ and suppose there are a closed subspace $E_<$ and a finite-dimensional subspace $E_>$ of E so that

$$E = E_< \oplus E_>, \ L(E_<) \subset E_<, \ L(E_>) \subset E_>.$$

Assume also that there are positive reals $b < 1$ and $a < b$ so that the spectra $\sigma_<$ and $\sigma_>$ of the induced maps

$$L_< : E_< \ni x \mapsto Lx \in E_< \text{ and } L_> : E_> \ni x \mapsto Lx \in E_>$$

are contained in $\{z \in \mathbb{C} : |z| < a\}$ and $\{z \in \mathbb{C} : b < |z|\}$, respectively.

Let $P_> : E \to E$ denote the projection onto $E_>$ along $E_<$, and set $P_< = \mathrm{id} - P_>$.

In this appendix we describe the proof of the following result on a smooth generalized center-unstable manifold.

Theorem III.1 *There exist open neighbourhoods $N_>$ of 0 in $E_>$, $N_<$ of 0 in $E_<$, N of p in U, and a C^1-map*

$$w : N_> \to E_<$$

with $w(N_>) \subset N_<$, $w(0) = 0$, and $Dw(0) = 0$ so that the shifted graph

$$W = p + \{z + w(z) : z \in N_>\}$$

satisfies

$$f(W \cap N) \subset W$$

and

$$\{x \in E : \text{There exists a trajectory } (x_n)_{-\infty}^{0} \text{ of } f \text{ in } p + N_> + N_< \text{ with } x_0 = x\} \subset W.$$

The proof is parallel to the proof of Theorem II.1. In the sequel we sketch the parts which have analogues in the proof of Theorem II.1, and give the details of the parts which are different.

1. We fix $\epsilon > 0$ with

$$a + \epsilon < b - \epsilon.$$

Next we find a norm $|\cdot|$ on E which is equivalent to the originally given one and satisfies

$$|x| = |P_> x| + |P_< x|,$$
$$|LP_< x| \leq (a+\epsilon)|P_< x|,$$
$$|LP_> x| \geq (b-\epsilon)|P_> x|$$

for all $x \in E$. Then we find a norm $|\cdot|_>$ on $E_>$ which is C^1-smooth on $E_> \setminus \{0\}$, and define

$$||x|| = \max\{|P_> x|_>, |P_< x|\}.$$

Let $E(\delta) = \{x \in E : ||x|| < \delta\}$.

2. Set $V = U - p$ and consider the transformed map

$$g^* : V \ni x \mapsto f(x+p) - p \in E$$

with fixed point 0 and $Dg^*(0) = L$. Let $r^* : V \to E$ be given by $r^*(x) = g^*(x) - Lx$ for $x \in V$. Extend r^* to a map $r : E \to E$ by $r(x) = 0$ for all $x \in E \setminus V$, and define $g = L + r$. Finally, for every $\delta > 0$, we define $r_\delta : E \to E$ by

$$r_\delta(x) = \rho\left(\frac{|P_> x|_>}{\delta}\right)\rho\left(\frac{|P_< x|}{\delta}\right)r(x),$$

and set $g_\delta = L + r_\delta$.

Fix $\delta_0 > 0$ so that $\overline{E(2\delta_0)} \subset V$. An analogue of Proposition II.2 permits to find $\delta_1 \in (0, \delta_0)$ and a nondecreasing function $\lambda : [0, \delta_1] \to [0, 1]$ with $\lim_{\delta\to 0+} \lambda(\delta) = 0 = \lambda(0)$ so that for each $\delta \in (0, \delta_1]$ and for all x, y in E,

$$|r_\delta(x)| \leq \delta\lambda(\delta)$$

and

$$|r_\delta(x) - r_\delta(y)| \leq \lambda(\delta)|x - y|.$$

It is convenient for every $\delta > 0$ to introduce the substitution operator

$$R_\delta : E^{-\mathbb{N}} \ni \xi \mapsto (r_\delta(x_n))_{-\infty}^0 \in E^{-\mathbb{N}}.$$

3. For every $\eta > 0$, let E_η^- denote the Banach space of all sequences $\xi = (x_n)_{-\infty}^0 \in E^{-\mathbb{N}}$ satisfying

$$\sup_{j\in -\mathbb{N}} |x_j|\eta^{-j} < \infty,$$

with the norm given by

$$|\xi|_\eta = \sup_{j\in -\mathbb{N}} |x_j|\eta^{-j}.$$

An analogue of Proposition II.3 guarantees that for every $z \in E_>$, $\phi = (f_n)_{-\infty}^0 \in E_\eta^-$, $\eta \in (a+\epsilon, b-\epsilon)$ and for every sequence $\xi = (x_n)_{-\infty}^0 \in E_\eta^-$ with

$$\begin{cases} x_{n+1} = Lx_n + f_n & \text{for all integers } n \leq -1, \\ P_> x_0 = z \end{cases} \tag{III.1}$$

we have

$$x_n = \sum_{j=-\infty}^{n-1} L_<^{n-j-1} P_< f_j - \sum_{j=n}^{-1} L_>^{n-j-1} P_> f_j + L_>^n z \qquad \text{for all integers } n \leq -1$$

and

$$x_0 = z + \sum_{j=-\infty}^{-1} L_<^{-j-1} P_< f_j.$$

Consider the map

$$K : \{\phi \in E^{-\mathbb{N}} : \phi \in E_\eta^- \text{ for some } \eta \in (a+\epsilon, b-\epsilon)\} \to E^{-\mathbb{N}}$$

given by

$$(K\phi)_n = \sum_{j=-\infty}^{n-1} L_<^{n-j-1} P_< f_j - \sum_{j=n}^{-1} L_>^{n-j-1} P_> f_j \qquad \text{for all integers } n \le -1$$

and

$$(K\phi)_0 = \sum_{j=-\infty}^{-1} L_<^{-j-1} P_< f_j.$$

Set $c(\eta) = \frac{1}{\eta-a-\epsilon} + \frac{1}{b-\epsilon-\eta}$, for every $\eta \in (a+\epsilon, b-\epsilon)$. We obtain $KE_\eta^- \subset E_\eta^-$ for all $\eta \in (a+\epsilon, b-\epsilon)$, and the linear operators $K_\eta : E_\eta^- \to E_\eta^-$ given by $K_\eta \phi = K\phi$ satisfy

$$|K_\eta \phi|_\eta \le c(\eta)|\phi|_\eta \qquad \text{for all } \phi \in E_\eta^-.$$

So $K_\eta \in L(E_\eta^-, E_\eta^-)$ and $|K_\eta| \le c(\eta)$. An analogue of Proposition II.4 shows that for all $\eta \in (a+\epsilon, b-\epsilon)$, $z \in E_>$, $\phi \in E_\eta^-$ the sequence

$$\xi = K_\eta \phi + (L_>^n z)_{-\infty}^0 \in E_\eta^-$$

solves (III.1).

4. For every $\eta \in (a+\epsilon, b-\epsilon)$, choose $\delta_\eta \in (0, \delta_1]$ such that

$$\lambda(\delta_\eta)c(\eta) < 1.$$

Let $\eta \in (a+\epsilon, b-\epsilon)$ and $\delta \in (0, \delta_\eta)$. Then R_δ induces a map $r_{\delta\eta} : E_\eta^- \ni \xi \mapsto R_\delta(\xi) \in E_\eta^-$ which is Lipschitz continuous with Lipschitz constant $\lambda(\delta)$.

The result in part 3 shows that for every $z \in E_>$ and $\xi \in E_\eta^-$, the problem

$$x_{n+1} = g_\delta(x_n) \quad \text{for all integers } n \le -1, \quad P_> x_0 = z$$

is equivalent to the fixed point equation

$$\xi = K_\eta(r_{\delta\eta}(\xi)) + (L_>^j z)_{-\infty}^0.$$

It can be verified that the map

$$T_{\delta\eta} : E_\eta^- \times E_> \to E_\eta^-$$

given by

$$T_{\delta\eta}(\xi, z) = K_\eta(r_{\delta\eta}(\xi)) + (L_>^j z)_{-\infty}^0$$

satisfies

$$|T_{\delta\eta}(\xi, z) - T_{\delta\eta}(\xi^*, z)|_\eta \le c(\eta)\lambda(\delta)|\xi - \xi^*|_\eta$$

for all ξ, ξ^* in E_η^- and for all $z \in E_>$. So for every $z \in E_>$ there is exactly one fixed point $\xi_{\delta\eta}(z) \in E_\eta^-$ of the contraction $T_{\delta\eta}(\cdot, z) : E_\eta^- \to E_\eta^-$. As $P_>(r_{\delta\eta}(z))_0 = z$, we conclude that $\xi \in E_\eta^-$ is a trajectory of g_δ with $P_> x_0 = z$ if and only if $\xi = \xi_{\delta\eta}(z)$.

We can also show that the map

$$\xi_{\delta\eta} : E_> \ni z \mapsto \xi_{\delta\eta}(z) \in E_\eta^-$$

is Lipschitz continuous. So the map

$$w_{\delta\eta} : E_> \ni z \mapsto P_<(\xi_{\delta\eta}(z))_0 \in E_<$$

is Lipschitz continuous. Clearly, $w_{\delta\eta}(0) = 0$.

5. For every $\eta \in (a+\epsilon, b-\epsilon)$ and $\delta \in (0, \delta_\eta)$, the graph

$$W_{\delta\eta} = \{z + w_{\delta\eta}(z) : z \in E_>\}$$

coincides with the set

$$\{x \in E : \text{There exists a trajectory } (x_n)_{-\infty}^0 \text{ of } g_\delta \text{ in } E_\eta^- \text{ with } x_0 = x\},$$

which is positively invariant with respect to g_δ. As $g_\delta(x) = g^*(x)$ for all $x \in E(\delta)$, we have

$$g^*(W_{\delta\eta} \cap E(\delta)) \subset W_{\delta\eta} \qquad \text{for such } \eta \text{ and } \delta.$$

Moreover, if $(x_j)_{-\infty}^0$ is given so that $x_j \in E(\delta)$ for all $j \in -\mathbb{N}$, then $(x_j)_{-\infty}^0 \in E_\eta^-$ due to the fact that $\eta < b - \epsilon < 1$. Therefore,

$$\{x \in E : \text{There exists a trajectory } (x_n)_{-\infty}^0 \text{ of } g^* \text{ in } (E(\delta))^{-\mathbb{N}} \text{ with } x_0 = x\} \subset W_{\delta\eta}.$$

6. We prove that if $0 < \delta < \delta_\eta$ and $\lambda(\delta) < \frac{1-a-\epsilon}{2}$, then for every $z \in E_>$ and for all integers $n \le 0$ we have $|P_<(\xi_{\delta\eta}(z))_n| < \frac{\delta}{2}$. In fact, the fixed point equation for $\xi_{\delta\eta}(z)$ implies that for all integers $n \le -1$ we have

$$\begin{aligned}
|P_<(\xi_{\delta\eta}(z))_n| &= \Big|\sum_{j=-\infty}^{n-1} L_<^{n-j-1} P_<[r_{\delta\eta}(\xi_{\delta\eta}(z))]_j\Big| \\
&\le \sum_{j=-\infty}^{n-1} (a+\epsilon)^{n-j-1} |r_\delta((\xi_{\delta\eta}(z))_j)| \\
&\le \sum_{j=-\infty}^{n-1} (a+\epsilon)^{n-j-1} \delta\lambda(\delta) \\
&\le \frac{1}{1-a-\epsilon}\delta\lambda(\delta) < \frac{\delta}{2}.
\end{aligned}$$

For $n = 0$, we have

$$\begin{aligned}
|P_<(\xi_{\delta\eta}(z))_0| &= \Big|\sum_{j=-\infty}^{-1} L_<^{-j-1} P_<[r_{\delta\eta}(\xi_{\delta\eta}(z))]_j\Big| \\
&\le \sum_{j=-\infty}^{-1} (a+\epsilon)^{-j-1} \delta\lambda(\delta) = \frac{1}{1-a-\epsilon}\delta\lambda(\delta) < \frac{\delta}{2}.
\end{aligned}$$

7. Let a bounded C^1-smooth map $q : \mathcal{O} \to E$ be given on an open convex subset $\mathcal{O} \subset E$ so that Dq is bounded. Consider the substitution operator

$$\hat{q} : \mathcal{O}^{-\mathbb{N}} \to E^{-\mathbb{N}}, \quad \hat{q}(\xi) = (q(x_n))_{-\infty}^0 \quad \text{for } \xi \in \mathcal{O}^{-\mathbb{N}}.$$

Assume that η and $\tilde{\eta}$ are given so that $0 < \tilde{\eta} < \eta < 1$. One can easily verify that $\hat{q}(\mathcal{O}^{-\mathbb{N}} \cap E_\eta^-) \subset E_{\tilde{\eta}}^-$. The estimate

$$|Dq(x_j)\hat{x}_j|\tilde{\eta}^{-j} \le \sup_{x\in\mathcal{O}} |Dq(x)||\hat{x}_j|\eta^{-j}\left(\frac{\tilde{\eta}}{\eta}\right)^{-j}$$

for all $\xi = (x_j)_{-\infty}^0 \in \mathcal{O}^{-\mathbb{N}}$ and $\hat{\xi} \in E_\eta^-$ shows that the linear maps

$$A_q(\xi) : E^{-\mathbb{N}} \ni \hat{\xi} \mapsto (Dq(x_j)\hat{x}_j)_{-\infty}^0 \in E^{-\mathbb{N}}, \qquad \xi \in \mathcal{O}^{-\mathbb{N}},$$

satisfy $A_q(\xi)E_\eta^- \subset E_{\tilde{\eta}}^-$ and define continuous linear maps

$$A(\xi)_{q\eta\tilde{\eta}} : E_\eta^- \to E_{\tilde{\eta}}^-.$$

We show that

(i) the map

$$A_{q\eta\tilde{\eta}} : \mathcal{O}^{-\mathbb{N}} \cap E_\eta^- \ni \xi \mapsto A(\xi)_{q\eta\tilde{\eta}} \in L(E_\eta^-, E_{\tilde{\eta}}^-)$$

is continuous.

Let $\xi \in \mathcal{O}^{-\mathbb{N}} \cap E_\eta^-$ be given and let $\tilde{\epsilon} > 0$. Choose an integer $j_0 \le 0$ such that for all integers $j \le j_0$,

$$2 \sup_{x\in\mathcal{O}} |Dq(x)| \left(\frac{\tilde{\eta}}{\eta}\right)^{-j} < \frac{\tilde{\epsilon}}{2}.$$

Next choose $\tilde{\delta} > 0$ such that for all $j \in \{j_0, j_0+1, \dots, 0\}$ and for all $x \in \mathcal{O}$ with $|x - x_j| < \tilde{\delta}\eta^j$,

$$|Dq(x) - Dq(x_j)| < \frac{\tilde{\epsilon}}{2} \left(\frac{\tilde{\eta}}{\eta}\right)^j.$$

Now consider $\hat{\xi} \in \mathcal{O}^{-\mathbb{N}} \cap E_\eta^-$ with $|\hat{\xi} - \xi|_\eta < \tilde{\delta}$ and consider $\tilde{\xi} \in E_\eta^-$ with $|\tilde{\xi}|_\eta \le 1$. For all integers $j \le j_0$,

$$|(Dq(\hat{x}_j) - Dq(x_j))\tilde{x}_j|\tilde{\eta}^{-j} \le 2 \sup_{x\in\mathcal{O}} |Dq(x)||\tilde{x}_j|\eta^{-j} \left(\frac{\tilde{\eta}}{\eta}\right)^{-j} < \frac{\tilde{\epsilon}}{2}.$$

For all $j \in \{j_0, j_0+1, \dots, 0\}$,

$$|(Dq(\hat{x}_j) - Dq(x_j))\tilde{x}_j|\tilde{\eta}^{-j} \le |Dq(\hat{x}_j) - Dq(x_j)||\tilde{x}_j|\eta^{-j} \left(\frac{\tilde{\eta}}{\eta}\right)^{-j} < \frac{\tilde{\epsilon}}{2}.$$

It follows that

$$|A_{q\eta\tilde{\eta}}(\hat{\xi}) - A_{q\eta\tilde{\eta}}(\xi)|_{L(E_\eta^-, E_{\tilde{\eta}}^-)} = \sup_{\tilde{\xi}\in E_\eta^-, |\tilde{\xi}|_\eta \le 1} \sup_{j\in -\mathbb{N}} |Dq(\hat{x}_j)\tilde{x}_j - Dq(x_j)\tilde{x}_j|\tilde{\eta}^{-j} \le \frac{\tilde{\epsilon}}{2} < \tilde{\epsilon},$$

and $A_{q\eta\tilde{\eta}}$ is continuous at ξ.

Arguments similar to those in the proof of assertion (iii) of Lemma II.6, and modifications according to the proof of (i) above prove that

(ii) for every $\tilde{\epsilon} > 0$ and $\xi \in \mathcal{O}^{-\mathbb{N}} \cap E_\eta^-$ there exists $\tilde{\delta} > 0$ so that for every $\tilde{\xi} \in \mathcal{O}^{-\mathbb{N}} \cap E_\eta^-$ with $|\tilde{\xi} - \xi|_\eta \le \tilde{\delta}$,

$$|\hat{q}(\tilde{\xi}) - \hat{q}(\xi) - A_{q\eta\tilde{\eta}}(\xi)[\tilde{\xi} - \xi]|_{\tilde{\eta}} \le \tilde{\epsilon}|\tilde{\xi} - \xi|_\eta.$$

8. Fix $\eta \in (a+\epsilon, b-\epsilon)$, $\tilde{\eta} \in (a+\epsilon, \eta)$ and $\delta > 0$ so that

$$\delta < \min\{\delta_{\tilde{\eta}}, \delta_\eta\}, \quad \lambda(\delta) < \frac{1-a-\epsilon}{2}.$$

We are going to employ Lemma II.8 to prove that $\xi_{\delta\tilde{\eta}}$ is C^1-smooth. Set $\Lambda = E_>$ and $P = \Lambda$.

Note that $r_\delta|_{\{x\in E: |P_< x| < \delta\}}$ is C^1-smooth. Therefore, the result in part 2 yields

$$\sup\{|Dr_\delta(x)| : |P_< x| < \delta\} \le \lambda(\delta).$$

An application of the result in part 7 to the open convex set

$$\mathcal{O} = \left\{x \in E : |P_< x| < \frac{\delta}{2}\right\}$$

and to the bounded C^1-map $q = r_\delta|_\mathcal{O}$ shows that the linear maps

$$A_{r_\delta}(\xi) : E^{-\mathbb{N}} \ni \hat{\xi} \mapsto (Dr_\delta(x_j)\hat{x}_j)_{-\infty}^0 \in E^{-\mathbb{N}}, \qquad \xi \in \mathcal{O}^{-\mathbb{N}},$$

induce a continuous map $A_{r_\delta \eta\tilde{\eta}}$ from the convex set

$$M = \left\{\xi \in E_\eta^- : |P_< x_j| < \frac{\delta}{2} \text{ for all } j \in -\mathbb{N}\right\} \subset E_\eta^-$$

into $L(E_\eta^-, E_{\tilde{\eta}}^-)$ and for every $\xi \in M$ and for every $\tilde{\epsilon} > 0$ there exists $\tilde{\delta} > 0$ such that

$$R_\delta(\xi) \in E_{\tilde{\eta}}^-,\ R_\delta(\tilde{\xi}) \in E_{\tilde{\eta}}^-, \text{ and } |R_\delta(\tilde{\xi}) - R_\delta(\xi) - A_{r_\delta\eta\tilde{\eta}}(\xi)(\tilde{\xi}-\xi)|_{\tilde{\eta}} \le \tilde{\epsilon}|\tilde{\xi}-\xi|_\eta \quad \text{(III.2)}$$

for all $\tilde{\xi} \in M$ with $|\tilde{\xi} - \xi|_\eta \le \tilde{\delta}$.

We now verify the hypotheses of Lemma II.7 with $Y = E_\eta^-$, $h = T_{\delta\eta}|_{Y\times P}$ and $\kappa = \lambda(\delta)\max\{c(\eta), c(\tilde{\eta})\} < 1$. The result in part 6 yields $\xi_{\delta\eta}(P) \subset M$.

Define $\Phi : P \to M$ by $\Phi(z) = \xi_{\delta\eta}(z)$ so that

$$h(\Phi(p), p) = \Phi(p) \qquad \text{for all } p \in P.$$

The map $h_0 = h|_{M\times P}$ satisfies

$$h_0(\xi, z) = T_{\delta\eta}(\xi, z) = K(R_\delta(\xi)) + (L_>^j z)_{-\infty}^0$$

which shows that for every $(\xi, z) \in M \times P$ the derivative $D_2 h_0(\xi, z)$ exists and is given by

$$D_2 h_0(\xi, z)\tilde{z} = (L_>^j \tilde{z}))_{-\infty}^0 \in E_\eta^-;$$

it is constant on $M \times P$ and therefore continuous.

We proceed to other hypotheses of Lemma II.8. Set $Y_1 = E_{\tilde{\eta}}^-$. Recall $E_\eta \subset E_{\tilde{\eta}}$ and $|\xi|_{\tilde{\eta}} \le |\xi|_\eta$ on E_η. Therefore, the map $j : Y \to Y_1$ given by $j(\xi) = \xi$ is linear, injective and continuous. The map $k = j \circ h_0$ is given by

$$k(\xi, z) = T_{\delta\tilde{\eta}}(\xi, z) = K(R_\delta(\xi)) + (L_>^j z)_{-\infty}^0,$$

and $A : M \times P \ni (\xi, z) \mapsto K_{\tilde{\eta}} \circ A_{r_\delta\eta\tilde{\eta}}(\xi) \in L(Y, Y_1)$ is continuous.

Let $\epsilon^* > 0$ and $(\xi, z) \in M \times P$ be given. Set

$$\tilde{\epsilon} = \frac{\epsilon^*}{1 + |K_{\tilde{\eta}}|}.$$

Consider $\tilde{\delta}$ so that (III.2) holds for all $\tilde{\xi} \in M$ with $|\tilde{\xi} - \xi|_\eta \le \delta$. For all such $\tilde{\xi} \in M$ we get

$$\begin{aligned} |k(\tilde{\xi}, z) - k(\xi, z) - A(\xi, z)(\tilde{\xi} - \xi)|_{\tilde{\eta}} &= |K_{\tilde{\eta}}[R_\delta(\tilde{\xi}) - R_\delta(\xi) - A_{r_\delta\eta\tilde{\eta}}(\xi)(\tilde{\xi} - \xi)]|_{\tilde{\eta}} \\ &\le |K_{\tilde{\eta}}|\tilde{\epsilon}|\tilde{\xi} - \xi|_\eta \le \epsilon^*|\tilde{\xi} - \xi|_\eta. \end{aligned}$$

So, hypothesis (ii) of Lemma II.8 is satisfied.

In order to define maps $h^{(1)}$, $h_1^{(1)}$ as in hypothesis (iii) of Lemma II.8, note first that for all $\xi \in M$, $\hat{\xi} \in Y = E_\eta^-$, $j \in -\mathbb{N}$, we have

$$|(A_{r_\delta}(\xi)\hat{\xi})_j|\eta^{-j} = |Dr_\delta(x_j)\hat{x}_j|\eta^{-j} \leq \lambda(\delta)|\hat{x}_j|\eta^{-j}.$$

Therefore, each $A_{r_\delta}(\xi), \xi \in M$ defines an element

$$A_{r_\delta \eta\eta}(\xi) \in L(Y,Y) \qquad \text{with } |A_{r_\delta \eta\eta}(\xi)| \leq \lambda(\delta).$$

Analogously, we obtain for each $\xi \in M$ a map

$$A_{r_\delta \tilde{\eta}\tilde{\eta}}(\xi) \in L(Y_1,Y_1) \qquad \text{with } |A_{r_\delta \tilde{\eta}\tilde{\eta}}(\xi)| \leq \lambda(\delta).$$

Define

$$h^{(1)} : M \times P \to L(Y,Y)$$

by $h^{(1)}(\xi,z) = K_\eta \circ A_{r_\delta \eta\eta}(\xi)$ and

$$h_1^{(1)} : M \times P \to L(Y_1,Y_1)$$

by $h_1^{(1)}(\xi,z) = K_{\tilde{\eta}} \circ A_{r_\delta \tilde{\eta}\tilde{\eta}}(\xi)$. For all $(\xi,z,\hat{\xi}) \in M \times P \times Y$, we obtain

$$A(\xi,z)\hat{\xi} = K(A_{r_\delta}(\xi)\hat{\xi}) = \begin{cases} jh^{(1)}(\xi,z)\hat{\xi}, \\ h_1^{(1)}(\xi,z)j\hat{\xi}, \end{cases}$$

and for all $(\xi,z) \in M \times P$,

$$\max\left\{|h^{(1)}(\xi,z)|, |h_1^{(1)}(\xi,z)|\right\} \leq \max\{c(\eta), c(\tilde{\eta})\}\lambda(\delta) = \kappa.$$

Hypothesis (iv) of Lemma II.8 is fulfilled because we have

$$(j \circ h^{(1)}(\xi,z))\hat{\xi} = (j \circ K_\eta \circ A_{r_\delta \eta\eta}(\xi))\hat{\xi} = (K_{\tilde{\eta}} \circ A_{r_\delta \eta\tilde{\eta}}(\xi))(\hat{\xi}) = A(\xi,z)\hat{\xi}$$

on $M \times P \times Y$, and $A : M \times P \to L(Y,Y_1)$ is continuous.

We saw earlier that $D_2 h_0$ is continuous. So Lemma II.8 yields that $j \circ \Phi = j \circ (\xi_{\delta\eta}|_P)$ is C^1-smooth. As $j \circ \Phi = \xi_{\delta\tilde{\eta}}$, we conclude that $\xi_{\delta\tilde{\eta}}$ is C^1-smooth, from which it follows that $w_{\delta\tilde{\eta}}$ is C^1-smooth. The fixed point equation for $D\xi_{\delta\tilde{\eta}}(0)$ according to Lemma II.8 and arguments as in Appendix II lead to $Dw_{\delta\tilde{\eta}}(0) = 0$.

9. We now choose convex open neighbourhoods $N_>$ of 0 in $E_>$ and $N_<$ of 0 in $E_<$ so that

$$N_> \subset \{z \in E_> : |z|_> < \delta\},$$
$$N_< \subset \{y \in E_< : |y| < \delta\},$$
$$N_> + N_< \subset E(\delta)(\subset V).$$

Recall $g^*(x) = g_\delta(x)$ on $E(\delta) \supset N_> + N_<$. Define $w : N_> \to E_<$ by $w(z) = w_{\delta\tilde{\eta}}(z)$. Then w is C^1-smooth with $w(0) = 0$ and $Dw(0) = 0$. Choose an open neighbourhood N of p in U so small that

$$N - p \subset N_> + N_<$$

and

$$P_> g^*(N-p) \subset N_>.$$

Consider

$$W = p + \{z + w(z) : z \in N_>\}.$$

For every $x \in W \cap N$, we get $x - p \in W_{\delta\tilde{\eta}} \cap E(\delta)$, and

$$f(x) - p = g^*(x - p) \in W_{\delta\tilde{\eta}} \cap (N_> + E_<) = W - p,$$

hence

$$f(W \cap N) \subset W.$$

Finally, if $x \in E$ is given so that there exists a trajectory $(x_n)_{-\infty}^0$ of f in $p + N_< + N_>$ such that $x_0 = x$, then $x_n - p \in E(\delta)$ for all $n \in -\mathbb{N}$ and $x_{n+1} - p = g^*(x_n - p)$ for all integers $n \leq -1$. The result in part 5 then shows that $x - p \in W_{\delta\tilde{\eta}}$. Using also $P_>(x - p) \in N_>$ we find $x - p = z + w(z)$ for some $z \in N_>$, or $x \in W$. This completes the proof. □

APPENDIX IV

Invariant cones close to neutrally stable fixed points with 1-dimensional center spaces

In this appendix, we assume that E is a Banach space over $\mathbb{R}$ and f is a C^1-map from an open neighbourhood U of 0 in E into E such that the following hold:

$$f(0) = 0. \tag{IV.1}$$

$$\text{1 is an isolated point of the spectrum } \sigma \text{ of } Df(0) \text{ with one-dimensional realified generalized eigenspace.} \tag{IV.2}$$

$$\text{There exists } \gamma < 1 \text{ such that } |\lambda| \le \gamma \text{ for all } \lambda \in \sigma \setminus \{1\}. \tag{IV.3}$$

Let $\chi_c \in E$ be an eigenvector of the eigenvalue 1 of $Df(0)$ and let Z be the realified generalized eigenspace of $Df(0)$ associated with the remaining spectrum points. We also assume that there exists a neighbourhood $V \subset U$ of 0 such that

$$f(V \cap Z) \subset Z, \quad f(V \cap \mathbb{R}\chi_c) \subset \mathbb{R}\chi_c. \tag{IV.4}$$

For a fixed $\alpha \in (\gamma, 1)$, we can apply Theorem I.1 to find a norm $||\cdot||_s$ which is equivalent to the originally given one in E such that the induced map $Df(0)_Z : Z \ni \chi \mapsto Df(0)\chi \in Z$ and the projections associated with the decomposition $E = Z \oplus \mathbb{R}\chi_c$ satisfy

$$||Df(0)_Z||_s < \alpha \tag{IV.5}$$

and

$$||\operatorname{Pr}_{\mathbb{R}\chi_c}||_s = ||\operatorname{Pr}_Z||_s = 1. \tag{IV.6}$$

We choose an open ball N_z in Z with respect to $||\cdot||_s$ and an open ball N_c in $\mathbb{R}\chi_c$ with respect to $||\cdot||_s$ such that $0 \in N_z + N_c \subset V$. The C^1-map

$$R : N_z + N_c \ni \rho \mapsto f(\rho) - Df(0)\rho \in E$$

satisfies

$$R(0) = 0,\ DR(0) = 0,\ \operatorname{Pr}_Z R(\operatorname{Pr}_{\mathbb{R}\chi_c} \rho) = 0,\ \operatorname{Pr}_{\mathbb{R}\chi_c} R(\operatorname{Pr}_Z \rho) = 0 \text{ for all } \rho \in N_z + N_c.$$

For every $\rho \in N_z + N_c$ we obtain

$$\begin{aligned} ||\operatorname{Pr}_{\mathbb{R}\chi_c} R(\rho)||_s &= ||\operatorname{Pr}_{\mathbb{R}\chi_c} R(\operatorname{Pr}_Z \rho) + \operatorname{Pr}_{\mathbb{R}\chi_c} \int_0^1 DR(\operatorname{Pr}_Z \rho + t \operatorname{Pr}_{\mathbb{R}\chi_c} \rho) \operatorname{Pr}_{\mathbb{R}\chi_c} \rho \, dt||_s \\ &= ||\operatorname{Pr}_{\mathbb{R}\chi_c} \int_0^1 DR(\operatorname{Pr}_Z \rho + t \operatorname{Pr}_{\mathbb{R}\chi_c} \rho) \operatorname{Pr}_{\mathbb{R}\chi_c} \rho \, dt||_s, \end{aligned}$$

and

$$\begin{aligned}
&|| \Pr_Z f(\rho)||_s \\
&\quad = || \Pr_Z Df(0)\rho + \Pr_Z R(\rho)||_s \\
&\quad = || \Pr_Z Df(0)\rho + \Pr_Z R(\Pr_{\mathbb{R}\chi_c} \rho) + \Pr_Z \int_0^1 DR(\Pr_{\mathbb{R}\chi_c} \rho + t \Pr_Z \rho) \Pr_Z \rho \, dt||_s \\
&\quad = || \Pr_Z Df(0)\rho + \Pr_Z \int_0^1 DR(\Pr_{\mathbb{R}\chi_c} \rho + t \Pr_Z \rho) \Pr_Z \rho \, dt||_s.
\end{aligned}$$

Thus, for a fixed $\hat{\epsilon} \in (0, \frac{1-\alpha}{2})$, we can choose an integer $n_0 > 0$ such that for all $\rho \in \frac{1}{n_0}(N_z + N_c)$ we have

$$|| \Pr_{\mathbb{R}\chi_c} R(\rho)||_s \le \hat{\epsilon} || \Pr_{\mathbb{R}\chi_c} \rho||_s \tag{IV.7}$$

and

$$|| \Pr_Z f(\rho)||_s \le (\alpha + \hat{\epsilon}) || \Pr_Z \rho||_s. \tag{IV.8}$$

We define the slope of vectors $\rho \in E$ with respect to the decomposition

$$E = \mathbb{R}\chi_c \oplus Z$$

and to the norm $|| \cdot ||_s$ by

$$\mathrm{sl}(\rho) = \begin{cases} \frac{|| \Pr_Z \rho||_s}{|| \Pr_{\mathbb{R}\chi_c} \rho||_s} & \text{if } \Pr_{\mathbb{R}\chi_c} \rho \neq 0, \\ \infty & \text{if } \Pr_{\mathbb{R}\chi_c} \rho = 0. \end{cases}$$

For $h > 0$, set

$$K_h^+ = \{\rho \in E : \mathrm{sl}(\rho) < h, \ \Pr_{\mathbb{R}\chi_c} \rho \in \mathbb{R}^+ \chi_c\}$$

and

$$K_h^- = \{\rho \in E : \mathrm{sl}(\rho) < h, \ \Pr_{\mathbb{R}\chi_c} \rho \in \mathbb{R}^- \chi_c\}.$$

We now derive the following result on *cone invariance.*

Proposition IV.1 *For every $h > 0$ there exists an integer $n_1 = n_1(h) \ge \max\{4, n_0\}$ such that for each $\chi \in \frac{1}{n_1}(N_c + N_z)$ we have*

$$f\left(\frac{1}{n_1}(N_z + N_c) \cap (\chi + K_h^+)\right) \subset f(\chi) + K_h^+,$$
$$f\left(\frac{1}{n_1}(N_z + N_c) \cap (\chi + K_h^-)\right) \subset f(\chi) + K_h^-.$$

Proof 1. For all χ, η in $N_z + N_c$ set

$$\begin{aligned}
R_D(\chi, \eta)(\eta - \chi) &= f(\eta) - f(\chi) - Df(\chi)(\eta - \chi) \\
&= \int_0^1 [Df(\chi + t(\eta - \chi)) - Df(\chi)](\eta - \chi) \, dt.
\end{aligned}$$

Then

$$f(\eta) - f(\chi) = Df(\chi)(\eta - \chi) + R_D(\chi, \eta)(\eta - \chi),$$

and

$$\begin{aligned} R_D(\chi,\eta)(\eta-\chi) - R(\eta-\chi) &= \int_0^1 [Df(\chi+t(\eta-\chi)) - Df(\chi)](\eta-\chi)\,dt \\ &\quad - \int_0^1 [Df(t(\eta-\chi)) - Df(0)](\eta-\chi)\,dt \\ &= \int_0^1 [Df(\chi+t(\eta-\chi)) - Df(t(\eta-\chi))](\eta-\chi)\,dt \\ &\quad + [Df(0) - Df(\chi)](\eta-\chi). \end{aligned}$$

2. Let $h>0$. Choose $\epsilon^*>0$ so small that

$$0 < \frac{\alpha h + 3(1+h)\epsilon^*}{1-3(1+h)\epsilon^*} < h, \quad 3(1+h)\epsilon^* < \frac{1}{2},$$

and choose an integer $n_1 \geq \max\{4, n_0\}$ so large that for all χ,η in $\frac{1}{n_1}(N_z+N_c)$ we have

$$\begin{aligned} &||Df(\chi) - Df(0)||_s \leq \epsilon^*, \\ &||R(\eta-\chi)||_s \leq \epsilon^* ||\eta-\chi||_s, \\ &||R_D(\chi,\eta)(\eta-\chi) - R(\eta-\chi)||_s \leq \epsilon^* ||\eta-\chi||_s. \end{aligned}$$

Then for all χ,η in $\frac{1}{n_1}(N_z+N_c)$ with $\mathrm{sl}(\eta-\chi)<h$ we have

$$\begin{aligned} &||\operatorname{Pr}_{\mathbb{R}\chi_c}[f(\eta)-f(\chi)]||_s \\ &\geq ||\operatorname{Pr}_{\mathbb{R}\chi_c} Df(\chi)(\eta-\chi)||_s - ||\operatorname{Pr}_{\mathbb{R}\chi_c} R_D(\chi,\eta)(\eta-\chi)||_s \\ &\geq ||\operatorname{Pr}_{\mathbb{R}\chi_c} Df(0)(\eta-\chi)||_s - ||\operatorname{Pr}_{\mathbb{R}\chi_c}[Df(\chi)-Df(0)](\eta-\chi)||_s \\ &\quad - ||\operatorname{Pr}_{\mathbb{R}\chi_c} R(\eta-\chi)||_s - ||\operatorname{Pr}_{\mathbb{R}\chi_c}[R_D(\chi,\eta)(\eta-\chi) - R(\eta-\chi)]||_s \\ &= ||Df(0)\operatorname{Pr}_{\mathbb{R}\chi_c}(\eta-\chi)||_s - ||\operatorname{Pr}_{\mathbb{R}\chi_c}[Df(\chi)-Df(0)](\eta-\chi)||_s \\ &\quad - ||\operatorname{Pr}_{\mathbb{R}\chi_c} R(\eta-\chi)||_s - ||\operatorname{Pr}_{\mathbb{R}\chi_c}[R_D(\chi,\eta)(\eta-\chi) - R(\eta-\chi)]||_s \\ &\geq ||\operatorname{Pr}_{\mathbb{R}\chi_c}(\eta-\chi)||_s - 3(1+h)\epsilon^* ||\operatorname{Pr}_{\mathbb{R}\chi_c}(\eta-\chi)||_s \\ &= [1-3(1+h)\epsilon^*]||\operatorname{Pr}_{\mathbb{R}\chi_c}(\eta-\chi)||_s > 0. \end{aligned}$$

Similarly,

$$\begin{aligned} &||\operatorname{Pr}_Z[f(\eta)-f(\chi)]||_s \\ &\leq ||\operatorname{Pr}_Z Df(\chi)(\eta-\chi)||_s + ||\operatorname{Pr}_Z R_D(\chi,\eta)(\eta-\chi)||_s \\ &\leq ||\operatorname{Pr}_Z Df(0)(\eta-\chi)||_s + ||\operatorname{Pr}_Z[Df(\chi)-Df(0)](\eta-\chi)||_s \\ &\quad + ||\operatorname{Pr}_Z R(\eta-\chi)||_s + ||\operatorname{Pr}_Z[R_D(\chi,\eta)(\eta-\chi) - R(\eta-\chi)]||_s \\ &= ||Df(0)\operatorname{Pr}_Z(\eta-\chi)||_s + ||\operatorname{Pr}_Z[Df(\chi)-Df(0)](\eta-\chi)||_s \\ &\quad + ||\operatorname{Pr}_Z R(\eta-\chi)||_s + ||\operatorname{Pr}_Z[R_D(\chi,\eta)(\eta-\chi) - R(\eta-\chi)]||_s \\ &\leq \alpha||\operatorname{Pr}_Z(\eta-\chi)||_s + 3(1+h)\epsilon^* ||\operatorname{Pr}_{\mathbb{R}\chi_c}(\eta-\chi)||_s \\ &\leq [\alpha h + 3(1+h)\epsilon^*]||\operatorname{Pr}_{\mathbb{R}\chi_c}(\eta-\chi)||_s. \end{aligned}$$

Hence

$$\mathrm{sl}[f(\eta)-f(\chi)] \leq \frac{\alpha h + 3(1+h)\epsilon^*}{1-3(1+h)\epsilon^*} < h.$$

3. We have

$$\begin{aligned}\operatorname{Pr}_{\mathbb{R}\chi_c}[f(\eta)-f(\chi)] &= \operatorname{Pr}_{\mathbb{R}\chi_c} Df(\chi)(\eta-\chi)+\operatorname{Pr}_{\mathbb{R}\chi_c} R_D(\chi,\eta)(\eta-\chi)\\ &\quad -\operatorname{Pr}_{\mathbb{R}\chi_c} Df(0)(\eta-\chi)+\operatorname{Pr}_{\mathbb{R}\chi_c} Df(0)(\eta-\chi)\\ &= \operatorname{Pr}_{\mathbb{R}\chi_c} Df(\chi)(\eta-\chi)+\operatorname{Pr}_{\mathbb{R}\chi_c} R_D(\chi,\eta)(\eta-\chi)\\ &\quad -\operatorname{Pr}_{\mathbb{R}\chi_c} Df(0)(\eta-\chi)+\operatorname{Pr}_{\mathbb{R}\chi_c}(\eta-\chi).\end{aligned}$$

As

$$\begin{aligned}&||\operatorname{Pr}_{\mathbb{R}\chi_c}[Df(\chi)-Df(0)](\eta-\chi)+\operatorname{Pr}_{\mathbb{R}\chi_c} R_D(\chi,\eta)(\eta-\chi)||_s\\ &\quad\le 3\epsilon^*(1+h)||\operatorname{Pr}_{\mathbb{R}\chi_c}(\eta-\chi)||_s\\ &\quad\le \frac{1}{2}||\operatorname{Pr}_{\mathbb{R}\chi_c}(\eta-\chi)||_s,\end{aligned}$$

we immediately obtain that $\operatorname{Pr}_{\mathbb{R}\chi_c}(\eta-\chi)\in\mathbb{R}^+\chi_c$ $(\in\mathbb{R}^-\chi_c)$ implies $\operatorname{Pr}_{\mathbb{R}\chi_c}[f(\eta)-f(\chi)]\in\mathbb{R}^+\chi_c$ $(\in\mathbb{R}^-\chi_c)$. □

The next two results deal with a trajectory $(\chi_j)_0^\infty$ of f in $(N_c+N_z)\setminus Z$ which converges to 0 as $j\to\infty$.

Proposition IV.2 *For every $h>0$ there exists $j_1\ge 0$ such that $\chi_j\in\frac{1}{n_1(h)}(N_z+N_c)\setminus\frac{1}{n_1(h)}N_z$ and $\operatorname{sl}(\chi_j)<h$ for all integers $j\ge j_1$.*

Proof For $\rho\in\frac{1}{n_0}(N_z+N_c)$ with $\operatorname{Pr}_{\mathbb{R}\chi_c}\rho\ne 0$, we apply (IV.7) and (IV.8) to get

$$\begin{aligned}||\operatorname{Pr}_Z f(\rho)||_s &\le (\alpha+\hat\epsilon)||\operatorname{Pr}_Z\rho||_s,\\ ||\operatorname{Pr}_{\mathbb{R}\chi_c} f(\rho)|| = ||Df(0)(\operatorname{Pr}_{\mathbb{R}\chi_c}\rho)&+\operatorname{Pr}_{\mathbb{R}\chi_c} R(\rho)||_s \ge (1-\hat\epsilon)||\operatorname{Pr}_{\mathbb{R}\chi_c}\rho||_s,\end{aligned}$$

and hence

$$\operatorname{sl}(f(\rho))\le\frac{\alpha+\hat\epsilon}{1-\hat\epsilon}\operatorname{sl}(\rho).$$

Thus the assertion follows since the choice of $\hat\epsilon$ implies that $(a+\hat\epsilon)/(1-\hat\epsilon)<1$. □

Fix $h>0$ and $j_1\ge 0$ as in the above Proposition. Then either $0\ne\operatorname{Pr}_{\mathbb{R}\chi_c}\chi_{j_1}\in\mathbb{R}^+\chi_c$ or $0\ne\operatorname{Pr}_{\mathbb{R}\chi_c}\chi_{j_1}\in\mathbb{R}^-\chi_c$.

Proposition IV.3 *If $\operatorname{Pr}_{\mathbb{R}\chi_c}\chi_{j_1}\in\mathbb{R}^+\chi_c$ $(\in\mathbb{R}^-\chi_c)$ then there exists $\delta>0$ such that for all $r\in(0,\delta)$ $(r\in(-\delta,0))$ the following holds.*

(i) $f(r\chi_c)=\hat r\chi_c$ with $0<\hat r<r$ $(r<\hat r<0)$.

(ii) The set $K_h^+\cap[r\chi_c+K_h^-]$ $(K_h^-\cap[r\chi_c+K_h^+])$ is positively invariant under f.

(iii) There exists $j_2=j_2(r)\ge j_1$ such that for all integers $j\ge j_2$, $\chi_j\in K_h^+\cap[r\chi_c+K_h^-]$ $(\in K_h^-\cap[r\chi_c+K_h^+])$.

Proof We give the proof in case $\operatorname{Pr}_{\mathbb{R}\chi_c}\chi_{j_1}\in\mathbb{R}^+\chi_c$. The other case can be dealt with analogously.

1. Proof of (i). Choose $\tilde\delta>0$ such that $(0,\tilde\delta)\chi_c\in\frac{1}{n_1}N_c$ and $\chi_{j_1}\in r\chi_c+K_h^+$ for all $r\in(0,\tilde\delta)$. Observe that for every $r>0$ and $\chi\in K_h^+\cap[r\chi_c+K_h^-]$, $\operatorname{Pr}_{\mathbb{R}\chi_c}\chi=\theta\chi_c$ with $0<\theta<r$ and hence $||\operatorname{Pr}_{\mathbb{R}\chi_c}\chi||_s\le r||\chi_c||_s$, which implies

$$||\chi||_s\le||\operatorname{Pr}_Z\chi||_s+||\operatorname{Pr}_{\mathbb{R}\chi_c}\chi||_s\le(1+h)r||\chi_c||_s.$$

It follows that there exists $\delta \in (0, \tilde{\delta})$ with

$$K_h^+ \cap [r\chi_c + K_h^-] \subset \frac{1}{n_1}[N_z + N_c] \qquad \text{for every } r \in (0, \delta).$$

Let $r \in (0, \delta)$. There exists $\hat{r} \neq 0$ such that $f(r\chi_c) = \hat{r}\chi_c$. By Proposition IV.1,

$$\hat{r}\chi_c = f(0 + r\chi_c) \in f(0) + K_h^+ = K_h^+$$

and hence $\hat{r} > 0$. We want to show $\hat{r} < r$. Assume $\hat{r} \geq r$. Proposition IV.1 yields

$$f\left(\frac{1}{n_1}(N_z + N_c) \cap [r\chi_c + K_h^+]\right) \subset \hat{r}\chi_c + K_h^+.$$

On the other hand, $\hat{r}\chi_c + K_h^+ \subset r\chi_c + K_h^+$ since $\eta \in \hat{r}\chi_c + K_h^+$ implies $\eta = \hat{r}\chi_c + \rho = r\chi_c + \rho + (\hat{r} - r)\chi_c$ for some $\rho \in K_h^+$, and using the assumption $\hat{r} \geq r$, we have

$$\begin{aligned} \mathrm{sl}(\rho + (\hat{r} - r)\chi_c) &= \frac{||\mathrm{Pr}_Z(\rho + (\hat{r} - r)\chi_c)||_s}{||\mathrm{Pr}_{\mathbb{R}\chi_c}(\rho + (\hat{r} - r)\chi_c)||_s} \\ &\leq \frac{||\mathrm{Pr}_Z \rho||_s}{||\mathrm{Pr}_{\mathbb{R}\chi_c} \rho||_s} = \mathrm{sl}(\rho) < h. \end{aligned}$$

Therefore, all χ_j with $j_1 \leq j \in \mathbb{N}$ belong to $r\chi_c + K_h^+$ which contradicts the fact that $\chi_j \to 0$ as $j \to \infty$.

2. Using Proposition IV.1 with $f(0) = 0$, $f(r\chi_c) = \hat{r}\chi_c$ and $0 \leq \hat{r} < r$ and using the fact that $\hat{r}\chi_c + K_h^- \subset r\chi_c + K_h^-$, one gets $f(K_h^+ \cap [r\chi_c + K_h^-]) \subset K_h^+ \cap [r\chi_c + K_h^-]$.

3. Proposition IV.1 gives $\mathrm{Pr}_{\mathbb{R}\chi_c} \chi_j \in \mathbb{R}^+\chi_c$ for all integers $j \geq j_1$. We next find an integer $n_2 \geq n_1$ so that $K_h^+ \cap \frac{1}{n_2}(N_z + N_c) \subset r\chi_c + K_h^-$. Since $\chi_j \to 0$ and $\mathrm{sl}(\chi_j) \to 0$ as $j \to \infty$, we find $j_2 \geq j_1$ so that $\chi_j \in K_h^+ \cap \frac{1}{n_2}(N_z + N_c)$ for all integers $j \geq j_2$. Therefore, $\chi_j \in K_h^+ \cap (r\chi_c + K_h^-)$ for all integers $j \geq j_2$. □

We now derive an application of Proposition IV.3 which is used in Chapter 10.

Consider a C^1-map g from an open neighbourhood V of 0 in a Banach space X over $\mathbb{R}$ into X with $g(0) = 0$. Let σ be the spectrum of $Dg(0)$ and assume that

- $\sigma_s = \sigma \cap \{z \in \mathbb{C} : |z| < 1\}$ is compact;
- $\sigma_c = \sigma \cap S_{\mathbb{C}}^1 = \{1\}$ and the realified generalized eigenspace of $1 \in \sigma$ is one-dimensional;
- $\sigma_u = \sigma \setminus (\sigma_s \cup \sigma_c)$ is nonempty and finite, and the associated realified generalized eigenspace Y is finite-dimensional.

Let $\chi_c \in X$ be an eigenvector of the eigenvalue 1 of $Dg(0)$, and let Z be the realified generalized eigenspace of $Dg(0)$ associated with σ_s. Then $X = \mathbb{R}\chi_c \oplus Z \oplus Y$. By Theorem II.1, there exist open neighbourhoods N_{cs} of 0 in $\mathbb{R}\chi_c \oplus Z$, N_u of 0 in Y, V_1 of 0 in V, and a C^1-map $w_{cs} : N_{cs} \to Y$ with $w_{cs}(0) = 0, Dw_{cs}(0) = 0$ and $w_{cs}(N_{cs}) \subset N_u$ so that the center-stable manifold

$$W^{cs} = \{x_{cs} + w_{cs}(x_{cs}) : x_{cs} \in N_{cs}\}$$

satisfies

$$g(W^{cs} \cap V_1) \subset W^{cs}, \quad \cap_{n=0}^{\infty} g^{-n}(N_{cs} + N_u) \subset W^{cs}.$$

W^{cs} is a C^1-submanifold of X. Consider the C^1-map $g^{cs} : W^{cs} \cap V_1 \to W^{cs}$ given by $g^{cs}(x) = g(x)$. Then $Dg^{cs}(0) : T_0 W^{cs} \to T_0 W^{cs}$ is given by the restriction of

$Dg(0)$ to $T_0W^{cs} = \mathbb{R}\chi_c \oplus Z$. We introduce center and stable manifolds of g^{cs}, as follows.

The projection $\mathrm{Pr}_{\mathbb{R}\chi_c\oplus Z}$ defines a C^1-diffeomorphism A from the C^1-submanifold W^{cs} of X onto the open subset N_{cs} of $\mathbb{R}\chi_c \oplus Z$, and $\tilde{N}_{cs} = A(W^{cs} \cap V_1)$ is an open neighbourhood of 0 in $\mathbb{R}\chi_c \oplus Z$. The map

$$\tilde{g}^{cs} : \tilde{N}_{cs} \ni x \mapsto A(g^{cs}(A^{-1}(x))) \in \mathbb{R}\chi_c \oplus Z$$

is C^1-smooth, with $D\tilde{g}^{cs}(0) = Dg^{cs}(0)$, $D\tilde{g}^{cs}(0)\chi_c = \chi_c$, $D\tilde{g}^{cs}(0)Z \subset Z$; the spectrum of the induced map

$$Z \ni x \mapsto D\tilde{g}^{cs}(0) \in Z$$

is $\sigma_s \subset \mathbb{C}_1$. So there exist an open neighbourhood U_c of 0 in $\mathbb{R}\chi_c$, an open neighbourhood U_s of 0 in Z and C^1-maps $w_c : U_c \to Z$ and $w_s : U_s \to \mathbb{R}\chi_c$ such that the sets

$$W^c = \{x_c + w_c(x_c) : x_c \in U_c\} \subset \tilde{N}_{cs},$$
$$W^s = \{x_s + w_s(x_s) : x_s \in U_s\} \subset \tilde{N}_{cs}$$

are a center and a stable manifold of $\tilde{g}^{cs}$ at the fixed point 0, in the sense of Theorems III.1 and I.2. The sets

$$W^{cs,c} = A^{-1}(W^c) \subset W^{cs} \cap V_1,$$
$$W^{cs,s} = A^{-1}(W^s) \subset W^{cs} \cap V_1$$

are called a center manifold and a stable manifold of g^{cs} at 0, respectively.

Proposition IV.4 *If there is a trajectory $(x_n)_0^\infty$ of g in $W^{cs} \setminus W^{cs,s}$ such that $x_n \to 0$ as $n \to \infty$, then there exist $k \in \mathbb{N}$ and a neighbourhood N of x_k in W^{cs} such that $g(N) \subset N$.*

Proof 1. Choose an open neighbourhood V_2 of 0 in V_1 such that $\mathrm{Pr}_{\mathbb{R}\chi_c\oplus Z}(W^{cs} \cap V_2) \subset U_c + U_s$. Define $B : W^{cs} \cap V_2 \to \mathbb{R}\chi_c \oplus Z$ by

$$B(x_c + x_s + w_{cs}(x_c + x_s)) = x_c - w_s(x_s) + x_s - w_c(x_c).$$

Clearly, B is C^1-smooth,

$$B(0) = 0, \text{ and } DB(0) : T_0(W^{cs} \cap V_2) \to \mathbb{R}\chi_c \oplus Z$$

is the identity map. Note also that for every $x_c + x_s \in \mathrm{Pr}_{\mathbb{R}\chi_c\oplus Z}(W^{cs} \cap V_2)$,

$$\begin{aligned} x_c + x_s + w_{cs}(x_c + x_s) \in W^{cs,c} &\iff x_c + x_s + w_{cs}(x_c + x_s) \in A^{-1}(W^c), \\ &\iff A(x_c + x_s + w_{cs}(x_c + x_s)) \in W^c, \\ &\iff x_c + x_s \in W^c, \\ &\iff x_s = w_c(x_c). \end{aligned}$$

It follows that for every $x_c + x_s \in \mathrm{Pr}_{\mathbb{R}\chi_c\oplus Z}(W^{cs} \cap V_2)$, $\mathrm{Pr}_Z\, B(x_c + x_s + w_{cs}(x_c + x_s)) = 0$ if and only if $x_c + x_s + w_{cs}(x_c + x_s) \in W^{cs,c}$. Similarly, we obtain for such points $\mathrm{Pr}_{\mathbb{R}\chi_c}\, B(x_c + x_s + w_{cs}(x_c + x_s)) = 0$ if and only if $x_c + x_s + w_{cs}(x_c + x_s) \in W^{cs,s}$. Using the Implicit Function Theorem, we find an open neighbourhood V_3 of 0 in V_2

and an open neighbourhood $\tilde{V}_3 \subset U_c + U_s$ such that B defines a C^1-diffeomorphism from $W^{cs} \cap V_3$ onto $\tilde{V}_3$. Let B^{-1} denote its inverse. Observe that we have

$$B(W^{cs,c} \cap V_3) = \tilde{V}_3 \cap \mathbb{R}\chi_c,$$
$$B(W^{cs,s} \cap V_3) = \tilde{V}_3 \cap Z.$$

Due to the local positive invariance of W^{cs} and the continuity of g, there exists an open neighbourhood V_4 of 0 in V_3 such that

$$g(V_4 \cap W^{cs}) \subset V_3 \cap W^{cs}.$$

Then $\tilde{V}_4 = B(V_4)$ is an open neighbourhood of 0 in $\tilde{V}_3$, with

$$B^{-1}(\tilde{V}_4) = W^{cs} \cap V_4.$$

The map

$$f : \tilde{V}_4 \ni x \mapsto B(g^{cs}(B^{-1}(x))) \in \tilde{V}_3$$

is C^1-smooth. We have $f(0) = 0$, so (IV.1) holds. From $DB(0) = \text{id}$ we infer $Df(0) = Dg^{cs}(0)$ which implies that (IV.2) and (IV.3) hold. Moreover,

$$f(\tilde{V}_4 \cap Z) = B(g^{cs}(B^{-1}(\tilde{V}_4 \cap Z))) \subset B(g^{cs}(W^{cs,s} \cap V_4)) \subset B(W^{cs,s} \cap V_3) = \tilde{V}_3 \cap Z \subset Z.$$

Similarly, $f(\tilde{V}_4 \cap \mathbb{R}\chi_c) \subset \mathbb{R}\chi_c$. Therefore, also (IV.4) is satisfied.

2. Since $x_j \in W^{cs} \setminus W^{cs,s}$ for all $j \in \mathbb{N}$ and $x_j \to 0$ as $j \to \infty$, we find $n_0 \in \mathbb{N}$ such that $x_j \in W^{cs} \cap V_4$ for $n_0 \leq j \in \mathbb{N}$. The points $\chi_j = B(x_j) \in \tilde{V}_4$, $n_0 \leq j \in \mathbb{N}$, satisfy $\chi_{j+1} = f(\chi_j)$ and $\chi_j \notin Z$, and we have $\chi_j \to 0$ as $j \to \infty$.

Choose $h > 0$. There exists $r_0 > 0$ such that for all $r \in (0, r_0)$,

$$K_h^+ \cap [r\chi_c + K_h^-] \subset \tilde{V}_4$$

and for all $r \in (-r_0, 0)$,

$$K_h^- \cap [r\chi_c + K_h^+] \subset \tilde{V}_4.$$

Proposition IV.3 guarantees that either there exist $r \in (0, r_0)$ with

$$f\left(K_h^+ \cap [r\chi_c + K_h^-]\right) \subset K_h^+ \cap [r\chi_c + K_h^-],$$

and $k \in \mathbb{N}$ with $\chi_j \in K_h^+ \cap [r\chi_c + K_h^-]$ for all integers $j \geq k$, or there exist $r \in (-r_0, 0)$ with

$$f\left(K_h^- \cap [r\chi_c + K_h^+]\right) \subset K_h^- \cap [r\chi_c + K_h^+]$$

and $k \in \mathbb{N}$ with $\chi_j \in K_h^- \cap [r\chi_c + K_h^+]$ for all integers $j \geq k$. In the first case,

$$N = B^{-1}\left(K_h^+ \cap [r\chi_c + K_h^-]\right)$$

is the desired neighbourhood of x_k in W^{cs}. In the second case, set

$$N = B^{-1}\left(K_h^- \cap [r\chi_c + K_h^+]\right).$$

□

APPENDIX V

Unstable sets of periodic orbits

Let E be a Banach space over $\mathbb{R}$ with norm $||\cdot||$. Consider a continuous semiflow $S : \mathbb{R}^+ \times E \to E$. Assume that $y : \mathbb{R} \to E$ is a periodic phase curve (i.e., $y(t+s) = S(t, y(s))$ for all $t \geq 0$ and $s \in \mathbb{R}$) which is not constant and has period $\omega > 0$. Let S be continuously differentiable on $(t_0, \infty) \times E$ for some $t_0 \in (0, \omega)$. Also let a closed linear subspace $H \subset E$ of codimension 1 be given so that $y'(\omega) \notin H$. Applying the Implicit Function Theorem as in Appendix I, one can show that there exist a convex bounded open neighborhood N of $y(0) = y(\omega)$ in E, $\epsilon \in (0, \omega)$, and a C^1-map $\tau : N \to (\omega - \epsilon, \omega + \epsilon)$ with the following properties:

(a) $\sup\{||D_1S(t,x)1|| : (t,x) \in (\omega - \epsilon, \omega + \epsilon) \times N\} < \infty$.

(b) $\tau(y(0)) = \omega$, and $S(\tau(x), x) \in y(0) + H$ for all $x \in N$.

(c) For all $(t,x) \in (\omega - \epsilon, \omega + \epsilon) \times N$ with $S(t,x) \in y(0) + H$, $t = \tau(x)$.

(d) $D_1S(\tau(x), x)1 \notin H$ for all $x \in N$.

(e) $\sup_{x \in N} ||D\tau(x)|| < \infty$.

We consider the Poincaré map $P : (y(0) + H) \cap N \ni x \mapsto S(\tau(x), x) \in y(0) + H$. Let σ^P denote the spectrum of $DP(y(0)) : H \to H$. We assume that the monodromy operator $D_2S(\omega, y(0)) : E \to E$ is compact, and denote its spectrum by σ^S. Recall that the sets

$$\sigma_s^S = \sigma^S \cap \{\mu \in \mathbb{C} : |\mu| < 1\}$$

and

$$\sigma_s^P = \sigma^P \cap \{\mu \in \mathbb{C} : |\mu| < 1\}$$

coincide, as well as $\sigma_u^S = \sigma^S \cap \{\mu \in \mathbb{C} : 1 < |\mu|\}$ and $\sigma_u^P = \sigma^P \cap \{\mu \in \mathbb{C} : 1 < |\mu|\}$. Set $\sigma_c^S = \sigma^S \cap S^1_{\mathbb{C}}$ and $\sigma_c^P = \sigma^P \cap S^1_{\mathbb{C}}$. The realified generalized eigenspaces H_s^P, H_c^P, H_u^P of $DP(y(0))$ associated with σ_s^P, σ_c^P, σ_u^P, respectively, satisfy

$$H = H_s^P \oplus H_c^P \oplus H_u^P,$$

and the realified generalized eigenspaces E_s^S, E_c^S, E_u^S of $D_2S(\omega, y(0))$ associated with σ_s^S, σ_c^S, σ_u^S, respectively, satisfy

$$E = E_s^S \oplus E_c^S \oplus E_u^S.$$

There is a constant $\lambda \in (0,1)$ so that

$$\lambda > \max\left\{ \sup_{\mu \in \sigma_s^S} |\mu|, \sup_{\mu \in \sigma_u^S} \frac{1}{|\mu|} \right\}.$$

Applying Theorem I.3 to $p = y(0)$, $U = (y(0) + H) \cap N$, $f = P$ and to $p = y(0)$, $U = N$, $f = S(\omega, \cdot)$, we find norms $||\cdot||_P$ and $||\cdot||_S$, which are equivalent to $||\cdot||_H$ and $||\cdot||$, respectively, constants α_P, α_S in $(0, \lambda)$, convex open neighbourhoods

$N_P^u \subset H \cap (N - y(0))$ of 0 in H, $N_S^u \subset N - y(0)$ of 0 in E, and open neighbourhoods $\widetilde{N_P^u} \subset N_P^u$ of 0 in H and $\widetilde{N_S^u} \subset N_S^u$ of 0 in E such that

$$\begin{aligned} W_P^u &= W^u(y(0), P, N_P^u) \\ &= \{x \in y(0) + N_P^u : \text{There is a trajectory } (x_n)_{-\infty}^0 \text{ of } P \text{ such that } x_0 = x, \\ &\qquad \lambda^n(x_n - y(0)) \in N_P^u \text{ for all } n \in -\mathbb{N}, \text{ and } \lambda^n(x_n - y(0)) \to 0 \text{ as } n \to -\infty\} \end{aligned}$$

and

$$\begin{aligned} W_S^u &= W^u(y(0), S(\omega,\cdot), N_S^u) \\ &= \{x \in y(0) + N_S^u : \text{There is a trajectory } (x_n)_{-\infty}^0 \text{ of } S(\omega,\cdot) \text{ such that } x_0 = x, \\ &\qquad \lambda^n(x_n - y(0)) \in N_S^u \text{ for all } n \in -\mathbb{N}, \text{ and } \lambda^n(x_n - y(0)) \to 0 \text{ as } n \to -\infty\} \end{aligned}$$

are C^1-submanifolds. Moreover, the mapping $W_P^u \cap (y(0) + \widetilde{N_P^u}) \ni x \mapsto P(x) \in y(0) + H$ defines a C^1-diffeomorphism P_u onto W_P^u with

$$||P_u^{-1}(x) - P_u^{-1}(y)||_P \leq \alpha_P ||x - y||_P \qquad \text{for all } x, y \text{ in } W_P^u,$$

and the mapping $W_S^u \cap (y(0) + \widetilde{N_S^u}) \ni x \mapsto S(\omega, x) \in E$ defines a C^1-diffeomorphism $S(\omega,\cdot)_u$ onto W_S^u with

$$||(S(\omega,\cdot)_u)^{-1}(x) - (S(\omega,\cdot)_u)^{-1}(y)||_S \leq \alpha_S ||x - y||_S \qquad \text{for all } x, y \text{ in } W_S^u.$$

We define the unstable set $W^u(y)$ of the periodic phase curve y by

$$W^u(y) = \bigcup_{t \geq 0} S(t, W_S^u).$$

The aim of this appendix is to show that

$$W^u(y) = \bigcup_{t \geq 0} S(t, W_P^u).$$

Proposition V.1 $W_P^u \subset W^u(y)$.

Proof 1. For every $x \in W_P^u$, there is a trajectory $(x_n)_{-\infty}^0$ of P such that $x_0 = x$, $\lambda^n(x_n - y(0)) \in N_P^u$ for all $n \in -\mathbb{N}$, and $\lambda^n(x_n - y(0)) \to 0$ as $n \to -\infty$. By the definition of $\widetilde{N_P^u}$, $x_n \in W_P^u \cap (y(0) + \widetilde{N_P^u})$ and $||x_n - y(0)||_P \leq \alpha_P ||x_{n+1} - y(0)||_P$ for all integers $n \leq -1$. Let $\tau_0 = 0$, $\tau_n = \tau(x_n)$ for all integers $n \leq -1$. Then $\tau_n \in (\omega - \epsilon, \omega + \epsilon)$ for all integers $n \leq -1$.

We claim that there exists $t^* \in \mathbb{R}$ such that $k\omega + \tau_{-1} + \cdots \tau_k \to t^*$ as $k \to -\infty$. To prove the claim, we use (e) and the equivalence of the norms $||\cdot||_P$ and $||\cdot||$ to find constants $K_1 > 0$ and $K_2 > 0$ such that

$$\begin{aligned} K_1 &\geq \sup_{x \in N} ||D\tau(x)||, \\ ||x|| &\leq K_2 ||x||_P \qquad \text{for all } x \in H. \end{aligned}$$

Then, we have

$$\begin{aligned} |\tau_n - \omega| &= |\tau(x_n) - \tau(y(0))| \leq K_1 ||x_n - y(0)|| \leq K_1 K_2 ||x_n - y(0)||_P \\ &\leq K_1 K_2 \alpha_P ||x_{n+1} - y(0)||_P \leq K_1 K_2 \alpha_P^2 ||x_{n+2} - y(0)||_P \leq \ldots \\ &\leq K_1 K_2 \alpha_P^{-n} ||x - y(0)||_P \end{aligned}$$

for all integers $n \le -1$. It follows that there exists $t^* \in \mathbb{R}$ such that the numbers

$$k\omega + \tau_{-1} + \ldots + \tau_k = [\tau_{-1} - \omega] + \ldots + [\tau_k - \omega], \qquad -1 \ge k \in -\mathbb{N},$$

are convergent to t^* as $k \to -\infty$, and for every integer $k \le -1$

$$\omega_k = k\omega - t^* + \tau_{-1} + \cdots + \tau_k$$

satisfies

$$|\omega_k| \le \sum_{-\infty}^{k-1} |\tau_n - \omega| \le \frac{K_1 K_2}{1-\alpha_P} \alpha_P^{-k+1} ||x - y(0)||_P.$$

2. Define $y^* : (-\infty, 0] \to E$ by

$$y^*(t) = S(t + \tau_0 + \tau_{-1} + \ldots + \tau_{k-1}, x_{k-1})$$

if $t \in (-\tau_0 - \tau_{-1} - \ldots - \tau_{k-1}, -\tau_0 - \tau_{-1} - \ldots - \tau_k]$ for some $k \in -\mathbb{N}$. As each τ_k with $k < 0$ is positive, the intervals involved in the above definition give a disjoint union of $(-\infty, 0]$. We claim that

$$y^*(t+s) = S(t, y^*(s)) \qquad \text{if } t \ge 0,\ s \le 0,\ t+s \le 0.$$

In fact, we can find integers $k \le 0$ and $m \le 0$ such that $s \in (-\tau_0 - \tau_{-1} - \ldots - \tau_{k-1}, -\tau_0 - \tau_{-1} - \ldots - \tau_k]$ and $t+s \in (-\tau_0 - \tau_{-1} - \ldots - \tau_{m-1}, -\tau_0 - \tau_{-1} - \ldots - \tau_m]$. As $t \ge 0$, we have $m \ge k$ and thus

$$x_{m-1} = S(\tau_{m-2}, x_{m-2}) = \ldots = S(\tau_{m-2} + \tau_{m-3} + \ldots + \tau_{k-1}, x_{k-1}).$$

Consequently,

$$\begin{aligned} y^*(t+s) &= S(t + s + \tau_0 + \ldots + \tau_{m-1}, x_{m-1}) \\ &= S(t + s + \tau_0 + \ldots + \tau_{m-1}, S(\tau_{m-2} + \ldots + \tau_{k-1}, x_{k-1})) \\ &= S(t + s + \tau_0 + \ldots + \tau_{k-1}, x_{k-1}) \\ &= S(t, S(s + \tau_0 + \tau_{-1} + \ldots + \tau_{k-1}, x_{k-1})) \\ &= S(t, y^*(s)). \end{aligned}$$

3. We claim that $\lambda^k[y^*(k\omega - t^*) - y(0)] \to 0$ as $k \to -\infty$. To verify this, we first note that

$$y^*(-\tau_0 - \tau_{-1} - \ldots - \tau_k) = x_k.$$

Note also that

$$k\omega - t^* = -\tau_0 - \tau_{-1} - \ldots - \tau_k + \omega_k = -\tau_0 - \tau_{-1} - \ldots - \tau_{k-1} + \omega_k + \tau_{k-1}.$$

Since $\tau_{k-1} \in (\omega - \epsilon, \omega + \epsilon)$ for all integers $k \le 0$ and $\omega_k \to 0$ as $k \to -\infty$, we must have $\omega_k + \tau_{k-1} > 0$ for large negative integers k. Hence, for large negative integers k, using the result in part 2 we have

$$\begin{aligned} &||y^*(k\omega - t^*) - y(0)||_S \\ &\quad \le ||y^*(k\omega - t^*) - x_k||_S + ||x_k - y(0)||_S \\ &\quad \le ||S(\tau_0 + \tau_{-1} + \ldots + \tau_{k-1} + k\omega - t^*, x_{k-1}) - S(\tau_{k-1}, x_{k-1})||_S + ||x_k - y(0)||_S \\ &\quad = ||S(\tau_{k-1} + \omega_k, x_{k-1}) - S(\tau_{k-1}, x_{k-1})||_S + ||x_k - y(0)||_S. \end{aligned}$$

From $x \in W_P^u$ it follows that $x_k \in N$ for all $k \in -\mathbb{N}$, and $x_k \to y(0)$ as $k \to -\infty$. Then $\tau_k = \tau(x_k) \to \tau(y(0)) = \omega$ as $k \to -\infty$. Using that $\omega_k \to 0$ as $k \to -\infty$,

$\omega_k + \tau_{k-1} \in (\omega - \epsilon, \omega + \epsilon)$ for all sufficiently large negative integers k. From (a) and from the equivalence of the norms we can find a constant K_3 so that

$$||S(t,x) - S(s,x)||_S \le K_3|t-s| \qquad \text{for all } (t,x),(s,x) \text{ in } (\omega-\epsilon,\omega+\epsilon)\times N.$$

Therefore, for large negative integers k, we have

$$\begin{aligned} ||y^*(k\omega - t^*) - y(0)||_S &\le K_3|\omega_k| + ||x_k - y(0)||_S \\ &\le K_3K_1K_2\frac{\alpha_P^{-k+1}}{1-\alpha_P}||x-y(0)||_P + ||x_k - y(0)||_S, \end{aligned}$$

and thereby

$$\lambda^k||y^*(k\omega - t^*) - y(0)||_S \le \frac{K_1K_2K_3\alpha_P}{1-\alpha_P}\left(\frac{\lambda}{\alpha_P}\right)^k ||x-y(0)||_P + \lambda^k||x_k - y(0)||_S.$$

The last estimate implies the assertion.

4. Choose $k \in -\mathbb{N}$ so that $-k\omega + t^* > 0$ and $\lambda^{k+n}[y^*((k+n)\omega - t^*) - y(0)] \in N_S^u$ for all integers $n \le 0$. The points $z_n = y^*(k\omega + n\omega - t^*)$, $n \in -\mathbb{N}$, satisfy $z_n = S(\omega, z_{n-1})$ and

$$\begin{aligned} \lambda^n[z_n - y(0)] &= \lambda^n[y^*(k\omega + n\omega - t^*) - y(0)] \\ &= \lambda^{-k}\lambda^{n+k}[y^*(k\omega + n\omega - t^*) - y(0)] \in \lambda^{-k}N_S^u \subset N_S^u. \end{aligned}$$

The above equality also implies $\lambda^n(z_n - y(0)) \to 0$ as $n \to -\infty$. Therefore, $z_0 = y^*(k\omega - t^*) \in W_S^u$ and thus

$$x = y^*(0) = S(-k\omega + t^*, z_0) \in S(-k\omega + t^*, W_S^u).$$

This implies $x \in \cup_{t\ge 0}S(t, W_S^u) = W^u(y)$. □

Proposition V.2 *There exists $\delta > 0$ so that $y(0) + E_\delta \subset N$ and*

$$S(\tau(x), x) \in W_P^u \qquad \textit{for all } x \in (y(0) + E_\delta)\cap W_S^u.$$

Proof 1. Let K_1, K_2, K_3, K_4 be positive constants such that

$$\left.\begin{aligned} &\text{(i)} && \sup_{x\in N}||D\tau(x)|| \le K_1, \\ &\text{(ii)} && K_3||x||_S \le ||x|| \le K_2||x||_S \quad \text{for all } x \in E, \\ &\text{(iii)} && ||S(t,x) - S(s,x)|| \le K_4|t-s| \quad \text{for all } (t,x),(s,x) \text{ in } (\omega-\epsilon,\omega+\epsilon)\times N. \end{aligned}\right\} \tag{V.1}$$

Choose $\delta > 0$ so small that the following holds:

$$\begin{aligned} &\text{(i)} && y(0) + E_{K_2(K_1K_2K_4K_3^{-1}+1)\delta} \subset N \cap (\widetilde{N_P^u} + y(0)), \\ &\text{(ii)} && \tau(y(0) + E_{K_2\delta}) \subset (\omega - \epsilon/2, \omega + \epsilon/2). \end{aligned} \tag{V.2}$$

Let $x \in (y(0) + E_\delta)\cap W_S^u$. Set $x^* = S(\tau(x), x)$. There exists a trajectory $(x_n)_{-\infty}^0$ of $S(\omega,\cdot)$ such that $x_0 = x$, $\lambda^n[x_n - y(0)] \in N_S^u$ for all integers $n \le 0$, and

$$\lambda^n[x_n - y(0)] \to 0 \qquad \text{as } n \to -\infty. \tag{V.3}$$

From the definition of $\widetilde{N_S^u}$, we have

$$x_n \in \widetilde{N_S^u} + y(0) \qquad \text{for all integers } n \le -1 \tag{V.4}$$

and

$$||x_n - y(0)||_S \leq \alpha_S ||x_{n+1} - y(0)||_S \leq \ldots \leq \alpha_S^{-n} ||x - y(0)||_S < \delta \quad \text{for all } n \in -\mathbb{N}. \tag{V.5}$$

This implies, by (ii) of (V.1), the estimate

$$||x_n - y(0)|| < K_2 \delta \qquad \text{for all } n \in -\mathbb{N}. \tag{V.6}$$

By (i) of (V.2), we have $x_n \in N$ and thus, $\tau(x_n)$ is defined for $n \leq 0$. (ii) of (V.2) implies that

$$\tau(x_n) \in (\omega - \epsilon/2, \omega + \epsilon/2) \qquad \text{for all } n \in -\mathbb{N}. \tag{V.7}$$

2. Define $z : (-\infty, \tau(x_0)] \to E$ by

$$z(t) = \begin{cases} S(t - (n-1)\omega, x_{n-1}) & \text{for } t \in ((n-1)\omega, n\omega] \text{ and } n \in -\mathbb{N}, \\ S(t, x) & \text{for } t \in (0, \tau(x_0)]. \end{cases}$$

Clearly,

$$z(n\omega) = x_n \qquad \text{for all } n \in -\mathbb{N} \tag{V.8}$$

and

$$z(\tau(x_0)) = S(\tau(x_0), x) = S(\tau(x), x) = x^*.$$

Using the same argument as that in part 2 of the proof for Proposition V.1, we have

$$z(t+s) = S(t, z(s)) \qquad \text{for all } t \geq 0 \text{ and } s \leq 0 \text{ with } t + s \leq \tau(x_0). \tag{V.9}$$

For each $n \in -\mathbb{N}$ set

$$t_n = \tau(x_{n-1}) + (n-1)\omega.$$

Then (V.8) and (V.9) imply

$$z(t_n) = S(\tau(x_{n-1}), x_{n-1}) \in y(0) + H \qquad \text{for all } n \in -\mathbb{N}. \tag{V.10}$$

3. We prove that $z(t_n) \in N \cap (\widetilde{N_P^u} + y(0))$ for all $n \in -\mathbb{N}$. Let $n \in -\mathbb{N}$. In fact, we have

$$\begin{aligned}
&||z(t_n) - y(0)||_S \\
&\quad \leq ||S(\tau(x_{n-1}), x_{n-1}) - x_n||_S + ||x_n - y(0)||_S \\
&\quad = ||S(\tau(x_{n-1}), x_{n-1}) - S(\omega, x_{n-1})||_S + ||x_n - y(0)||_S \\
&\quad \leq \frac{K_4}{K_3} |\tau(x_{n-1}) - \omega| + ||x_n - y(0)||_S && \text{(by (ii) and (iii) of (V.1))} \\
&\quad = \frac{K_4}{K_3} |\tau(x_{n-1}) - \tau(y(0))| + ||x_n - y(0)||_S \\
&\quad \leq \frac{K_4}{K_3} K_1 ||x_{n-1} - y(0)|| + ||x_n - y(0)||_S && \text{(by (i) of (V.1))} \\
&\quad \leq \frac{K_1 K_2 K_4}{K_3} ||x_{n-1} - y(0)||_S + ||x_n - y(0)||_S && \text{(by (ii) of (V.1))} \\
&\quad < (K_1 K_2 K_4 K_3^{-1} + 1)\delta && \text{(by (V.5))}.
\end{aligned}$$

Thus,

$$||z(t_n) - y(0)|| < K_2 (K_1 K_2 K_4 K_3^{-1} + 1)\delta,$$

and (V.2) gives $z(t_n) \in N \cap (\widetilde{N_P^u} + y(0))$.

4. We prove that $P(z(t_{n-1})) = z(t_n)$ for all $n \in -\mathbb{N}$. Let $n \in -\mathbb{N}$. Note that (V.7) implies

$$|\tau(x_{n-1}) - \tau(x_{n-2})| < \epsilon.$$

Therefore,

$$t_n - t_{n-1} = \tau(x_{n-1}) - \tau(x_{n-2}) + \omega \in (\omega - \epsilon, \omega + \epsilon).$$

On the other hand, we have

$$z(t_n) = S(t_n - t_{n-1}, z(t_{n-1})),$$

and by (V.10), $z(t_n) \in y(0) + H$. Therefore, by (c) of the properties of τ, we must have

$$t_n - t_{n-1} = \tau(z(t_{n-1}))$$

and hence

$$z(t_n) = P(z(t_{n-1})).$$

5. We prove $\lambda^n[z(t_n) - y(0)] \to 0$ as $n \to -\infty$. In fact, using the estimate in part 3 and (V.3), we obtain

$$||\lambda^n[z(t_n) - y(0)]||_S \leq \frac{K_1K_2K_4}{K_3}\lambda\lambda^{n-1}||x_{n-1} - y(0)||_S + \lambda^n||x_n - y(0)||_S \text{ for } n \in -\mathbb{N}$$

which implies the assertion.

6. The conclusion in part 5 enables us to find a $k \in -\mathbb{N}$ such that $z(t_k) \in W^u_P$. In case $k < 0$, the relation $z(t_k) \in \widetilde{N^u_P} + y(0)$ due to part 3 gives

$$z(t_{k+1}) = P(z(t_k)) \in W^u_P.$$

Repeating the above argument, we get $z(t_0) \in W^u_P \cap (\widetilde{N^u_P} + y(0))$.

Note that $|\tau(x_0) - \tau(x_{-1})| < \epsilon$ implies

$$\tau(x_0) - t_0 = \tau(x_0) - \tau(x_{-1}) + \omega \in (\omega - \epsilon, \omega + \epsilon),$$

and, in particular, $\tau(x_0) - t_0 > 0$. Therefore, using property (V.9), we have

$$x^* = z(\tau(x_0)) = S(\tau(x_0) - t_0, z(t_0)) = S(\tau(x_0) - \tau(x_{-1}) + \omega, z(t_0)).$$

Since $z(t_0) \in W^u_P \cap (\widetilde{N^u_P} + y(0)) \subset N$ and $\tau(x_0) - \tau(x_{-1}) + \omega \in (\omega - \epsilon, \omega + \epsilon)$, property (c) of τ implies that

$$\tau(x_0) - \tau(x_{-1}) + \omega = \tau(z(t_0))$$

and hence

$$x^* = S(\tau(z(t_0)), z(t_0)) = P(z(t_0)).$$

As $z(t_0) \in W^u_P \cap (\widetilde{N^u_P} + y(0))$, we conclude that $x^* \in W^u_P$. □

Theorem V.3 $W^u(y) = \cup_{t \geq 0} S(t, W^u_P)$.

Proof Proposition V.1 implies $\cup_{t\geq 0}S(t, W^u_P) \subset \cup_{t\geq 0}S(t, W^u_S) = W^u(y)$. Now consider $x \in W^u(y)$. Then there exist $t \geq 0$ and $x^* \in W^u_S$ such that $x = S(t, x^*)$. So, there is a trajectory $(x^*_n)^0_{-\infty}$ of $S(\omega, \cdot)$ with $x^*_0 = x^*$ and $\lambda^n[x^*_n - y(0)] \to 0$ as $n \to -\infty$. Choose $\delta > 0$ according to Proposition V.2. There is $n \in -\mathbb{N}$ such that $t - n\omega > \omega + \epsilon$ and $x^*_n \in (y(0) + E_\delta) \cap W^u_S$. Therefore, $S(\tau(x^*_n), x^*_n) \in W^u_P$ and thus $x = S(t - n\omega - \tau(x^*_n), S(\tau(x^*_n), x^*_n) \in S(t - n\omega - \tau(x^*_n), W^u_P) \subset \cup_{t\geq 0}S(t, W^u_P)$. □

APPENDIX VI

A discrete Lyapunov functional and a-priori estimates

In this appendix we introduce a discrete, integer-valued Lyapunov functional, specializing and modifying work begun by Mallet-Paret **[42]**. A more general version, which requires more complicated proofs, is contained in **[43, 44]**. The functional counts sign changes of segments x_t along solutions to certain delay differential equations.

Define a map $\mathrm{sc} : C \setminus \{0\} \to \mathbb{N} \cup \{\infty\}$ as follows:

$$\mathrm{sc}(\phi) = 0 \qquad \text{if either } \phi \geq 0 \text{ or } \phi \leq 0;$$

otherwise,

$$\mathrm{sc}(\phi) = \sup\bigl\{k \in \mathbb{N} \setminus \{0\} : \text{There exists a strictly increasing finite sequence } (s^i)_0^k \text{ in } [-1, 0] \text{ with } \phi(s^{i-1})\phi(s^i) < 0 \text{ for all } i \in \{1, 2, \dots, k\}\bigr\}.$$

Using the above sign change function, we define a functional $V : C \setminus \{0\} \to 2\mathbb{N}$ by

$$V(\phi) = \begin{cases} \mathrm{sc}(\phi) & \text{if } \mathrm{sc}(\phi) \text{ is even or } \infty, \\ \mathrm{sc}(\phi) + 1 & \text{if } \mathrm{sc}(\phi) \text{ is odd.} \end{cases}$$

Useful for the evaluation of V is the set

$$\begin{aligned} R = \{\phi \in C^1 : \, & \phi(0) \neq 0 \text{ or } \dot{\phi}(0)\phi(-1) > 0, \\ & \phi(-1) \neq 0 \text{ or } \dot{\phi}(-1)\phi(0) < 0, \\ & \text{all zeros of } \phi \text{ in } (-1, 0) \text{ are simple}\}. \end{aligned}$$

The basic properties of V are contained in the next two lemmas. The proofs here are simpler for our special case than those in Mallet-Paret and Sell **[43, 44]** for the general case.

Lemma VI.1

(i) For every $\phi \in C \setminus \{0\}$ and for every sequence $(\phi_n)_0^\infty$ in $C \setminus \{0\}$ with $\phi_n \to \phi$ as $n \to \infty$,

$$V(\phi) \leq \liminf_{n \to \infty} V(\phi_n).$$

(ii) For every $\phi \in R$ and for every sequence $(\phi_n)_0^\infty$ in $C^1 \setminus \{0\}$ with $||\phi_n - \phi||_{C^1} \to 0$ as $n \to \infty$,

$$V(\phi) = \lim_{n \to \infty} V(\phi_n) < \infty.$$

Proof 1. Proof of (i). The case $V(\phi) = 0$ is obvious. If $V(\phi) = \infty$ then $\phi = \lim_{n\to\infty} \phi_n$ easily implies that $V(\phi_n) \to \infty$ as $n \to \infty$, and thus $V(\phi) = \infty = \liminf_{n\to\infty} V(\phi_n)$. Assume that $0 < V(\phi) < \infty$. Choose $k \in \mathbb{N}$ and reals $s^0, s^1, \dots, s^k$ in $[-1, 0]$ so that $k = \mathrm{sc}(\phi)$, $-1 \le s^0 < s^1 < \dots < s^k \le 0$ and $\phi(s^{i-1})\phi(s^i) < 0$ for all $i \in \{1, \dots, k\}$. Let $\delta = \min\{|\phi(s^i)| : i \in \{0, 1, \dots, k\}\}$. Select $n_0 \in \mathbb{N}$ such that $||\phi_n - \phi|| < \delta$ for all integers $n \ge n_0$. It follows that $\phi_n(s^{i-1})\phi_n(s^i) < 0$ for all integers $n \ge n_0$ and $i \in \{1, \dots, k\}$. Consequently, $\mathrm{sc}(\phi_n) \ge k = \mathrm{sc}(\phi)$ for all integers $n \ge n_0$, and hence $V(\phi) \le \liminf_{n\to\infty} V(\phi_n)$.

2. Proof of (ii). Let $\phi \in R$. If ϕ has no zeros then $V(\phi) = 0$, and it is easy to see that for n sufficiently large ϕ_n has no zero; it follows that $V(\phi_n) = 0$. Otherwise the number of zeros of ϕ is finite, and they are all simple. Suppose the zeros are given by a strictly increasing finite sequence $(z^i)_0^k$, $k \in \mathbb{N}$. For $\epsilon > 0$ and $i \in \{0, \dots, k\}$, let $N(z^i, \epsilon)$ denote the open ϵ-neighbourhood of z^i in $[-1, 0]$. We can choose $\epsilon \in (0, 1)$ so small that the intervals $N(z^0, \epsilon), N(z^1, \epsilon), \dots, N(z^k, \epsilon)$ are pairwise disjoint, and

$$
\begin{aligned}
|\phi(s)| &> \epsilon^2 \quad \text{for all } s \in [-1, 0] \setminus \cup_{i=0}^k N(z^i, \epsilon),\\
|\dot\phi(s)| &> \epsilon \quad \text{for all } s \in \cup_{i=0}^k N(z^i, \epsilon).
\end{aligned}
$$

In case $-1 < z^0$ we can achieve $-1 \notin N(z^0, \epsilon)$, and in case $z^k < 0$ we can achieve $0 \notin N(z^k, \epsilon)$. Select $n_0 \in \mathbb{N}$ so that $||\phi_n - \phi||_{C^1} < \epsilon^2$ for all integers $n \ge n_0$. Let $n \in \mathbb{N}$ with $n \ge n_0$ be given. Then

$$
\begin{aligned}
\phi_n(s)\phi(s) &> 0 \qquad \text{for all } s \in [-1, 0] \setminus \cup_{i=0}^k N(z^i, \epsilon),\\
\dot\phi_n(s)\dot\phi(s) &> 0 \qquad \text{for all } s \in \cup_{i=0}^k N(z^i, \epsilon).
\end{aligned}
$$

Hence, in the case $-1 < z^0 \le z^k < 0$ it is easy to see that $\mathrm{sc}(\phi_n) = \mathrm{sc}(\phi)$. Now assume $z^0 = -1$. Then $z^k \ne 0$ because of $\phi \in R$. It follows that either $\mathrm{sc}(\phi_n) = \mathrm{sc}(\phi)$ or $\mathrm{sc}(\phi_n) = \mathrm{sc}(\phi) + 1$. From $\dot\phi(-1)\phi(0) < 0$ and the simplicity of the zeros of ϕ it follows that $\mathrm{sc}(\phi)$ is odd. Consequently, $V(\phi_n) = V(\phi)$. The case $z^k = 0$ is analogous. □

Assume that $I \subseteq \mathbb{R}$ is an interval, $\mu \ge 0$, $b : I \to \mathbb{R}$ is a positive continuous function, $z : I + [-1, 0] \to \mathbb{R}$ is continuous, $z|_I$ is differentiable, and

$$
\dot z(t) = -\mu z(t) + b(t) z(t-1) \tag{VI.1}
$$

for all $t > \inf I$ in I. Then the function

$$
y : I + [-1, 0] \ni t \mapsto e^{\mu t} z(t) \in \mathbb{R}
$$

satisfies

$$
\dot y(t) = e^{\mu} b(t) y(t-1) \tag{VI.2}
$$

for all $t > \inf I$ in I. From the definition of y it is obvious that for all $t \in I$ with $z_t \in C \setminus \{0\}$, $y_t \in C \setminus \{0\}$ and $\mathrm{sc}(z_t) = \mathrm{sc}(y_t)$, $V(z_t) = V(y_t)$. Using (VI.2) it is easy to see that if $z(t) \ne 0$ for some $t \in I + [-1, 0]$, then $z_t \ne 0$ and $y_t \ne 0$ for all $t \in I$. Let us show that

$$
z_t \in R \qquad \text{for all } t \in I \text{ with } t - 1 \in I \text{ and } y_t \in R.
$$

To see this, let $t \in I$ with $t - 1 \in I$ and $y_t \in R$ be given. Then we have $z_t \in C^1$ and

$$
\dot y(s) = e^{\mu s}(\mu z(s) + \dot z(s)) \tag{VI.3}
$$

for $s \in (t-1,t]$; at $s=t-1$ the analogue of (VI.3) for the right derivatives $\overset{+}{y}(t-1)$ and $\overset{+}{z}(t-1)$ holds. Assume that $s \in (t-1,t)$ and $z(s)=0$. From $y_t \in R$ and $y(s)=e^{\mu s}z(s)$, it follows that $y(s)=0$ and $\dot{y}(s) \neq 0$. Hence (VI.3) implies $\dot{z}(s) \neq 0$. Now assume that $z(t)=0$. Then $y_t \in R$ and $y(t)=e^{\mu t}z(t)$ imply $y(t)=0$ and $\dot{y}(t)y(t-1)>0$. Hence, (VI.3) with $s=t$ yields $\dot{y}(t)\dot{z}(t)>0$. Using equations (VI.1), (VI.2) and the facts that $z(t)=0$ and $b(t)>0$, it follows that $y(t-1)z(t-1)>0$. Therefore, $\dot{z}(t)z(t-1)>0$. Finally, assume that $z(t-1)=0$. Then $y(t-1)=0$ follows. From $y_t \in R$, we obtain $\overset{+}{y}(t-1)y(t)<0$. This fact and $y(t)=e^{\mu t}z(t)$ imply $y(t)z(t)>0$. Using (VI.3) with $s=t-1$, it follows that $\overset{+}{y}(t-1)\overset{+}{z}(t-1)>0$. From the relations $y(t)z(t)>0$, $\overset{+}{y}(t-1)y(t)<0$ and $\overset{+}{y}(t-1)\overset{+}{z}(t-1)>0$, we conclude that $\overset{+}{z}(t-1)z(t)<0$. Consequently, $z_t \in R$.

Lemma VI.2 *Assume that I is an interval in $\mathbb{R}$, $\mu \geq 0$, $b: I \to \mathbb{R}$ is a positive continuous function, $z: I+[-1,0] \to \mathbb{R}$ is continuous, $z|_I$ is differentiable, and (VI.1) holds for all $t > \inf I$ in I. Assume also $z(t) \neq 0$ for some $t \in I+[-1,0]$. Then*

(i) for all t_1, t_2 in I with $t_1 < t_2$, $V(z_{t_1}) \geq V(z_{t_2})$;

(ii) for all $t \in I$ with $t-2 \in I$ and $z(t)=z(t-1)=0$, either $V(z_t)=\infty$ or $V(z_{t-2}) > V(z_t)$;

(iii) for all $t \in I$ with $t-3 \in I$ and $V(z_{t-3}) = V(z_t) < \infty$, $z_t \in R$.

Proof 1. Consider $y: I+[-1,0] \ni t \mapsto e^{\mu t}z(t) \in \mathbb{R}$. By the remarks preceding Lemma VI.2, it is enough to show (i),(ii) and (iii) for y instead of z.

2. Proof of (i) for y instead of z. We claim that it suffices to show that for all $t \in I$ there exists $\epsilon_0 = \epsilon_0(t) > 0$ such that for all $\epsilon \in [0,\epsilon_0]$ with $t+\epsilon \in I$,

$$V(y_t) \geq V(y_{t+\epsilon}). \tag{VI.4}$$

Indeed, let t_1, t_2 in I with $t_1 < t_2$ be given and assume that for every $t \in I$ there is $\epsilon_0 = \epsilon_0(t) > 0$ so that for all $\epsilon \in [0,\epsilon_0]$ with $t+\epsilon \in I$ we have (VI.4). Define $t^* = \sup\{s \in [t_1,t_2] : V(y_{t_1}) \geq V(y_u) \text{ for all } t_1 \leq u \leq s\}$. Then $t_1 < t^* \leq t_2$. From the definition of t^* it follows that there is a sequence $(s_n)_0^\infty$ in $[t_1,t^*]$ so that $s_n \to t^*$ as $n \to \infty$ and $V(y_{t_1}) \geq V(y_{s_n})$ for all $n \in \mathbb{N}$. Clearly, $y_{s_n} \to y_{t^*}$ as $n \to \infty$. Then Lemma VI.1(i) yields $V(y_{t_1}) \geq V(y_{t^*})$. If $t^* < t_2$ then there is $\epsilon_0(t^*) \in (0, t_2-t^*]$ so that $V(y_{t_1}) \geq V(y_{t^*}) \geq V(y_{t^*+\epsilon})$ for all $\epsilon \in [0,\epsilon_0(t^*)]$. This contradicts the definition of t^*. Consequently, $t^* = t_2$, and thus the claim holds.

Let $t \in I$. We prove that there is $\epsilon_0 = \epsilon_0(t) > 0$ such that for all $\epsilon \in [0,\epsilon_0]$ with $t+\epsilon \in I$ we have (VI.4). The cases $V(y_t)=\infty$ and $y(t) \neq 0$ are obvious. Assume that $V(y_t) < \infty$ and $y(t)=0$. From $V(y_t)<\infty$ it follows that there is $\epsilon_0 \in (0,1)$ such that y does not change sign on $[t-1, t-1+\epsilon_0]$. Assume that $y(s) \geq 0$ for all $s \in [t-1,t-1+\epsilon_0]$. From equation (VI.2) and $b>0$, it follows that $\dot{y}(s) \geq 0$ and, by $y(t)=0$, $y(s) \geq 0$ for $s \in [t,t+\epsilon_0] \cap I$. If $y(s)=0$ for all $s \in [t,t+\epsilon_0] \cap I$, then (VI.4) holds for all $\epsilon \in [0,\epsilon_0]$ with $t+\epsilon \in I$. Assume that $y(\hat{s})>0$ for some $\hat{s} \in [t,t+\epsilon_0] \cap I$. Then, by (VI.2) and $b>0$, there is $\bar{s} \in [t-1,t-1+\epsilon_0]$ such that $y(\bar{s})>0$. From $V(y_t)<\infty$ it also follows that there exists $\delta \in (0,1)$ such that y is not identically zero on $[t-\delta,t]$ and either $y(s) \geq 0$ or $y(s) \leq 0$ for all $s \in [t-\delta,t]$. If $y(s) \geq 0$ on $[t-\delta,t]$, then y does not change sign on $[t-\delta,t+\epsilon_0] \cap I$, and thus (VI.4) holds for all $\epsilon \in [0,\epsilon_0]$ with $t+\epsilon \in I$. If $y(s) \leq 0$ on $[t-\delta,t]$, then it follows that $\mathrm{sc}(y_t)$ is odd and $V(y_t) = \mathrm{sc}(y_t)+1$. Since y has at most one sign change on

$[t-\delta, t+\epsilon] \cap I$ for all $\epsilon \in [0, \epsilon_0]$, we obtain that $\mathrm{sc}(y_{t+\epsilon}) \le \mathrm{sc}(y_t)+1$ for all $\epsilon \in [0, \epsilon_0]$ with $t+\epsilon \in I$. Hence, using the definition of V, (VI.4) follows for all $\epsilon \in [0, \epsilon_0]$ with $t+\epsilon \in I$. The proof in case $y \le 0$ on $[t-1, t-1+\epsilon_0]$ is analogous.

3. Proof of (ii) for y instead of z. Let $t \in I$ with $t-2 \in I$, $y(t) = y(t-1) = 0$, and $V(y_t) < \infty$ be given. Choose $k \in \mathbb{N}$ and $(s^i)_0^{k+2}$ such that $k = \mathrm{sc}(y_t)$,

$$t-1 = s^0 < s^1 < \ldots < s^{k+1} < s^{k+2} = t,$$

and, in case $k > 0$

$$y(s^i)y(s^{i+1}) < 0 \qquad \text{for all } i \in \{1, 2, \ldots, k\},$$

and in case $k = 0$ the number s^1 is chosen such that $y(s^1) \neq 0$. Applying the mean value theorem to each interval $[s^i, s^{i+1}]$, and using equation (VI.2) and $y(s^0) = y(s^{k+2}) = 0$, we get a sequence $(\bar{s}^i)_0^{k+1}$ such that

$$t-2 < \bar{s}^0 < \bar{s}^1 < \ldots < \bar{s}^k < \bar{s}^{k+1} < t-1$$

and

$$y(\bar{s}^i)y(\bar{s}^{i+1}) < 0 \qquad \text{for all } i \in \{0, 1, \ldots, k\}.$$

Therefore, $\mathrm{sc}(y_{t-1}) \ge k+1$. If k is even then, using also (i), we find

$$V(y_{t-2}) \ge V(y_{t-1}) \ge \mathrm{sc}(y_{t-1}) \ge k+1 > k = V(y_t).$$

Assume that k is odd. Then $y(s^1)$ and $y(s^{k+1})$ have different signs. It follows that $y(\bar{s}^0)$ and $y(\bar{s}^{k+1})$ have the same sign. Using $y(t-1) = 0$, the mean value theorem on $[\bar{s}^{k+1}, t-1]$, and (VI.2), it follows that there is $s^* \in (\bar{s}^{k+1}, t-1)$ so that $y(s^*-1)y(\bar{s}^0) < 0$. Consequently, $\mathrm{sc}(y_{s^*}) \ge k+2$. Then (i), $s^* \in (t-2, t-1)$ and the definition of V combined imply that

$$V(y_{t-2}) \ge V(y_{s^*}) \ge \mathrm{sc}(y_{s^*}) \ge k+2 > k+1 = V(y_t).$$

4. Proof of (iii) for y instead of z. Let $t \in I$ with $t-3 \in I$ and $V(y_{t-3}) = V(y_t) < \infty$ be given. (i) implies that $V(y_{s-2}) = V(y_s) < \infty$ for all $s \in [t-1, t]$. From (ii) it follows that $(y(s), y(s-1)) \neq (0,0)$ for all $s \in [t-1, t]$. Hence, by using equation (VI.2), we obtain $(y(s), \dot{y}(s)) \neq (0,0)$ for all $s \in [t-1, t]$. In particular, $y(t) = 0$ implies $\dot{y}(t) \neq 0$, and moreover $\dot{y}(t)y(t-1) > 0$ since $b(t) > 0$ in equation (VI.2). Assume $y(t-1) = 0$. Then $y(t) \neq 0$ and $\dot{y}(t-1) \neq 0$. Suppose that $y(t)\dot{y}(t-1) > 0$. Then it is easy to see that $\mathrm{sc}(y_t)$ is even. Let $k = \mathrm{sc}(y_t)$ and choose $(s^i)_0^{k+1}$ such that

$$t-1 = s^0 < s^1 < \ldots < s^k < s^{k+1} = t$$

and

$$y(s^i)y(s^{i+1}) < 0 \qquad \text{for all } i \in \{1, 2, \ldots, k\}.$$

Applying the mean value theorem on the intervals $[s^i, s^{i+1}]$ for all $i \in \{0, 1, \ldots, k\}$ and using equation (VI.2), we get $(\bar{s}^i)_0^k$ such that

$$\bar{s}^i + 1 \in (s^i, s^{i+1}) \qquad \text{for all } i \in \{0, 1, \ldots, k\},$$

$$t-2 < \bar{s}^0 < \bar{s}^1 < \ldots < \bar{s}^{k-1} < \bar{s}^k < t-1$$

and

$$y(\bar{s}^i)y(\bar{s}^{i+1}) < 0 \qquad \text{for all } i \in \{0, 1, \ldots, k-1\}.$$

If $\dot{y}(t-1) > 0$, then $y(t) > 0$, and it follows that $y(\bar{s}^k) > 0$. From $\dot{y}(t-1) > 0$ and $y(t-1) = 0$, we obtain that there exists $\bar{s}^{k+1} \in (\bar{s}^k, t-1)$ such that $y(\bar{s}^{k+1}) < 0$. Consequently, $\mathrm{sc}(y_{t-1}) \geq k+1$ and hence $V(y_{t-1}) \geq k+2 > k = V(y_t)$. Using (i), it follows that $V(y_{t-3}) \geq V(y_{t-1}) > V(y_t)$, which is a contradiction. Similarly, the case $\dot{y}(t-1) < 0$ leads to a contradiction. Therefore, $y(t-1) = 0$ implies $y(t)\dot{y}(t-1) < 0$. Thus, $y_t \in R$. □

The next result shows that the Lyapunov functional V can be effectively used to show that solutions of a linear delay differential equation with feedback conditions can not decay too fast at ∞. We should mention that estimates of this type go back to Walther **[61]** for slowly oscillatory solutions, and to Mallet-Paret **[42]** for solutions of more general oscillatory behaviours. See also Cao **[16]** and Arino **[3]** for further developments and simplifications. The proof below follows Cao **[16]** and Arino **[3]**.

Lemma VI.3 *Assume that $\mu \geq 0$ and b_0, b_1 are positive constants with $b_0 \leq b_1$. There exists $k_0 = k_0(\mu, b_0, b_1) > 0$ such that if $t_0 \in \mathbb{R}$, if $b : [t_0 - 5, t_0] \to \mathbb{R}$ is a continuous function with*

$$b_0 \leq b(t) \leq b_1 \qquad \textit{for all } t \in [t_0 - 5, t_0],$$

and if $z : [t_0 - 6, t_0] \to \mathbb{R}$ is continuous, differentiable for $t_0 - 5 < t \leq t_0$, satisfies (VI.1) for $t_0 - 5 < t \leq t_0$, and has the property $z_{t_0-5} \neq 0$, and if

$$V(z_{t_0-5}) \leq 2,$$

then

$$||z_{t_0-1}|| \leq k_0 ||z_{t_0}||.$$

Proof 1. Let $\hat{b}_0 > 0$, $\hat{b}_1 \geq \hat{b}_0$ be given. We show that there exists a constant $\hat{k}_0 > 0$ such that for every $t_0 \in \mathbb{R}$, for every continuous function $\hat{b} : [t_0 - 5, t_0] \to \mathbb{R}$ with $\hat{b}_0 \leq \hat{b}(t) \leq \hat{b}_1$ for all $t \in [t_0 - 5, t_0]$, and for every continuous $y : [t_0 - 6, t_0] \to \mathbb{R}$ which is differentiable and satisfies

$$\dot{y}(t) = \hat{b}(t) y(t-1) \qquad \text{for } t_0 - 5 < t \leq t_0,$$

and has the property $y_{t_0-5} \neq 0$ and $V(y_{t_0-5}) \leq 2$, we have

$$||y_{t_0-1}|| \leq \hat{k}_0 ||y_{t_0}||.$$

As a first step to this we prove that for every $\delta \in (0, \min\{1, 1/\hat{b}_0\})$ there exist a constant $c > 1$ so that for every $t_0, \hat{b}, y$ as above and for every closed interval $I \subset [t_0 - 5, t_0 - 1]$ of length δ,

$$\min_{s \in I} |y(s)| \leq c ||y_{t_0}||. \tag{VI.5}$$

Define real functions c_1, c_2, c_3, c_4 on $(0, \infty) \times (0, \infty)$ by

$$c_1(\tilde{\delta}, \tilde{b}_0) = \frac{2}{\tilde{b}_0 \tilde{\delta}} \quad \text{and} \quad c_i(\tilde{\delta}, \tilde{b}_0) = \frac{12 c_{i-1}(\tilde{\delta}/6, \tilde{b}_0)}{\tilde{b}_0 \tilde{\delta}} \quad \text{for } i \in \{2, 3, 4\}.$$

Let $\hat{b}_0 > 0$, $\hat{b}_1 \geq \hat{b}_0$, $\delta \in (0, \min\{1, 1/\hat{b}_0\})$ be given. We begin with the case $I \subset [t_0 - 2, t_0 - 1]$, that is, $I = [s^1 - 1, s^2 - 1]$ with $t_0 - 1 \leq s^1 < s^2 \leq t_0$ and $s^2 - s^1 = \delta$. Integrating the differential equation for y on $[s^1, s^2]$, we get

$$2||y_{t_0}|| \geq |y(s^2) - y(s^1)| = |\int_{s^1}^{s^2} \hat{b}(s) y(s-1)\, ds| = |\int_I \hat{b}(s-1) y(s)\, ds| \geq \hat{b}_0 \delta \min_{s \in I} |y(s)|.$$

Hence

$$\min_{s \in I} |y(s)| \leq c_1(\delta, \hat{b}_0) ||y_{t_0}||.$$

Now consider the case $I \subset [t_0 - 3, t_0 - 1]$. If the length of $I \cap [t_0 - 2, t_0 - 1]$ is $\frac{\delta}{2}$ or larger then inequality (VI.5) holds with $c = c_1(\frac{\delta}{2}, \hat{b}_0)$. Assume the length of $I \cap [t_0 - 3, t_0 - 2]$ is larger than $\frac{\delta}{2}$. Let t^1, t^2 in $[t_0 - 2, t_0 - 1]$ be given such that $[t^1 - 1, t^2 - 1] \subset I \cap [t_0 - 3, t_0 - 2]$ and $t^2 - t^1 = \frac{\delta}{2}$. Considering the intervals $[t^1, t^1 + \frac{\delta}{6}]$ and $[t^2 - \frac{\delta}{6}, t^2]$ in $[t_0 - 2, t_0 - 1]$, the first part of the proof gives that

$$\min_{s \in [t^1, t^1 + \frac{\delta}{6}]} |y(s)| \leq c_1(\frac{\delta}{6}, \hat{b}_0) ||y_{t_0}||, \qquad \min_{s \in [t^2 - \frac{\delta}{6}, t^2]} |y(s)| \leq c_1(\frac{\delta}{6}, \hat{b}_0) ||y_{t_0}||.$$

By the mean value theorem, there exists $t^* \in [t^1, t^2]$ such that

$$|\dot{y}(t^*)| \leq \frac{12 c_1(\frac{\delta}{6}, \hat{b}_0)}{\delta} ||y_{t_0}||.$$

The differential equation for y then implies

$$|y(t^* - 1)| \leq \frac{|\dot{y}(t^*)|}{\hat{b}_0} \leq \frac{12 c_1(\frac{\delta}{6}, \hat{b}_0)}{\hat{b}_0 \delta} ||y_{t_0}||.$$

Since $t^* - 1 \in I$ and $c_1(\frac{\delta}{2}, \hat{b}_0) < \frac{12 c_1(\frac{\delta}{6}, \hat{b}_0)}{\hat{b}_0 \delta}$, inequality (VI.5) is satisfied with $c = c_2(\delta, \hat{b}_0)$.

Repeating the above argument twice, we obtain that $c = c_4(\delta, \hat{b}_0)$ is an appropriate constant for any $I \subset [t_0 - 5, t_0 - 1]$.

2. We complete the proof of the first statement in part 1. Let $\hat{b}_0 > 0$ and $\hat{b}_1 \geq \hat{b}_0$ be given. Choose $\delta \in (0, \min\{\frac{1}{6}, \frac{1}{\hat{b}_1}\})$. By the result in part 1 there is a constant $c = c(\delta, \hat{b}_0) > 1$ such that

$$\min_{s \in I} |y(s)| \leq c ||y_{t_0}||$$

for every interval $I \subset [t_0 - 5, t_0 - 1]$ of length δ. Choose $\hat{k}_0 = \hat{k}_0(\hat{b}_0, \hat{b}_1) > c$ such that

$$\frac{\hat{k}_0 - c}{\hat{b}_1 \delta} > c, \qquad \frac{\hat{k}_0 - c}{\hat{b}_1 \delta} > \hat{k}_0,$$

and assume

$$||y_{t_0 - 1}|| > \hat{k}_0 ||y_{t_0}||$$

for $t_0, \hat{b}_0, y$ as above. Then there exists $t^* \in [t_0 - 2, t_0 - 1)$ with

$$|y(t^*)| > \hat{k}_0 ||y_{t_0}||.$$

We have

$$\min_{s \in [t^* - \delta, t^*]} |y(s)| \leq c ||y_{t_0}||, \qquad \min_{s \in [t^*, t^* + \delta]} |y(s)| \leq c ||y_{t_0}||.$$

The first inequality follows from the result in part 1, as well as the second inequality in case $t^* + \delta \le t_0 - 1$. In case $t_0 - 1 < t^* + \delta$ the second inequality clearly holds since $c > 1$. The mean value theorem implies the existence of $s^1 \in (t^* - \delta, t^*)$ and $s^2 \in (t^*, t^* + \delta)$ such that

$$|\dot{y}(s^i)| \ge \frac{\hat{k}_0 - c}{\delta}||y_{t_0}|| \qquad \text{for } i \in \{1,2\}, \quad \text{and} \quad \dot{y}(s^1)\dot{y}(s^2) < 0.$$

Without loss of generality, we may assume $\dot{y}(s^1) > 0$, since $-y$ is also a solution of the differential equation for y. From this equation we get

$$|y(s^i - 1)| \ge \frac{|\dot{y}(s^i)|}{\hat{b}_1} \ge \frac{\hat{k}_0 - c}{\hat{b}_1 \delta}||y_{t_0}|| > \hat{k}_0||y_{t_0}|| \qquad \text{for } i \in \{1,2\}$$

and

$$y(s^1 - 1) > 0, \quad y(s^2 - 1) < 0.$$

A second application of the result in part 1 yields

$$\min_{s\in[s^1-1-\delta, s^1-1]} |y(s)| \le c||y_{t_0}||, \qquad \min_{s\in[s^2-1, s^2-1+\delta]} |y(s)| \le c||y_{t_0}||.$$

The mean value theorem gives

$$\dot{y}(u^1) \ge \frac{\hat{k}_0 - c}{\delta}||y_{t_0}||, \;\; \dot{y}(u^0) \le -\frac{\hat{k}_0}{\delta}||y_{t_0}||, \;\; \dot{y}(u^2) \ge \frac{\hat{k}_0 - c}{\delta}||y_{t_0}||$$

for some $u^1 \in (s^1 - 1 - \delta, s^1 - 1)$, $u^2 \in (s^2 - 1, s^2 - 1 + \delta)$ and $u^0 \in (s^1 - 1, s^2 - 1)$. By the differential equation for y,

$$y(u^i - 1) \ge \frac{\hat{k}_0 - c}{\hat{b}_1 \delta}||y_{t_0}|| > c||y_{t_0}|| \qquad \text{for } i \in \{1,2\},$$

$$y(u^0 - 1) < -\frac{\hat{k}_0}{\delta \hat{b}_1}||y_{t_0}|| \le -\hat{k}_0||y_{t_0}||.$$

Applying the result in part 1 to the intervals $[u^1 - 1 - \delta, u^1 - 1]$ and $[u^2 - 1, u^2 - 1 + \delta]$ and using the mean value theorem, we can show that there exists a $w^1 \in (u^1 - 1 - \delta, u^1 - 1)$ such that $\dot{y}(w^1) > 0$; there exists a $w^2 \in (u^1 - 1, u^0 - 1)$ such that $\dot{y}(w^2) < 0$; there exists a $w^3 \in (u^0 - 1, u^2 - 1)$ such that $\dot{y}(w^3) > 0$; there exists a $w^4 \in (u^2 - 1, u^2 - 1 + \delta)$ such that $\dot{y}(w^4) < 0$. As $w^1 > u^1 - 1 - \delta \ge s^1 - 2 - 2\delta \ge t^* - 2 - 3\delta \ge t_0 - 4 - 3\delta > t_0 - 5$ and $w^1 < w^2 < w^3 < w^4$, we may use the differential equation for y to get the reals v^1, v^2, v^3, v^4 in $(u^1 - 2 - \delta, u^2 - 2 + \delta)$ such that $v^i = w^i - 1$, $i \in \{1,2,3,4\}$, and

$$y(v^i)y(v^{i+1}) < 0 \qquad \text{for all } i \in \{1,2,3\}.$$

We have

$$v^4 - v^1 \le u^2 - u^1 + 2\delta \le s^2 - s^1 + 4\delta \le 6\delta < 1, \quad v^1 = w^1 - 1 > t_0 - 6$$

and

$$v^4 = w^4 - 1 \ge u^2 - 2 \ge s^2 - 3 \ge t^* - 3 \ge t_0 - 5.$$

It follows that

$$V(y_{v^4}) > 2.$$

Then the monotonicity of V implies $V(y_{t_0 - 5}) \ge V(y_{v^4}) > 2$, a contradiction to the assumption.

3. Now let $\mu \geq 0$, $b_0 > 0$, $b_1 \geq b_0$ be given. Set $\hat{b}_0 = e^{\mu} b_0$, $\hat{b}_1 = e^{\mu} b_1$. Choose $\hat{k}_0 > 0$ according to the statement at the beginning of part 1. Consider $k_0 = e^{2\mu}\hat{k}_0$. Let t_0, b, z be given as in Lemma. Consider $\hat{b} : [t_0 - 5, t_0] \ni t \mapsto e^{\mu} b(t) \in \mathbb{R}$ and $y : [t_0 - 6, t_0] \ni t \mapsto e^{\mu t} z(t) \in \mathbb{R}$. We obtain

$$||y_{t_0-1}|| \leq \hat{k}_0 ||y_{t_0}||,$$

and it follows that

$$\begin{aligned} ||z_{t_0-1}|| &= \max_{s\in[-1,0]} |y(t_0 - 1 + s)e^{-\mu(t_0-1+s)}| \\ &\leq e^{-\mu(t_0-2)} ||y_{t_0-1}|| \leq e^{-\mu(t_0-2)} \hat{k}_0 ||y_{t_0}|| \\ &= e^{-\mu(t_0-2)} \hat{k}_0 \max_{s\in[-1,0]} |e^{\mu(t_0+s)} z(t_0 + s)| \\ &\leq e^{2\mu} \hat{k}_0 ||z_{t_0}|| = k_0 ||z_{t_0}||. \end{aligned}$$

□

In the remaining part of this appendix we prove three technical results. The first one gives conditions under which solutions of a sequence of equations have a convergent subsequence whose limit satisfies equation (VI.1).

Lemma VI.4 *Let intervals $I = (-\infty, t_0]$ with $t_0 \in \mathbb{R}$, or $I = [t_0, \infty)$ with $t_0 \in \mathbb{R}$ be given, or consider $I = \mathbb{R}$ and $t_0 = 0$. In the first, second and third of the cases set $I_1 = (-\infty, t_0] = I_2$, $I_1 = [t_0 + 1, \infty)$ and $I_2 = [t_0 + 2, \infty)$, $I_1 = \mathbb{R} = I_2$, respectively. Let sequences of continuous real functions b^n on I_1 and z^n on I_0, $n \in \mathbb{N}$, be given so that all restrictions $z^n|_{I_1}$ are differentiable. Let $\mu \geq 0$, $c > 0$, and a further continuous real function b on I_1 be given so that*

$$b^n \to b \qquad \text{as } n \to \infty \text{ uniformly on compact subsets of } I_1, \tag{VI.6}$$

$$\dot{z}^n(t) = -\mu z^n(t) + b^n(t) z^n(t-1) \qquad \text{for all } t > \inf I_1 \text{ in } I_1 \text{ and all } n \in \mathbb{N}, \tag{VI.7}$$

and

$$|z^n(t)| \leq c e^{c|t-t_0|} \qquad \text{for all } t \in I_0 \text{ and all } n \in \mathbb{N}. \tag{VI.8}$$

Then there exist a continuous function $z : I_1 \to \mathbb{R}$ so that $z|_{I_2}$ is differentiable, and a subsequence $(z^{n_k})_{k=0}^{\infty}$ of $(z^n)_{n=0}^{\infty}$ such that

$$\begin{aligned} z^{n_k} &\to z \qquad \text{as } k \to \infty \text{ uniformly on compact subsets in } I_1, \\ \dot{z}^{n_k} &\to \dot{z} \qquad \text{as } k \to \infty \text{ uniformly on compact subsets in } I_2, \end{aligned}$$

and

$$\dot{z}(t) = -\mu z(t) + b(t) z(t-1) \qquad \text{for all } t > \inf I_2 \text{ in } I_2.$$

Proof We give a proof only for the case $I_0 = (-\infty, t_0]$, $t_0 \in \mathbb{R}$, since the other cases are analogous. Recall that in case $I_0 = (-\infty, t_0]$ we have $I_1 = I_2 = (-\infty, t_0]$.

For each $k \in \mathbb{N}$, (VI.6) and (VI.8) imply that there are positive numbers B_k and Z_k such that

$$|b^n(t)| \leq B_k, \ |z^n(t)| \leq Z_k \qquad \text{for all } t \in [t_0 - k, t_0] \text{ and } n \in \mathbb{N}.$$

From (VI.7) it follows that

$$|\dot{z}^n(t)| \leq (\mu + B_k) Z_{k+1} \qquad \text{for all } t \in [t_0 - k, t_0] \text{ and } n \in \mathbb{N}.$$

Then the Arzèla–Ascoli Theorem implies that for each $k \in \mathbb{N}$ every subsequence of $(z^n)_0^\infty$ has a subsequence converging uniformly on $[t_0 - k, t_0]$. By using this property of $(z^n)_0^\infty$, we construct sequences $(z^{i,n})_{n=0}^\infty$, $i \in \mathbb{N}$, such that for every integer $i \geq 1$, $(z^{i,n})_{n=0}^\infty$ is a subsequence of $(z^{i-1,n})_{n=0}^\infty$ which converges uniformly on $[t_0 - i, t_0]$. Then the diagonal sequence $(z^{n,n})_{n=0}^\infty$ is a subsequence $(z^{n_k})_{k=0}^\infty$ of $(z^n)_{n=0}^\infty$ which converges pointwise to a continuous function $z : (-\infty, t_0] \to \mathbb{R}$. For every $i \in \mathbb{N}$ the convergence is uniform on $[t_0 - i, t_0]$. Considering (VI.6) and (VI.7) for the subsequence $(n_k)_{k=0}^\infty$, we find that

$$\dot{z}^{n_k} \to -\mu z + bz(\cdot - 1) \qquad \text{as } k \to \infty$$

uniformly on $[t_0 - i, t_0]$ for every $i \in \mathbb{N}$. It follows that z is differentiable and

$$\dot{z}(t) = -\mu z(t) + b(t)z(t-1)$$

for all $t \in (-\infty, t_0]$. □

The next two results can be used to verify condition (VI.8) in Lemma VI.4.

Lemma VI.5 *Let $\mu \geq 0$, $b_0 > 0$ and $b_1 \geq b_0$. There exists $c = c(\mu, b_0, b_1) > 0$ such that if $t_0 \in \mathbb{R}$ and $b : (-\infty, t_0] \to \mathbb{R}$ is a continuous function with*

$$b_0 \leq b(t) \leq b_1 \qquad \textit{for all } t \leq t_0,$$

and $z : (-\infty, t_0] \to \mathbb{R}$ is differentiable and satisfies (VI.1) for all $t \leq t_0$, and if $z(t) \neq 0$ for some $t \leq t_0$, and

$$V(z_t) \leq 2 \qquad \textit{for all } t \leq t_0,$$

then

$$||z_t|| \leq ce^{c(t_0 - t)}||z_{t_0}|| \qquad \textit{for all } t \leq t_0.$$

Proof Lemma VI.3 implies that there exists $k_0 = k_0(\mu, b_0, b_1) > 0$ with

$$||z_{t-1}|| \leq k_0||z_t|| \qquad \text{for all } t \leq t_0. \tag{VI.9}$$

Define $c = 1 + k_0$. Let $t \in [t_0 - (n+1), t_0 - n]$ for some $n \in \mathbb{N}$. It follows from (VI.9) that

$$||z_t|| \leq k_0||z_{t+1}|| \leq k_0^2||z_{t+2}|| \leq \ldots \leq k_0^n||z_{t+n}|| \leq (1 + k_0)^n||z_{t+n}||.$$

From $t+n \in [t_0 - 1, t_0]$ and (VI.9) with $t = t_0$, we obtain $||z_{t+n}|| \leq ||z_{t_0}|| + ||z_{t_0-1}|| \leq (1 + k_0)||z_{t_0}||$. Consequently,

$$||z_t|| \leq (1 + k_0)^{n+1}||z_{t_0}||.$$

As $t \in [t_0 - (n+1), t_0 - n]$ implies $n \leq t_0 - t$, and $\log(1 + k_0) < 1 + k_0$, we have $(1 + k_0)^n = e^{n \log(1+k_0)} \leq e^{c(t_0 - t)}$. Therefore,

$$||z_t|| \leq (1 + k_0)(1 + k_0)^n||z_{t_0}|| \leq ce^{c(t_0 - t)}||z_{t_0}|| \qquad \text{for all } t \leq t_0.$$

□

A bound for solutions $z : [t_0 - 1, \infty) \to \mathbb{R}$, $t_0 \in \mathbb{R}$, of (VI.1) can be obtained from the Gronwall Lemma.

Lemma VI.6 *Assume that $\mu \geq 0$, $t_0 \in \mathbb{R}$, $b_0 > 0$, and $b : [t_0, \infty) \to \mathbb{R}$ is a continuous function with $|b(t)| \leq b_0$ for all $t \geq t_0$, and $z : [t_0 - 1, \infty) \to \mathbb{R}$ is a solution of equation (VI.1). Then*

$$||z_t|| \leq ||z_{t_0}|| e^{b_0(t-t_0)} \qquad \textit{for all } t \geq t_0.$$

Proof Multiplying equation (VI.1) with $e^{\mu t}$ we obtain

$$\frac{d}{dt}(e^{\mu t} z(t)) = e^{\mu t} b(t) z(t-1) \ \text{ for } \ t > t_0.$$

Integrating and multiplying with $e^{-\mu t}$ we get

$$|z(t)| \leq e^{\mu(t_0 - t)} |z(t_0)| + \int_{t_0}^{t} e^{\mu(s-t)} |b(s)||z(s-1)| \, ds, \qquad t \geq t_0.$$

Using $\mu \geq 0$ it is easily obtained that

$$||z_t|| \leq ||z_{t_0}|| + \int_{t_0}^{t} b_0 ||z_s|| \, ds, \qquad t \geq t_0.$$

Hence the Gronwall Lemma implies the desired inequality. □

APPENDIX VII

Floquet multipliers for a class of linear periodic delay differential equations

Let $\mu \geq 0$ and let a continuous periodic strictly positive function $a : \mathbb{R} \to \mathbb{R}$ be given with a period $\omega > 1$. Let $C_{\mathbb{C}}$ denote the Banach space over $\mathbb{C}$ of complex-valued continuous functions ϕ on $[-1, 0]$, with $||\phi|| = \max_{-1\leq t\leq 0} |\phi(t)|$. All functions ϕ in C and in $C_{\mathbb{C}}$ define solutions x^{ϕ} on $[-1, \infty)$ of the initial value problem

$$\dot{x}(t) = -\mu x(t) + a(t)x(t-1) \tag{VII.1}$$

and

$$x_0 = \phi,$$

with values in $\mathbb{R}$ and in $\mathbb{C}$, respectively; these solutions are given by the variation-of-constants formula

$$x(t) = e^{-\mu(t-n)}x(n) + \int_n^t e^{-\mu(t-s)}a(s)x(s-1)ds, \qquad n \in \mathbb{N},\ n \leq t \leq n+1.$$

The period maps $M : C \ni \phi \mapsto x_\omega^\phi \in C$ and $M_{\mathbb{C}} : C_{\mathbb{C}} \ni \phi \mapsto x_\omega^\phi \in C_{\mathbb{C}}$ are called monodromy operators. We have

$$(M_{\mathbb{C}}\phi)(t) = M(\operatorname{Re}\phi)(t) + iM(\operatorname{Im}\phi)(t) \qquad \text{for all } \phi \in C_{\mathbb{C}} \text{ and } t \in [-1, 0]$$

and

$$M^j\phi = x_{j\omega}^\phi \qquad \text{for all } \phi \in C,\ j \in \mathbb{N}.$$

This means that the asymptotic behaviour of solutions of equation (VII.1) is governed by the behaviour of M and its iterates.

M and $M_{\mathbb{C}}$ are linear, continuous, compact and have the same spectrum σ. (Recall that according to our convention in Chapter 1 the spectrum of M is the spectrum of its complexification; the operator $M_{\mathbb{C}}$ is conjugate to this complexification by a topological isomorphism.) Every point $\lambda \in \sigma \setminus \{0\}$ is an eigenvalue of finite multiplicity, and is isolated in σ. These eigenvalues in $\sigma \setminus \{0\}$ are called Floquet multipliers.

Incidentally, we observe that $0 \in \sigma$.

A complex number $\lambda \neq 0$ is a Floquet multiplier if and only if $\bar{\lambda}$ is a Floquet multiplier. For a given spectral subset $\sigma_0 \subset \sigma$, i.e., σ_0 and $\sigma \setminus \sigma_0$ are both nonempty and compact, with $\sigma_0 = \bar{\sigma}_0$ the realified generalized eigenspace of M associated with σ_0 coincides with the set of elements $\phi \in C$ given by real-valued elements in the generalized eigenspace $M_{\mathbb{C}}$ associated with σ_0.

First we need a real, positive eigenvalue of M with an eigenvector in the interior of the cone $K = \{\phi \in C : 0 \leq \phi\}$. Using the variation-of-constants formula above, one easily verifies the inclusions

$$M(K) \subset K, \quad M^2(K \setminus \{0\}) \subset \overset{\circ}{K}$$

and the fact that for every $\phi \in K$ the function

$$[-1, \infty) \ni t \mapsto e^{\mu t} x^{\phi}(t) \in \mathbb{R}$$

is increasing on $[0, \infty)$. In particular,

$$x^{\phi}(t) \geq e^{-\mu t} \phi(0) \qquad \text{for all } \phi \in K \text{ and } t \geq 0.$$

Proposition VII.1

(i) There exist $\lambda > 0$ and $\phi \in \overset{\circ}{K}$ with $M\phi = \lambda\phi$.

(ii) For every $\lambda > 0$ so that there exists $\phi \in \overset{\circ}{K}$ with $M\phi = \lambda\phi$ the realified generalized eigenspace $C_{<\lambda}$ of M given by the spectral subset $\{z \in \sigma : |z| < \lambda\}$ satisfies

$$C_{<\lambda} \cap (K \cup (-K)) = \{0\}.$$

Proof 1. The set

$$A = \{\phi \in K : \phi(0) = 1, [-1, 0] \ni t \mapsto e^{\mu t}\phi(t) \in \mathbb{R} \text{ is nondecreasing}\}$$

is closed, bounded, and convex. For $\phi \in A$ we have $x^{\phi}_{\omega} \in K$, $x^{\phi}_{\omega}(0) = x^{\phi}(\omega) \geq e^{-\mu\omega}$, and $[-1, 0] \ni t \mapsto e^{\mu t} x^{\phi}_{\omega}(t) \in \mathbb{R}$ is nondecreasing. Also, there exists a bound $c \geq 0$ for all $|x^{\phi}(t)|$ with $\phi \in A$ and $\omega - 2 \leq t \leq \omega$. It follows that the continuous map

$$T : A \ni \phi \mapsto \frac{1}{x^{\phi}(\omega)} M\phi \in C$$

has range in A and that the derivatives

$$\frac{d}{dt} T(\phi)(t) = \frac{1}{x^{\phi}(\omega)} \dot{x}^{\phi}(\omega + t) \qquad \text{for all } \phi \in A \text{ and } t \in [-1, 0]$$

are bounded. So the Arzèla–Ascoli Theorem implies that $\overline{T(A)} \subset A$ is compact. By Schauder's Fixed Point Theorem, there is a fixed point ϕ of T. So,

$$M\phi = x^{\phi}(\omega)\phi.$$

Clearly, $x^{\phi}(\omega) > 0$ and $\phi = (x^{\phi}(\omega))^{-2} M^2(\phi) \in \overset{\circ}{K}$.

2. Suppose $C_{<\lambda}$ contains an element $\phi \neq 0$ in $K \cup (-K)$. Using positive invariance and linearity, we find an element $\psi \in \overset{\circ}{K} \cap C_{<\lambda}$ so that $[-1, 0] \ni t \mapsto e^{\mu t}\psi(t) \in \mathbb{R}$ is increasing and $\psi(0) = 1$. Consequently, the set

$$B = \{\phi \in C_{<\lambda} \cap K : \phi(0) = 1, \ [-1, 0] \ni t \mapsto e^{\mu t}\phi(t) \in \mathbb{R} \ \text{ is nondecreasing}\}$$

is nonempty, closed, bounded, and convex. Arguing as in part 1, we find $\rho > 0$ and $\psi \in B \subset C_{<\lambda}$ so that $M\psi = \rho\psi$ and $\psi \in \overset{\circ}{K}$. Necessarily, $\rho < \lambda$. Choose $c > 0$ with $c\phi \ll \psi$. Then

$$\lambda^j c\phi = M^j(c\phi) \leq M^j \psi = \rho^j \psi \qquad \text{for all } j \in \mathbb{N},$$

which leads to a contradiction. □

We should mention that alternatively one can obtain Proposition VII.1 (and much more which we do not need here) from the Krein-Rutman Theorem applied to M^2 (see, for example, Theorem 19.3 in Deimling **[21]**).

Observe that in terms of the Lyapunov functional in Appendix VI,

$$K \cap (-K) \setminus \{0\} = V^{-1}(0),$$

so that we have an eigenvector in $V^{-1}(0)$ for some eigenvalue $\lambda > 0$, and

$$C_{<\lambda} \cap V^{-1}(0) = \emptyset.$$

In the sequel, we prove a stronger version of such relations between realified generalized eigenspaces of M and oscillation frequencies, with

$$T = V^{-1}(\{0, 2\})$$

in place of $V^{-1}(0)$. Note that using Lemma VI.1(i), we can show that

$$\overline{T} = T \cup \{0\}.$$

The approach below follows Chow, Diekmann and Mallet-Paret **[18]**, Chow and Walther **[19]** and Lani-Wayda and Walther **[41]**, but is more complicated. Closely related results can also be obtained from the more general work of Mallet-Paret and Sell **[43]**, which requires a somewhat larger apparatus.

We begin with three simple results which have nothing to do with Floquet multipliers.

Lemma VII.2 *Let $L \subset C$ be a linear space. If there exists $n \in \mathbb{N}$ such that each $\phi \in L \setminus \{0\}$ has at most n zeros then* $\dim L \leq n + 1$.

Proof Suppose $\phi_1, \dots, \phi_{n+2}$ are linearly independent elements of L. There exists $c \in \mathbb{R}^{n+2} \setminus \{0\}$ with

$$\sum_{k=1}^{n+2} c_k \phi_k \Big(-1 + \frac{j}{n}\Big) = 0$$

for all $j \in \{0, \dots, n\}$. It follows that $\phi = \sum_{k=1}^{n+2} c_k \phi_k \in L \setminus \{0\}$ has $n+1$ zeros $-1 + \frac{j}{n}$, $0 \leq j \leq n$, a contradiction. □

Let X denote the space of all continuous functions $x : \mathbb{R} \to \mathbb{R}$, equipped with the topology of uniform convergence on compact sets. Set

$$J = \{x \in X : x_t \in T \quad \text{for all } t \in \mathbb{R}\}.$$

Proposition VII.3 *Let $x \in J$ be a solution of equation (VII.1). Then each segment x_t, $t \in \mathbb{R}$, has at most two zeros.*

Proof The function $y : \mathbb{R} \ni t \mapsto e^{\mu t} x(t) \in \mathbb{R}$ satisfies

$$\dot{y}(t) = e^{\mu} a(t) y(t-1)$$

for all $t \in \mathbb{R}$, the zeroset of x coincides with the zeroset of y, and $V(x_t) = V(y_t)$ for all $t \in \mathbb{R}$. (i) and (iii) of Lemma VI.2 show that the function $\mathbb{R} \ni t \mapsto V(y_t) \in \{0, 2\}$ is constant on some interval $(-\infty, t_0]$, $y_t \in R$ for all $t \leq t_0$, and in particular all zeros of y in $(-\infty, t_0]$ are simple. The equation for y implies that if y has no zero on an interval $[s-1, s]$, then y has no zero on $[s, \infty)$. If $z^0 \neq z^1$ are zeros of y then, by the mean value theorem, $\dot{y}$ has a zero between z^0 and z^1, and, by the equation for y, y has a zero between $z^0 - 1$ and $z^1 - 1$. These properties and $V(x_t) \leq 2$ for all $t \in \mathbb{R}$ combined imply that the zeroset of y is either empty or given by a strictly increasing sequence

$$(z_n)_{n \in I}$$

where either $I = \mathbb{Z}$ or $I = -\mathbb{N}$,

$$0 < z_n - z_{n-1} \leq 1 \qquad \text{for all } n \in I,$$

and there exists $n_0 \in I$ such that

$$z_n - z_{n-2} > 1 \qquad \text{for all integers } n \leq n_0.$$

It suffices to show that, in case of nonempty zeroset of y, $z_n - z_{n-2} > 1$ for all $n \in I$. We prove this by induction. Assume $z_n - z_{n-2} > 1$. We saw that y has a zero z in $(z_n - 1, z_{n+1} - 1)$. From $z_n - z_{n-2} > 1$ and $z_{n+1} - z_n \leq 1$ it follows that $z = z_{n-1}$. Consequently, $z_{n+1} - z_{n-1} > 1$. □

Proposition VII.4 $\overline{J} = J \cup \{0\}$.

Proof We have $0 \in \overline{J}$ since we can find a sequence of positive constant functions which converges to 0. Conversely, suppose $x \in \overline{J} \setminus (J \cup \{0\})$. Then $2 < V(x_t)$ for some $t \in \mathbb{R}$. Then there are points $s_1 < s_2 < s_3 < s_4$ in $[t-1, t]$ with $\operatorname{sign} x(s_j) \operatorname{sign} x(s_{j+1}) < 0$ for $j \in \{1, 2, 3\}$. There is a sequence $(x_n)_0^\infty$ in J with $x_n \to x$ as $n \to \infty$. It follows that for n sufficiently large and for $j \in \{1, 2, 3\}$, $\operatorname{sign} x_n(s_j) = \operatorname{sign} x(s_j)$. Therefore, $2 < V((x_n)_t)$, which is a contradiction. □

Next we follow Section 3 of Chow and Walther **[19]** which collects some simple conclusions from general facts of Floquet theory for retarded functional differential equations in Chapter 8 of Hale **[27]** or Chapter 14 of Diekmann, Verduyn Lunel, van Gils and Walther **[22]**.

For $0 \neq \lambda \in \sigma$, let $m = m(\lambda)$ denote the dimension of the generalized eigenspace

$$G(\lambda) = \bigcup_{k \in \mathbb{N}} \ker(M_{\mathbb{C}} - \lambda \operatorname{id})^k$$

of $M_{\mathbb{C}}$ associated with λ and let $d = d(\lambda)$ denote the dimension of the eigenspace

$$E(\lambda) = \ker(M_{\mathbb{C}} - \lambda \operatorname{id}).$$

Let us recall from Chow and Walther **[19]** how to obtain linear subspaces $\mathcal{E}(\lambda)$ and $\mathcal{G}(\lambda)$ of X which consist of solutions of equation (VII.1) whose segments x_0 form the realified eigenspace

$$E_{\mathbb{R}}(\lambda) = \{\phi \in C \colon \text{There exists } \psi \in E(\lambda) + E(\overline{\lambda}) \text{ with } \phi(t) = \psi(t) \text{ for all } t \in [-1, 0]\}$$

and the realified generalized eigenspace

$$G_{\mathbb{R}}(\lambda) = \{\phi \in C : \text{There exists } \psi \in G(\lambda) + G(\overline{\lambda}) \text{ with } \phi(t) = \psi(t) \text{ for all } t \in [-1, 0]\},$$

respectively. There is a basis $\phi_1, \ldots, \phi_m$ of $G(\lambda)$ so that $\phi_1, \ldots, \phi_d$ is a basis of $E(\lambda)$. Define the square matrix $\widetilde{M} \in \mathbb{C}^{m \times m}$ by

$$(M_{\mathbb{C}}\phi_1, \ldots, M_{\mathbb{C}}\phi_m) = (\phi_1, \ldots, \phi_m)\widetilde{M}.$$

Let I denote the unit matrix, with columns $e^1, \dots, e^m$ in $\mathbb{C}^m$. Then $N = \widetilde{M} - \lambda I$ is nilpotent, and if κ is the minimal positive integer with

$$\ker(M_{\mathbb{C}} - \lambda\,\mathrm{id})^{\kappa} = \ker(M_{\mathbb{C}} - \lambda\,\mathrm{id})^{\kappa+1},$$

then

$$N^{\kappa} = 0 \neq N^{\kappa-1}.$$

The only eigenvalue of $\widetilde{M}$ is λ. The unit vectors $e^1, \dots, e^d$ span the eigenspace of $\widetilde{M}$ associated with λ. Choose $\chi \in \mathbb{C}$ with $e^{\omega\chi} = \lambda$. Then $\operatorname{Re}\chi = (\ln|\lambda|)/\omega$ is uniquely determined by λ. Set

$$\widetilde{N} = \ln\left(I + \frac{1}{\lambda}N\right) = \sum_{l=1}^{\kappa-1} \frac{(-1)^{l+1}}{l}\left(\frac{1}{\lambda}N\right)^l \qquad \text{if } \kappa > 1$$

and

$$\widetilde{N} = 0 \qquad \text{if } \kappa = 1,$$

and

$$B = \chi I + \frac{1}{\omega}\widetilde{N}.$$

It follows that

$$\widetilde{M} = e^{\omega B} = e^{\omega\chi}e^{\widetilde{N}}, \quad \widetilde{N}^{\kappa} = 0 \neq \widetilde{N}^{\kappa-1}.$$

The only eigenvalue of B is χ, and the eigenspaces of $\widetilde{M}$ and B associated with λ and χ respectively coincide.

For $t \geq 0$, consider the row vector $\tilde{p}_t$ whose kth component, $k \in \{1, \dots, m\}$, is the segment y_t of the solution $y : [-1, \infty) \to \mathbb{C}$ of equation (VII.1) starting at the kth component of the row vector $[-1, 0] \ni s \mapsto (\phi_1(s), \dots, \phi_m(s))e^{-tB} \in (\mathbb{C}^m)^{tr}$. One has $\tilde{p}_{t+\omega} = \tilde{p}_t$ for all $t \geq 0$. Let $y^{(k)} : [-1, \infty) \to \mathbb{C}$ denote the solutions of equation (VII.1) with $y_0^{(k)} = \phi_k$, $k \in \{1, \dots, m\}$. By linearity,

$$\tilde{p}_t = (y_t^{(1)}, \dots, y_t^{(m)})e^{-tB} \qquad \text{for each } t \geq 0.$$

Now it becomes obvious that the map $\tilde{p} : (0, \infty) \ni t \mapsto \tilde{p}_t(0) \in (\mathbb{C}^m)^{tr}$ is continuously differentiable. Let p denote the periodic extension of $\tilde{p}$ to $\mathbb{R}$. Clearly, p is C^1-smooth. For every $c \in \mathbb{C}^m$, the function

$$y^c : \mathbb{R} \ni t \mapsto p(t)e^{tB}c \in \mathbb{C}$$

is a solution of equation (VII.1) with $y_0^c = (\phi_1, \dots, \phi_m)c$.

Let Y denote the complex vector space of continuous functions $\mathbb{R} \to \mathbb{C}$ equipped with the topology of uniform convergence on compact subsets of $\mathbb{R}$. Consider the subspaces

$$\begin{aligned} \mathcal{G}_{\mathbb{C}}(\lambda) &= \{y^c \in Y : c \in \mathbb{C}^m\}, \\ \mathcal{E}_{\mathbb{C}}(\lambda) &= \{y^c \in Y : c_{d+1} = \ldots = c_m = 0\} = \{y^c \in Y : y_0^c \in E(\lambda)\}. \end{aligned}$$

$\mathcal{G}_{\mathbb{C}}(\lambda)$ and $\mathcal{E}_{\mathbb{C}}(\lambda)$ have bases given by $c \in \{e_1, \dots, e_m\}$ and $c \in \{e_1, \dots, e_d\}$, respectively. So

$$\dim \mathcal{G}_{\mathbb{C}}(\lambda) = m, \quad \dim \mathcal{E}_{\mathbb{C}}(\lambda) = d.$$

Moreover, for $\tilde{\lambda} \in \sigma \setminus \{\lambda\}$ with $\tilde{\lambda} \neq 0$, $\mathcal{G}_{\mathbb{C}}(\tilde{\lambda}) \cap \mathcal{G}_{\mathbb{C}}(\lambda) = \{0\}$.

For functions $y^c \in \mathcal{E}_{\mathbb{C}}(\lambda)$, we obtain

$$y^c(t) = e^{\chi t}p(t)c \qquad \text{for all } t \in \mathbb{R}.$$

Consequently,

$$y^c(t) = e^{\beta t} q(t)$$

with $\beta = (\ln|\lambda|)/\omega$ and $q : \mathbb{R} \ni t \mapsto e^{it \operatorname{Im} \chi} p(t)c \in \mathbb{C}$. q is C^1-smooth and is a finite sum of products of periodic functions. It follows that q is almost periodic. Namely, for every $\epsilon > 0$ there exists $L > 0$ so that every interval of length L contains $\theta \in \mathbb{R}$ with

$$|q(t+\theta) - q(t)| < \epsilon \qquad \text{for all } t \in \mathbb{R}.$$

Similarly, $\dot{q}$ is almost periodic.

If $y = y^c \in \mathcal{G}_{\mathbb{C}}(\lambda)$, then

$$y^c(t) = e^{\beta t} \sum_{l=0}^{\kappa-1} q_l(t) t^l \qquad \text{for all } t \in \mathbb{R}$$

with β given as before and functions

$$q_l : \mathbb{R} \in t \mapsto p(t) e^{it \operatorname{Im} \chi} \frac{1}{\omega^l l!} \widetilde{N}^l c \in \mathbb{C}$$

for $l \in \{0, \dots, \kappa - 1\}$. Each q_l is C^1-smooth and almost periodic, with an almost periodic derivative. The coefficient $\tilde{y} : \mathbb{R} \ni t \mapsto e^{\beta t} q_{\kappa-1}(t) \in \mathbb{C}$ belongs to $\mathcal{E}_{\mathbb{C}}(\lambda)$ since $\widetilde{N}^{\kappa-1} c$ is an eigenvector of B and $\widetilde{M}$ or zero. For $\widetilde{N}^{\kappa-1} c \neq 0$, we have $\tilde{y} \neq 0$. It follows that the set

$$\{y \in \mathcal{G}_{\mathbb{C}}(\lambda) : \tilde{y} \neq 0\}$$

is dense in $\mathcal{G}_{\mathbb{C}}(\lambda)$.

For $0 \neq \lambda \in \sigma$ with $\operatorname{Im} \lambda \geq 0$, let

$$\mathcal{E}(\lambda) = \{x \in X : \text{There exists } y \in \mathcal{E}_{\mathbb{C}}(\lambda) + \mathcal{E}_{\mathbb{C}}(\overline{\lambda}) \text{ with } x(t) = y(t) \text{ for all } t \in \mathbb{R}\},$$
$$\mathcal{G}(\lambda) = \{x \in X : \text{There exists } y \in \mathcal{G}_{\mathbb{C}}(\lambda) + \mathcal{G}_{\mathbb{C}}(\overline{\lambda}) \text{ with } x(t) = y(t) \text{ for all } t \in \mathbb{R}\}.$$

Define $\beta : (0, \infty) \to \mathbb{R}$ by $\beta(\rho) = \ln \rho/\omega$. For $\rho > 0$ with $\sigma \cap \rho S^1_{\mathbb{C}} \neq 0$, define

$$\mathcal{E}_\rho = \bigoplus_{\lambda \in \sigma, \rho = |\lambda|, 0 \leq \operatorname{Im} \lambda} \mathcal{E}(\lambda),$$

$$\mathcal{G}_\rho = \bigoplus_{\lambda \in \sigma, \rho = |\lambda|, 0 \leq \operatorname{Im} \lambda} \mathcal{G}(\lambda).$$

These spaces of real-valued solutions of equation (VII.1) have the following properties:

(P1) $\dim \mathcal{G}(\lambda) = m(\lambda)$ for $\lambda \in \sigma \cap \mathbb{R}$; $\dim \mathcal{G}(\lambda) = 2m(\lambda)$ for $\lambda \in \sigma$ with $\operatorname{Im} \lambda > 0$.

(P2) For each $x \in \mathcal{E}_\rho$ there is an almost periodic C^1-function $q : \mathbb{R} \to \mathbb{R}$ with almost periodic derivative so that

$$x(t) = e^{\beta(\rho) t} q(t) \qquad \text{for all } t \in \mathbb{R}.$$

(P3) There is an integer $\kappa = \kappa(\rho) \geq 1$ so that for every $x \in \mathcal{G}_\rho$ there exist almost periodic real-valued C^1-functions $q_0, \dots, q_{k-1}$ with almost periodic derivatives such that

$$x(t) = e^{\beta(\rho) t} \sum_{l=0}^{\kappa-1} q_l(t) t^l \qquad \text{for all } t \in \mathbb{R},$$

and

$$\tilde{x} : \mathbb{R} \ni t \mapsto e^{\beta(\rho)t} q_{\kappa-1}(t) \in \mathbb{R}$$

belongs to $\mathcal{E}_\rho$.

(P4) The set $\{x \in \mathcal{G}_\rho : \tilde{x} \neq 0\}$ is dense in $\mathcal{G}_\rho$.

The next four results are variants of results in Section 4 of Chow and Walther **[19]**, but we should mention that the proof of Proposition VII.8 is considerably more complicated than that for the corresponding result in Chow and Walther **[19]**.

Proposition VII.5 *Let $\rho > 0$. If $x \in \mathcal{E}_\rho$ and $x_t \in T$ for some $t \in \mathbb{R}$ then $x \in J$.*

Proof We have $V(x_s) \leq 2$ for all $s \geq t$. Suppose $2 < V(x_s)$ for some $s < t$. The function $q : \mathbb{R} \ni s \mapsto e^{-\beta(\rho)s}x(s) \in \mathbb{R}$ is almost periodic, and there are $s_1 < s_2 < s_3 < s_4$ in $[s-1, s]$ with $q(s_j)q(s_{j+1}) < 0$ for $j \in \{1, 2, 3\}$. Using the almost periodicity we find $p \geq t - s$ so that $q(s_j + p)q(s_{j+1} + p) < 0$ for $j \in \{1, 2, 3\}$. It follows that $V(x_{s+p}) \geq 3$, which is a contradiction to the fact that V is decreasing along the solution x of equation (VII.1). □

For $\rho > 0$ with $\sigma \cap \rho S^1_{\mathbb{C}} \neq \emptyset$, let E_ρ denote the direct sum

$$E_\rho = \bigoplus_{\lambda \in \sigma, |\lambda| = \rho, \operatorname{Im} \lambda \geq 0} E_{\mathbb{R}}(\lambda).$$

Proposition VII.6 *Let $\rho > 0$ with $\sigma \cap \rho S^1_{\mathbb{C}} \neq \emptyset$ be given. If $E_\rho \cap T \neq \emptyset$ then*

$$\bigoplus_{\rho^* \geq \rho} \mathcal{G}_{\rho^*} \subset \overline{J}.$$

Proof 1. We show $\mathcal{E}_\rho \subset \overline{J}$. Let $y \in \mathcal{E}_\rho$. The hypothesis implies that for some $x \in \mathcal{E}_\rho$, $x_0 \in T$. Proposition VII.5 yields $x \in J$.

If $0 = \alpha y + (1 - \alpha)x$ for some $\alpha \in [0, 1]$, then $\alpha \neq 0$ (since $x \neq 0$) and

$$y = \frac{\alpha - 1}{\alpha} x \in \frac{\alpha - 1}{\alpha} J \subset \overline{J}.$$

So, it remains to consider the case $0 \notin x + [0, 1](y - x)$. Set

$$\alpha_0 = \sup\{\alpha \in [0, 1] : x + \alpha(y - x) \in J\}$$

and

$$z = x + \alpha_0(y - x).$$

We claim that $z \in J$. This is clearly the case if $\alpha_0 = 0$ since $x \in J$. If $0 < \alpha_0 \leq 1$, then there is a sequence of points α_j, $j \in \mathbb{N}$, in $[0, 1]$ with $\alpha_j \to \alpha_0$ as $j \to \infty$. The functions $x^{(j)} = x + \alpha_j(y - x)$ converge in X to z as $j \to \infty$. Hence $z \in \overline{J}$. We have $z \neq 0$. Therefore, Proposition VII.4 gives $z \in J$.

We now claim that $\alpha_0 = 1$ from which it follows that $y = z \in J$. In fact, if $\alpha_0 < 1$ then the functions $x_{(j)} = x + (\alpha_0 + \frac{1}{j})(y - x)$, $\frac{1}{1-\alpha_0} \leq j \in \mathbb{N}$, and their derivatives converge in X to z and $\dot{z}$, respectively, as $j \to \infty$. (i) and (iii) of Lemma VI.2 and $z \in J$ combined imply that there exists $t \in \mathbb{R}$ so that $V(z_s) = V(z_t)$ for all $s \in (-\infty, t]$, and $z_t \in R$. Using (ii) of Lemma VI.1, we obtain $V((x_{(j)})_t) =$

$V(z_t) \leq 2$ for some integer $j \geq \frac{1}{1-\alpha_0}$. Proposition VII.5 gives $x_{(j)} \in J$, which is a contradiction to the definition of α_0 as a supremum.

2. We prove $\mathcal{G}_\rho \subset \overline{J}$. Let $y \in \mathcal{G}_\rho$. There is a sequence of functions $x_{(j)}$, $j \in \mathbb{N}$, in $\mathcal{G}_\rho$ with $x_{(j)} \to y$ in X as $j \to \infty$, and $0 \neq \tilde{x}_{(j)} \in \mathcal{E}_\rho$ for all $j \in \mathbb{N}$. In view of Proposition VII.4, it is sufficient to show that each $x = x_{(j)}$, $j \in \mathbb{N}$, belongs to J. Set $\beta = \beta(\rho)$, $\kappa = \kappa(\rho)$. There are almost periodic continuously differentiable functions q_l, $l \in \{0, \ldots, \kappa-1\}$, with bounded derivatives so that

$$x(t) = e^{\beta t} \sum_{l=0}^{\kappa-1} t^l q_l(t) \qquad \text{for all } t \in \mathbb{R},$$

and

$$z : \mathbb{R} \in t \mapsto e^{\beta t} q_{\kappa-1}(t) \in \mathbb{R}$$

belongs to $\mathcal{E}_\rho \setminus \{0\} \subset J$ (the inclusion is due to the result in part 1). All q_l are bounded, due to almost periodicity. Set $q = q_{\kappa-1}$. In case $\kappa = 1$, $x = z \in J$ and we are done. So let $\kappa > 1$ in the sequel. (i) of Lemma VI.2 and $V(z_t) \leq 2$ on $\mathbb{R}$ imply that there exists $t_0 < 0$ with $V(z_t) = V(z_{t_0})$ on $(-\infty, t_0]$. Choose $t_1 < t_0$ with $q(t_1) \neq 0$. The almost periodicity of q permits to find a sequence $(t_n)_0^\infty$ in $(-\infty, t_0]$ with $t_n \to -\infty$ as $n \to \infty$ so that $|q(t_1)|/2 \leq ||q_{t_n}||$ for all $n \in \mathbb{N}$, and $||q_{t_n}|| \to r$ as $n \to \infty$ for some $r > 0$.

For every $n \in \mathbb{N}$ consider the solution

$$w^n : \mathbb{R} \ni t \mapsto \frac{1}{||z_{t_n}||} z(t_n + t) \in \mathbb{R}$$

of the equation

$$\dot{w}(s) = -\mu w(s) + a(t_n + s) w(s-1).$$

There are integers k_n with $s_n = t_n + k_n \omega \in [0, \omega]$ and a subsequence $(s_{n_j})_0^\infty$ which converges to some $s_\infty \in [0, \omega]$. The uniform continuity of the periodic function a yields

$$\sup_{t \in \mathbb{R}} |a(t_{n_j} + t) - a(s_\infty + t)| \to 0 \qquad \text{as } j \to \infty.$$

Using Lemmas VI.5 and VI.6 with $b(t) = a(t_n + t)$ for $t \leq 0$, and Lemma VI.4 with $I_0 = I_1 = I_2 = (-\infty, 2]$, $b(t) = a(s_\infty + t)$ and $b^j(t) = a(t_{n_j} + t)$ for $t \in I_1$, we find a subsequence of functions $\hat{w}^k = w^{n_{j_k}}$ and a C^1-smooth function $w : (-\infty, 2] \to \mathbb{R}$ such that $\hat{w}^k$ and $\dot{\hat{w}}^k$ converge uniformly on every bounded interval in $(-\infty, 2]$ to w and $\dot{w}$ as $j \to \infty$, and

$$\dot{w}(s) = -\mu w(s) + (a_\infty + s) w(s-1) \qquad \text{for all } s \in (-\infty, 2].$$

By $||w_0|| = 1$ we have $w_s \neq 0$ for all $s \leq 2$.

Set $u_k = t_{n_{j_k}}$ for $k \in \mathbb{N}$. It follows that for every $s \leq 2$,

$$V(w_s) \leq \liminf_{k \to \infty} V(||z_{u_k}||^{-1} z_{u_k + s}) \leq 2.$$

2.1. We claim that $V(w_s) = V(w_0) \leq 2$ on $(-\infty, 0]$ and $w_0 \in R$. The equality is obvious from the last estimate if $V(z_s) = 0$ on $(-\infty, t_0]$. So it remains to consider the case $V(z_s) = 2$ on $(-\infty, t_0]$. Then $V(w_0) = 2$, for otherwise $V(w_0) = 0$ implies that w_2 has no zero and thus implies that z_{u_k+2} has no zero for k sufficiently large, a contradiction to $V(z_{u_k+2}) = 2$. (i) of Lemma VI.2 and the last estimate give $V(w_s) = 2$ on $(-\infty, 0]$. Finally, (iii) of Lemma VI.2 yields $w_0 \in R$.

2.2. Next, consider the C^1-functions $\hat{q}_k = ||q_{u_k}||^{-1}q_{u_k}$ for $k \in \mathbb{N}$. We claim that there exist a subsequence $(\hat{q}_{k_l})_0^\infty$ and a constant $c \neq 0$ so that

$$\sup_{t\in[-1,0]} \left|\hat{q}_{k_l}(t) - ce^{-\beta t}w(t)\right| \to 0$$

and

$$\sup_{t\in[-1,0]} \left|\dot{\hat{q}}_{k_l}(t) - ce^{-\beta t}(-\beta w(t) + \dot{w}(t))\right| \to 0$$

as $l \to \infty$.

In fact, the equality $z(s) = e^{\beta s}q(s)$ on $\mathbb{R}$ yields

$$||z_s|| = e^{\beta s} \max_{t\in[-1,0]} \left|e^{\beta t}q_s(t)\right| \leq e^{\beta s}\max\{e^{-\beta}, 1\}||q_s||.$$

It follows that there is a subsequence $(k_l)_0^\infty$ of $(k)_0^\infty$ so that the numbers

$$\frac{||z_{u_{k_l}}||}{||q_{u_{k_l}}||}e^{-\beta u_{k_l}}$$

converge to some $c \in \mathbb{R}$ as $l \to \infty$. Using the equalities

$$\frac{q_s(t)}{||q_s||} = \frac{||z_s||}{||q_s||}e^{-\beta s}e^{-\beta t}\frac{z_s(t)}{||z_s||} \qquad \text{for all } t \in [-1,0]$$

and

$$\frac{\dot{q}_s(t)}{||q_s||} = \frac{||z_s||}{||q_s||}e^{-\beta s}e^{-\beta t}(-\beta z_s(t) + \dot{z}_s(t))\frac{1}{||z_s||} \qquad \text{for all } t \in [-1,0]$$

with $s = u_{k_l}$ and the convergence properties of the points $||z_{u_k}||^{-1}z_{u_k}$, we obtain the required convergence property of $\hat{q}_{k_l}$. Observe $||\hat{q}_{k_l}|| = 1$ for all $l \in \mathbb{N}$. This yields $c \neq 0$.

2.3. Consider $\phi : [-1,0] \ni t \mapsto ce^{-\beta t}w(t) \in \mathbb{R}$. We claim that $V(\phi) = V(w_0)$ and $\phi \in R$. The equality follows from the fact that the zeros of ϕ and w_0 coincide. The equation $\dot{\phi}(s) = ce^{-\beta s}(-\beta w(s) + \dot{w}(s))$, $-1 \leq s \leq 0$, and the simplicity of the zeros of w_0 combined imply that all zeros of ϕ are simple. Moreover, if $\phi(0) = 0$ then $w(0) = 0$. Assume $\phi(0) = 0$. As $w_0 \in R$, we must have $\dot{w}(0)w(-1) > 0$. But $\dot{\phi}(0) = c\dot{w}(0)$ and $\phi(-1) = ce^{\beta}w(-1)$. It follows that $\dot{\phi}(0)\phi(-1) > 0$. In case $\phi(-1) = 0$ we find analogously that $\dot{\phi}(-1)\phi(0) < 0$. It follows that $\phi \in R$.

2.4. The boundedness of the functions q_l and their derivatives and the fact that $\lim_{k\to\infty} ||q_{u_k}|| = r > 0$ combined imply that the C^1-functions

$$r_l : [-1,0] \ni s \mapsto \frac{1}{||q_{u_{k_l}}||}\sum_{i=0}^{\kappa-2}(u_{k_l} + s)^{i-(\kappa-1)}q_i(u_{k_l} + s) \in \mathbb{R}$$

and their derivatives converge to 0 uniformly on $[-1,0]$ as $l \to \infty$. It follows that the C^1-functions

$$\chi_l = \frac{1}{||q_{u_{k_l}}||}q_{u_{k_l}} + r_l$$

and their derivatives $\dot{\chi}_l$ converge to ϕ and $\dot{\phi}$, respectively, uniformly on $[-1,0]$ as $l \to \infty$. We apply (ii) of Lemma VI.1 to obtain

$$V(\phi) = \lim_{l\to\infty} V(\chi_l), \quad \text{or } V(\chi_l) = V(\phi) \quad \text{for sufficiently large } l.$$

$V(\chi_l) = V(\phi)$ gives $V(x_{u_{k_l}}) = V(\phi)$. Using (i) of Lemma VI.1 we deduce $V(x_s) \le 2$ on $\mathbb{R}$, which means $x \in J$.

3. Let $v \in \oplus_{\rho^* \ge \rho} \mathcal{G}_{\rho^*}$. Then $v = y + \hat{v}$ with $y \in \mathcal{G}_\rho$ and $\hat{v} \in \oplus_{\rho^* > \rho} \mathcal{G}_{\rho^*}$. There is a sequence $(x^j)_0^\infty$ in $\mathcal{G}_\rho$ with $0 \ne \tilde{x}^j \in \mathcal{E}_\rho$ for all $j \in \mathbb{N}$ and $x^j \to y$ in X as $j \to \infty$. In view of Proposition VII.4, it is sufficient to show that for every $j \in \mathbb{N}$, $x^j + \hat{v}$ belongs to J. Let $j \in \mathbb{N}$. Set $x = x^j$, $\beta = \beta(\rho)$, $\kappa = \kappa(\rho)$. There are almost periodic continuously differentiable functions q_l, $l \in \{0, \dots, \kappa - 1\}$, with bounded derivatives so that

$$x(t) = e^{\beta t} \sum_{l=0}^{\kappa-1} t^l q_l(t) \qquad \text{on } \mathbb{R},$$

and

$$z : \mathbb{R} \ni t \mapsto e^{\beta t} q_{\kappa-1}(t) \in \mathbb{R}$$

belongs to $\mathcal{E}_\rho \setminus \{0\} \subset J$. We have

$$(x + \hat{v})(t) = e^{\beta t} t^{\kappa-1} \Big(\sum_{l=0}^{\kappa-1} t^{l-(\kappa-1)} q_l(t) + e^{-\beta t} t^{1-\kappa} \hat{v}(t) \Big) \qquad \text{for all } t < 0.$$

Since $\beta < \beta(\rho^*)$ for all $\rho^* > \rho$ with $\sigma \cap \rho^* S^1_{\mathbb{C}} \ne \emptyset$, we know that the C^1-function $\hat{r} : (-\infty, 0) \ni t \mapsto e^{-\beta t} t^{1-\kappa} \hat{v}(t) \in \mathbb{R}$ and its derivative $\dot{\hat{r}}$ converge to 0 as $t \to -\infty$.

Set $q = q_{\kappa-1}$. We can proceed as in part 2 with either $r_l + \frac{1}{||q_{u_{k_l}}||} \hat{r}_{u_{k_l}}$ in place of r_l (if $\kappa > 1$) or $\frac{1}{||q_{u_{k_l}}||} \hat{r}_{u_{k_l}}$ in place of r_l (if $\kappa = 1$) to show that $x + \hat{v} \in J$. □

Corollary VII.7 *Let $\rho > 0$ with $\sigma \cap \rho S^1_{\mathbb{C}} \ne \emptyset$ and $E_\rho \cap T \ne \emptyset$ be given. Then*

$$\sum_{\lambda \in \sigma, |\lambda| \ge \rho} m(\lambda) \le 3.$$

Proof Using Propositions VII.6, VII.4 and VII.3, we conclude that each non-zero $\phi \in \oplus_{\lambda \in \sigma, |\lambda| \ge \rho, \operatorname{Im} \lambda \ge 0} G_{\mathbb{R}}(\lambda)$ has at most 2 zeros. Applying Lemma VII.2 and observing

$$\sum_{\lambda \in \sigma, |\lambda| \ge \rho} m(\lambda) = \sum_{\lambda \in \sigma, |\lambda| \ge \rho, \operatorname{Im} \lambda \ge 0} \dim G_{\mathbb{R}}(\lambda) = \dim \bigoplus_{\lambda \in \sigma, |\lambda| \ge \rho, \operatorname{Im} \lambda \ge 0} G_{\mathbb{R}}(\lambda),$$

we obtain the desired inequality. □

We need a stronger version of Proposition VII.6.

Proposition VII.8 *Let $\rho > 0$ with $\sigma \cap \rho S^1_{\mathbb{C}} \ne \emptyset$ be given. If there exists $y \in \mathcal{G}_\rho$ with $y_t \in T$ for some $t \in \mathbb{R}$, then $\oplus_{\rho^* \ge \rho} \mathcal{G}_{\rho^*} \subset \overline{J}$.*

Proof 1. In view of Proposition VII.6 it is sufficient to show that $E_\rho \cap T \ne \emptyset$.

There exists a sequence of elements $y^n \in \mathcal{G}_\rho$ with $\tilde{y}^n \ne 0$ and $y^n \to y$ in X as $n \to \infty$. Using equation (VII.1) we obtain $\dot{y}^n \to \dot{y}$ in X as $n \to \infty$. (i) and (iii) of Lemma VI.2 yield a number $s \ge t$ so that $y_s \in R$; (ii) of Lemma VI.1 gives an integer j with $V(y^j_s) = V(y_s) \le 2$. Set $x = y_j$, $\beta = \beta(\rho)$, $\kappa = \kappa(\rho)$. There exist almost periodic real-valued C^1-functions $q_0, \dots, q_{\kappa-1}$ with bounded derivatives so that

$$x(t) = e^{\beta t} \sum_{j=0}^{\kappa-1} t^j q_j(t) \qquad \text{on } \mathbb{R},$$

and

$$\tilde{x} : \mathbb{R} \ni t \mapsto e^{\beta t} q_{\kappa-1}(t) \in \mathbb{R}$$

belongs to $\mathcal{E}_\rho \setminus \{0\}$. Set $q = q_{\kappa-1}$. The almost periodicity of q permits to find a strictly increasing sequence $(t_n)_0^\infty$ with $t_n \to \infty$, and $t_0 \geq s+7$ so that the numbers $||q_{t_n}|| > 0$ are bounded below by some constant $c > 0$. There are a subsequence of points $s_m = t_{n_m}$, $m \in \mathbb{N}$, and integers j_m so that the points $s_m + j_m\omega$, $m \in \mathbb{N}$, belong to $[0, \omega]$ and converge to some s_ω as $m \to \infty$. Observe that for every $m \in \mathbb{N}$ the function

$$x^{(m)} : \mathbb{R} \ni t \mapsto \frac{1}{||x_{s_m}||} x(s_m + t) \in \mathbb{R}$$

is a solution of the equation

$$\dot{x}(t) = -\mu x(t) + a(s_m + t)x(t-1).$$

The coefficients $\mathbb{R} \ni t \mapsto a(s_m + t) \in \mathbb{R}$ converge uniformly on $\mathbb{R}$ to $\mathbb{R} \ni t \mapsto a(s_\omega + t) \in \mathbb{R}$ as $n \to \infty$. Using Lemma VI.6 with $b(t) = a(s_m + t)$ for $t \geq 0$ and Lemma VI.4 with $I_0 = [0, \infty)$, $I_1 = [1, \infty)$, $I_2 = [2, \infty)$, $b^m(t) = a(s_m + t)$ and $b(t) = a(s_\omega + t)$ for $t \in I_1$, we find a sequence of integers $u_n = s_{m_n}$, $n \in \mathbb{N}$, and a continuous function $w : [1, \infty) \to \mathbb{R}$ which is differentiable on $(2, \infty)$ such that the functions

$$x^n : [1, \infty) \ni t \mapsto \frac{x(u_n + t)}{||x_{u_n}||} \in \mathbb{R}$$

converge uniformly on every bounded interval in $[1, \infty)$ to w, and the derivatives $\dot{x}^n$ converge uniformly on every bounded interval of $(2, \infty)$ to $\dot{w}$, and

$$\dot{w}(t) = -\mu w(t) + a(s_\omega + t)w(t-1) \qquad \text{for } t > 2.$$

An application of Lemma VI.3 yields a positive lower bound for the quotients $\frac{||x_{s_m+2}||}{||x_{s_m}||}$, $m \in \mathbb{N}$, which in turn gives $w_2 \neq 0$. Next, (i) of Lemma VI.1 yields $V(w_2) \leq 2$. Using (i) and (iii) of Lemma VI.2, we obtain some $\tau_0 \geq 3$ with $w_\tau \in R \cap T$ for all $\tau \geq \tau_0$. The almost periodicity of q permits to find $\tau \geq \tau_0$ with

$$||q_{u_n+\tau}|| \geq \frac{c}{2} \qquad \text{for all } n \in \mathbb{N}.$$

2. Set $\phi = w_\tau$. Then

$$\frac{1}{||x_{u_n}||} x_{u_n+\tau} \to \phi, \quad \frac{1}{||x_{u_n}||} \dot{x}_{u_n+\tau} \to \dot{\phi} \qquad \text{as } n \to \infty.$$

For $n \in \mathbb{N}$, define $\phi_n \in C$ by

$$\phi_n(s) = e^{-\beta(u_n+\tau+s)}(u_n + \tau + s)^{1-\kappa} x(u_n + \tau + s).$$

We claim that there exist a constant $c_0 > 0$ and a strictly increasing sequence $(n_j)_0^\infty$ so that the functions

$$\frac{1}{||\phi_{n_j}||} \phi_{n_j} \text{ converge uniformly to } [-1, 0] \ni t \mapsto c_0 e^{-\beta t} \phi(t) \in \mathbb{R}$$

and their derivatives

$$\frac{1}{||\phi_{n_j}||} \dot{\phi}_{n_j} \text{ converge uniformly to } [-1, 0] \ni t \mapsto c_0 e^{-\beta t}(-\beta\phi(t) + \dot{\phi}(t)) \in \mathbb{R}$$

as $j \to \infty$. To prove the claim, we first note that for every $n \in \mathbb{N}$ and $t \in \mathbb{R}$,

$$\frac{\phi_{n_j}(t)}{||\phi_{n_j}||} = \frac{e^{-\beta t} x_{u_n+\tau}(t)}{||x_{u_n}||} \left(\frac{u_n + \tau}{u_n + \tau + t} \right)^{\kappa - 1} \frac{||x_{u_n}||}{\max_{s \in [-1,0]} |e^{-\beta s}(\frac{u_n+\tau}{u_n+\tau+s})^{\kappa-1} x_{u_n+\tau}(s)|}.$$

For $n \to \infty$, the functions $[-1,0] \ni t \mapsto \frac{e^{-\beta t} x_{u_n+\tau}(t)}{||x_{u_n}||} \in \mathbb{R}$ converge uniformly to $[-1,0] \ni t \mapsto e^{-\beta t}\phi(t) \in \mathbb{R}$, and the functions $[-1,0] \ni t \mapsto (\frac{u_n+\tau}{u_n+\tau+t})^{\kappa-1} \in \mathbb{R}$ converge uniformly to $[-1,0] \ni t \mapsto 1 \in \mathbb{R}$. We show that the third factor is bounded. For all $s \in [-1,0]$,

$$e^{-\beta s}(\frac{u_n + \tau}{u_n + \tau + s})^{\kappa-1} |x(u_n + \tau + s)| \geq \min\{e^{\beta}, 1\} |x_{u_n+\tau}(s)|.$$

Thus the norm in the denominator is bounded from below by

$$\min\{e^{\beta}, 1\} ||x_{u_n+\tau}||.$$

Let j_τ be the smallest integer in $[\tau, \infty)$. Repeated application of Lemma VI.3 yields a constant $c^* > 0$ so that

$$||x_{u_n}|| \leq c^* ||x_{u_n+j_\tau}|| \qquad \text{for all } n \in \mathbb{N}.$$

As $j_\tau - \tau \in [0,1]$, we can use Lemma VI.6 to get a constant $c^{**} > 0$ so that

$$||x_{u_n+j_\tau}|| \leq c^{**} ||x_{u_n+\tau}|| \qquad \text{for all } n \in \mathbb{N}.$$

Combining all estimates above, we obtain the asserted boundedness of the third terms, and thus there is a strictly increasing subsequence $(n_j)_{j=0}^{\infty}$ along which the third factor converges to a constant $c_0 \geq 0$. It follows that the functions $\frac{1}{||\phi_{n_j}||}\phi_{n_j}$ converge to $[-1,0] \ni s \mapsto c_0 e^{-\beta s}\phi(s) \in \mathbb{R}$ as $j \to \infty$. We have $c_0 > 0$ since the limit has norm 1. Taking derivatives of $\chi_n = \frac{1}{||\phi_n||}\phi_n$, we get

$$\dot{\chi}_n(s) = -\beta\chi_n(s) + \frac{1-\kappa}{u_n + \tau + s}\chi_n(s) + \frac{1}{||\phi_n||} e^{-\beta(u_n+\tau+s)} (u_n+\tau+s)^{1-\kappa} \dot{x}(u_n+\tau+s)$$

for all $n \in \mathbb{N}$ and $s \in [-1,0]$. The functions $-\beta\chi_{n_j}$ converge to $[-1,0] \ni s \mapsto -c_0\beta e^{-\beta s}\phi(s) \in \mathbb{R}$ as $j \to \infty$. The functions on $[-1,0]$ defined by the second terms on the right hand side of the last equation tend to 0 uniformly on $[-1,0]$ as $j \to \infty$ since $||\chi_n|| = 1$. The functions defined by the third terms can be treated as the functions $\frac{1}{||\phi_n||}\phi_n$ above so that they converge to the function $[-1,0] \ni s \mapsto c_0 e^{-\beta s} \lim_{n\to\infty} \frac{1}{||x_{u_n}||} \dot{x}_{u_n+\tau}(s) \in \mathbb{R}$, which equals $[-1,0] \ni s \mapsto c_0 e^{-\beta s}\dot{\phi}(s) \in \mathbb{R}$.

3. The function $\psi : [-1,0] \ni s \mapsto c_0 e^{-\beta s}\phi(s) \in \mathbb{R}$ is continuously differentiable with

$$\dot{\psi}(s) = c_0 e^{-\beta s}(-\beta\phi(s) + \dot{\phi}(s)) \qquad \text{for all } s \in [-1,0].$$

We claim that $\psi \in R$. This is true since $\phi \in R$, the zeros of ψ and ϕ coincide, $\text{sign} \circ \psi = \text{sign} \circ \phi$, and since for every zero z of ϕ and ψ, $\text{sign}(\dot{\psi}(z)) = \text{sign}(\dot{\phi}(z))$.

We also observe that $\phi \in T$ implies $\psi \in T$.

4. The boundedness of q and its derivative implies that there is a strictly increasing sequence of reals $v_k = u_{n_{j_k}} + \tau$, $k \in \mathbb{N}$, so that q_{v_k} tends to some $r \in C$ as $k \to \infty$, with $||r|| \geq c/2$.

In case $\kappa > 1$, the boundedness of the functions q_l and their derivatives implies that for $k \to \infty$ the segments

$$\xi_k : [-1,0] \ni s \mapsto \sum_{i=0}^{\kappa-2} (v_k + s)^{i-(\kappa-1)} q_i(v_k + s) \in \mathbb{R}$$

and their derivatives converge to zero uniformly on $[-1,0]$. It follows that the reals

$$\max_{s\in[-1,0]} \left| e^{-\beta(v_k+s)} (v_k + s)^{1-\kappa} x(v_k + s) \right|$$

converge to $||r|| > 0$ as $k \to \infty$. Consequently, the segments

$$\eta_l : [-1,0] \ni s \mapsto e^{-\beta(v_k+s)} (v_k + s)^{1-\kappa} x(v_k + s) \in \mathbb{R}$$

and their derivatives converge for $k \to \infty$ in C to $||r||\psi$ and its derivative, respectively. Using, in case $\kappa > 1$, the convergence property of the segments ξ_k we conclude that the segments q_{v_k} and their derivatives converge for $k \to \infty$ in C to $||r||\psi \in R \cap T$ and its derivatives, respectively.

Now, we can apply (ii) of Lemma VI.1 and find integers $k \in \mathbb{N}$ with

$$2 \geq V(||r||\psi) = V(q_{v_k}) = V(\tilde{x}_{v_k}).$$

Choose $m \in \mathbb{N}$ with $v_k \leq m\omega$. Then $\tilde{x}_{m\omega} \neq 0$. By (i) of Lemma VI.2,

$$2 \geq V(\tilde{x}_{m\omega}) = V(M^m \tilde{x}_0), \quad \text{or } M^m \tilde{x}_0 \in T.$$

On the other hand, $M^m \tilde{x}_0 \in E_\rho$, since $\tilde{x}_0 \in E_\rho$. Therefore, $T \cap E_\rho \neq \emptyset$. □

For $\rho > 0$, let $C_{\leq\rho}$ and $C_{\rho<}$ denote the realified generalized eigenspaces of M associated with the spectral sets $\{\lambda \in \sigma : |\lambda| \leq \rho\}$ and $\{\lambda \in \sigma : \rho < |\lambda|\}$, respectively. We let $C_{\rho<} = \{0\}$ if $\{\lambda \in \sigma : \rho < |\lambda|\} = \emptyset$. Note that

$$C_{\rho<} = \bigoplus_{\lambda\in\sigma, \rho<|\lambda|, \operatorname{Im}\lambda\geq 0} G_{\mathbb{R}}(\lambda)$$

in case there exists $\lambda \in \sigma$ with $\rho < |\lambda|$.

Theorem VII.9 *The set $A(\sigma, C\backslash T) = \{r \in (0,\infty) : \sigma \cap r S^1_{\mathbb{C}} \neq \emptyset,\ C_{\leq r} \cap T = \emptyset\}$ is nonempty and has a maximum. Let $\rho = \max A(\sigma, C\backslash T)$. We have $\sigma \cap (\rho, \infty) \neq \emptyset$, and $C_{\leq\rho} \cap T = \emptyset$, $C_{\rho<} \subset \overline{T}$, $\dim C_{\rho<} \leq 3$.*

Proof 1. We claim that there exists $c > 0$ so that for every solution $y : [-1,\infty) \to \mathbb{R}$ of equation (VII.1) and for every $t \geq 0$ with $y_t \neq 0$ and $V(y_t) \leq 2$ there is an integer $j \geq 0$ with

$$||M^l(M^j y_0)|| \geq c^l ||M^j y_0|| \qquad \text{for all } l \in \mathbb{N}.$$

To prove the claim, we let $k \in \mathbb{N}$ denote the smallest integer in $[\omega, \infty)$. By Lemma VI.3 there exists $c_0 > 0$ so that for every solution $x : [-1,\infty) \to \mathbb{R}$ of equation (VII.1) and for every $s \geq 0$ with $V(x_\theta) \leq 2$ for $\theta \in [s,\infty)$, we have

$$c_0 ||x_{s+4}|| \leq ||x_{s+4+1}||.$$

Hence, $c_0^k ||x_{s+4}|| \leq ||x_{s+4+k}||$. Integration of the differential equation on $[s+4+\omega, s+4+k]$ yields

$$||x_{s+4+k}|| \leq (1 + \max a) ||x_{s+4+\omega}||$$

so that for $c = c_0^k(1+\max a)^{-1}$, we get

$$c||x_{s+4}|| \leq ||x_{s+4+\omega}||.$$

Let t and y be given as in the claim. Choose an integer j_0 with $j_0\omega \geq t$. Assertion (i) of Lemma VI.2 implies $V(y_\theta) \leq 2$ for $\theta \in [j_0\omega, \infty)$. Recall $\omega > 1$. So for every $l \in \mathbb{N}$ we obtain

$$c||y_{(j_0+4)\omega+l\omega}|| \leq ||y_{(j_0+4)\omega+(l+1)\omega}||.$$

It follows that

$$c^l||M^{j_0+4}y_0|| \leq ||M^l(M^{j_0+4}y_0)|| \qquad \text{for all } l \in \mathbb{N}.$$

2. Lemma I.1 ensures that for every $\rho > 0$ there exist a norm $||\cdot||_\rho$ on C, which is equivalent to $||\cdot||$, and $\epsilon(\rho) \in (0,\rho)$ so that

$$||M\phi||_\rho \leq (\rho+\epsilon(\rho))||\phi||_\rho \qquad \text{for all } \phi \in C_{\leq\rho},$$

and

$$(\rho+2\epsilon(\rho))||\phi|| \leq ||M\phi||_\rho \qquad \text{for all } \phi \in C_{\rho<}.$$

We now claim that for $0 < \rho < \frac{c}{2}$, $C_{\leq\rho} \cap T = \emptyset$. Suppose that there exists $\phi \in C_{\leq\rho} \cap T$ for some $\rho \in (0, \frac{c}{2})$. Using the result of part 1 and the positive invariance of $C_{\leq\rho}$ and T under M, we obtain an element $\psi \in C_{\leq\rho} \cap T$ and a constant $c > 0$ so that

$$||M^l\psi|| \geq c^l||\psi|| \qquad \text{for all } l \in \mathbb{N}.$$

Combining this with the estimate for $||M\phi||_\rho$ on $C_{\leq\rho}$, we find

$$2\rho \geq \rho + \epsilon(\rho) \geq c,$$

a contradiction to $\rho < \frac{c}{2}$.

3. So the set

$$A(\sigma, C\setminus T) = \{r \in (0,\infty) : \sigma \cap rS^1_{\mathbb{C}} \neq \emptyset,\ C_{\leq r} \cap T = \emptyset\}$$

is nonempty. It has a maximum because $\sigma \setminus \{0\}$ consists of isolated points. Let $\rho = \max A(\sigma, C\setminus T)$. By Proposition VII.1, there exists $\lambda \in \sigma \cap (\rho,\infty)$. Set

$$r = \min\{|\lambda| : \lambda \in \sigma \text{ and } \rho < |\lambda|\}$$

and

$$G_r = \bigoplus_{\lambda\in\sigma,|\lambda|=r,0\leq \operatorname{Im}\lambda} G_{\mathbb{R}}(\lambda).$$

G_r is the realified generalized eigenspace of M associated with the finite spectral set $\sigma \cap rS^1_{\mathbb{C}}$, and we have

$$C_{\leq r} = C_{\leq\rho} \oplus G_r.$$

We claim that $G_r \cap T \neq \emptyset$. The definition of ρ implies that there is an element $\phi \in (C_{\leq r} \cap T)\setminus C_{\leq\rho}$. Then $\phi = \chi + \psi$ with $\chi \in C_{\leq\rho}$ and $0 \neq \psi \in G_r$. The relations $MG_r \subset G_r$ and $\dim G_r < \infty$ imply that a subsequence of the normed iterates $\psi_j = \frac{1}{||M^j\psi||}M^j\psi$, $j \in \mathbb{N}$, converges to some $\beta \in G_r \setminus \{0\}$. Consider the normed iterates $\phi_j = \frac{1}{||M^j\phi||}M^j\phi \in T$, $j \in \mathbb{N}$.

The case $\phi \neq \psi$. Then $M^j(\phi-\psi) \neq 0$ for all $j \in \mathbb{N}$, and we have the estimate

$$
\begin{aligned}
&||\phi_j - \psi_j|| \\
&= \left\| \frac{||M^j(\phi-\psi)||}{||M^j\phi||} \frac{1}{||M^j(\phi-\psi)||} M^j(\phi-\psi) + \left(\frac{||M^j\psi||}{||M^j\phi||} - 1\right) \frac{1}{||M^j\psi||} M^j\psi \right\| \\
&\leq 2\frac{||M^j(\phi-\psi)||}{||M^j\phi||} \leq \frac{2}{\left|\frac{||M^j\psi||}{||M^j(\phi-\psi)||} - 1\right|}.
\end{aligned}
$$

Moreover, $\phi - \psi \in C_{\leq\rho}$ implies

$$||M^j(\phi-\psi)||_\rho \leq (\rho+\epsilon(\rho))^j ||\phi-\psi||_\rho,$$

and $\psi \in G_r \subset C_{\rho<}$ implies

$$||M^j\psi||_\rho \geq (\rho+2\epsilon(\rho))^j ||\psi||_\rho.$$

Therefore,

$$\frac{||M^j\psi||_\rho}{||M^j(\phi-\psi)||_\rho} \geq \left(\frac{\rho+2\epsilon(\rho)}{\rho+\epsilon(\rho)}\right)^j \frac{||\psi||_\rho}{||\phi-\psi||_\rho} \to \infty \qquad \text{as } j \to \infty.$$

Thus, the above estimate and the equivalence of the norm $||\cdot||_\rho$ with $||\cdot||$ yield $\phi_j - \psi_j \to 0$ as $j \to \infty$. So there is a subsequence of $(\phi_j)_0^\infty$ which converges to β, too. It follows that $\beta \in \overline{T}$, and we obtain $\beta \in G_r \cap T$.

The case $\psi = \phi$. Then $\psi_j \in T$ for all $j \in \mathbb{N}$. Hence $\beta \in \overline{T}$, and we obtain $\beta \in G_r \cap T$ as before.

This verifies the claim that $G_r \cap T \neq \emptyset$. We can now apply Proposition VII.8 to $\mathcal{G}_r$ and obtain

$$\bigoplus_{\rho^* \geq r} \mathcal{G}_{\rho^*} \subset \overline{J},$$

which gives

$$E_r \subset C_{\rho<} \subset \overline{T}.$$

By Corollary VII.7, $\dim C_{\rho<} \leq 3$. □

APPENDIX VIII

Some results from topology

We need the following variant of the Jordan Curve Theorem.

Theorem VIII.1 *(Schoenfliess Theorem)*

(i) If $f: S^1 \to \mathbb{R}^2$ is continuous and injective then there is a homeomorphism $h: D^2 \to f(S^1) \cup \operatorname{int} f(S^1)$ with $h(x) = f(x)$ for all $x \in S^1$.

(ii) If $f: S^1 \to S^2$ is continuous and injective then $S^2 \setminus f(S^1)$ has two connected components K_1 and K_2, and there exist homeomorphisms $h_1: D^2 \to f(S^1) \cup K_1$, $h_2: D^2 \to f(S^1) \cup K_2$ with $h_1(x) = h_2(x) = f(x)$ for all $x \in S^1$.

See Schoenfliess **[53]** or Theorems X.40.13 and X.40.14 in Rinow **[52]**.

A higher-dimensional analogue of the Jordan Curve Theorem can be stated as follows.

Theorem VIII.2 *(Jordan–Brouwer Separation Theorem) If $f: S^{n-1} \to \mathbb{R}^n$, $n \geq 2$, is continuous and injective then $\mathbb{R}^n \setminus f(S^{n-1})$ has exactly two connected components, one bounded and the other unbounded.*

See, for example, page 234 of Bredon **[14]**.

In the situation of Theorem VIII.2 we denote the bounded and unbounded components of $\mathbb{R}^n \setminus f(S^{n-1})$ by $\operatorname{int} f(S^{n-1})$ and $\operatorname{ext} f(S^{n-1})$, respectively.

In case $n = 2$ in the situation of Theorem VIII.2, $\operatorname{int} f(S^1)$ is homeomorphic to the open unit disk in $\mathbb{R}^2$ by Theorem VIII.1(i). But for $n \geq 3$, $\operatorname{int} f(S^{n-1})$ is not necessarily homeomorphic to the open unit ball in $\mathbb{R}^n$, as the example of Alexander's horned sphere shows (see, for example, Hocking and Young **[34]**). Additional properties of $\operatorname{int} f(S^{n-1})$ are required to prove that it is homeomorphic to the open unit ball in $\mathbb{R}^n$. We shall use a result of Bing **[12]** which gives a sufficient condition to guarantee that in case $n = 3$ the set $\operatorname{int} f(S^2)$ is homeomorphic to the open unit ball in $\mathbb{R}^3$.

The diameter $\operatorname{diam}(X)$ of a set $X \subset \mathbb{R}^3$ is defined by $\operatorname{diam}(X) = \sup\{|x-y|_{\mathbb{R}^3} : x \in X,\ y \in X\}$. A set $X \subset \mathbb{R}^3$ is called uniformly locally 1-connected if for every $\epsilon > 0$ there exists $\delta > 0$ such that every continuous map $a: \partial([0,1]\times[0,1]) \to X$ with $\operatorname{diam}\big(a(\partial([0,1]\times[0,1]))\big) < \delta$ can be extended to a continuous map $b: [0,1]\times[0,1] \to X$ with $\operatorname{diam}\big(b([0,1]\times[0,1])\big) < \epsilon$.

Theorem VIII.3 *If the set $A \subset \mathbb{R}^3$ is homeomorphic to S^2 and $\operatorname{int} A$ is uniformly locally 1-connected then there is a homeomorphism from $A \cup \operatorname{int} A$ onto D^3 which sends A onto S^2.*

See Theorem XVIII.5.C in Bing **[12]**.

Bibliography

[1] Abraham, R. and Robbin, J., *Transversal Mappings and Flows,* Benjamin, New York, 1967.

[2] Ammar, Y., *Eine dreidimensionale invariante Mannigfaltigkeit für autonome Differentialgleichungen mit Verzögerung,* Dissertation, Universität München, 1993.

[3] Arino, O., *A note on "The Discrete Lyapunov Function ...",* J. Differential Equations **104** (1993), 169–181.

[4] Arino, O. and Krisztin, T., *The 2-dimensional attractor of a differential equation with state-dependent delay,* in preparation.

[5] Arino, O. and Séguier, P., *Existence of oscillating solutions for certain differential equations with delay,* Functional Differential Equations and Approximation of Fixed Points (H.-O. Peitgen, H.-O. Walther, eds.), Lecture Notes in Math., Vol. 730, pp. 46–64, Springer-Verlag, New York, 1979.

[6] Arino, O. and Séguier, P., *Contribution à l'étude des comportements des solutions d'équation différentielle à retard par des méthodes de monotonie et de bifurcation,* Thèses d'Etat, Bordeaux I, Pau, 1980.

[7] Arndt, M., *Repräsentation räumlicher und zeitlicher Stetigkeit durch Synchronisation neuronaler Signale,* Dissertation, Universität Marburg, Fachbereich Physik, 1983.

[8] Bélair, J., *Stability in a model of a delayed neural network,* J. Dynamics and Differential Equations **5** (1993), 607–623.

[9] Bélair, J., Campbell, S. A. and van den Driessche, P., *Frustration, stability, and delay-induced oscillations in a neural network model,* SIAM J. Appl. Math. **56** (1996), 245–255.

[10] Bélair, J. and Dufour, S., *Stability in a three-dimensional system of delay-differential equations,* Canad. Appl. Math. Quarterly **4** (1996), 135–156.

[11] Bellman, R. and Cooke, K. L., *Differential-Difference Equations,* Academic Press, New York, 1963.

[12] Bing, R. H., *The Geometric Topology of 3-Manifolds,* Amer. Math. Soc. Colloq. Publ., Vol. 40, Providence, RI, 1983.

[13] Braun, H., *Personal communication.*

[14] Bredon, G. E., *Topology and Geometry,* Springer-Verlag, New York, 1995.

[15] Burton, T. A. and Hatvani, L., *Stability theorems for nonautonomous functional differential equations by Lyapunov functionals,* Tôhoku Math. J. **41** (1989), 65–104.

[16] Cao, Y., *The discrete Lyapunov function for scalar delay differential equations,* J. Differential Equations **87** (1990), 365–390.

[17] Chen, X.-Y., Hale, J. K. and Tan, B., *Invariant foliations for C^1 semigroups in Banach spaces,* J. Differential Equations **139** (1997), 283–318.

[18] Chow, S.-N., Diekmann, O. and Mallet-Paret, J., *Stability, multiplicity and global continuation of symmetric periodic solutions of a nonlinear Volterra integral equation,* Japan J. Appl. Math. **2** (1985), 433–469.

[19] Chow, S.-N. and Walther, H.-O., *Characteristic multipliers and stability of symmetric periodic solutions of $\dot{x}(t) = g(x(t-1))$,* Trans. Amer. Math. Soc. **307** (1988), 127–142.

[20] Cohen, M. A. and Grossberg, S., *Absolute stability of global pattern formation and parallel memory storage by competitive neural networks,* IEEE Trans. SMC. **13** (1983), 815–826.

[21] Deimling, K., *Nonlinear Functional Analysis,* Springer-Verlag, Berlin, 1985.

[22] Diekmann, O., van Gils, S. A., Verduyn Lunel, S. M. and Walther, H.-O., *Delay Equations, Functional-, Complex-, and Nonlinear Analysis,* Springer-Verlag, New York, 1995.

[23] Dieudonné, J., *Foundations of Modern Analysis,* Academic Press, New York, 1960.

[24] van den Driessche, P. and Zou, X. F., *Global attractivity in delayed Hopfield neural network models,* SIAM J. Math. Anal., to appear.

[25] Fiedler, F. and Mallet-Paret, J., *Connections between Morse sets for delay differential equations,* J. reine angew. Math. **397** (1989), 23–41.

[26] Győri, I., *Connections between compartmental systems with pipes and integrodifferential equations,* Mathematical Modelling **7** (1987), 1215–1238.

[27] Hale, J. K., *Theory of Functional Differential Equations,* Springer-Verlag, New York, 1977.

[28] Hale, J. K., *Asymptotic Behavior of Dissipative Systems,* Amer. Math. Soc., Providence, RI, 1988.

[29] Hale, J. K. and Lin, X. B., *Symbolic dynamics and nonlinear semiflows,* Annali Mat. Pura Appl. **144** (1986), 693–709.

[30] Hale, J. K. and Verduyn Lunel, S. M., *Introduction to Functional Differential Equations,* Springer-Verlag, New York, 1993.

[31] Hatvani, L. and Krisztin, T., *Asymptotic stability for a differential-difference equation containing terms with and without delay,* Acta Sci. Math. (Szeged) **60** (1995), 371–384.

[32] Herz, A. V. M., *Global Analysis of recurrent neural networks,* in Models of Neural Networks, Vol. 3 (E. Domany, J. L. van Hemmen and K. Schulten, eds.), Springer-Verlag, New York, 1994.

[33] Hirsch, M., *Stability and convergence in strongly monotone dynamical systems,* J. reine angew. Math. **383** (1988), 1–53.

[34] Hocking, J. G. and Young, G. S., *Topology,* Addison-Wesley, Dover, 1988.

[35] Hopfield, J. J., *Neural networks and physical systems with emergent collective computational abilities,* Proc. Natl. Acad. Sci. **79** (1982), 2554–2558.

[36] Hopfield, J. J., *Neurons with graded response have collective computational properties like two-stage neurons,* Proc. Natl. Acad. Sci. **81** (1984), 3088–3092.

[37] Iooss, G., *Bifurcation of Maps and Applications,* North-Holland, Amsterdam, 1979.

[38] Krisztin, T., *On the convergence of solutions of functional differential equations,* Acta Sci. Math. **43** (1981), 45–54.

[39] Krisztin, T., *Convergence of solutions of a nonlinear integrodifferential equation arising in compartmental systems,* Acta Sci. Math. **47** (1984), 471–485.

[40] Krisztin, T., *An invariance principle of Lyapunov–Razumikhin type and compartmental systems,* in Proc. of the First World Congress of Nonlinear Analysts (V. Lakshmikantham, ed.), pp. 1371–1379, Walther de Gruyter, Berlin, 1996.

[41] Lani-Wayda, B. and Walther, H.-O., *Chaotic motion generated by delayed negative feedback, Part I: a transversality criterion,* Differential and Integral Equations **8** (1995), 1407–1452.

[42] Mallet-Paret, J., *Morse decompositions for differential delay equations,* J. Differential Equations **72** (1988), 270–315.

[43] Mallet-Paret, J. and Sell, G., *Systems of differential delay equations: Floquet multipliers and discrete Lyapunov functions,* J. Differential Equations **125** (1996), 385–440.

[44] Mallet-Paret, J. and Sell, G., *The Poincaré–Bendixson theorem for monotone cyclic feedback systems with delay,* J. Differential Equations **125** (1996), 441–489.

[45] Mallet-Paret, J. and Walther, H.-O., *Rapid oscillations are rare in scalar systems governed by monotone negative feedback with a time delay,* Math. Inst., University of Giessen, 1994, preprint.

[46] Marcus, C. M. and Westervelt, R. M., *Stability of analog neural networks with delay,* Phys. Rev. A **39** (1989), 347–356.

[47] Morita, M., *Associative memory with non-monotone dynamics,* Neural Networks **6** (1993), 115-126.

[48] Neugebauer, A., *Invariante Mannigfaltigkeiten und Neigungslemmata für Abbildungen in Banachräumen,* Diploma thesis, Universität München, 1988.

[49] Olien, L. and Bélair, J., *Bifurcations, stability, and monotonicity properties of a delayed neural network model,* Physica D. **102** (1997), 349–363.

[50] Pakdaman, K., Malta, C. P., Grotta-Ragazzo, C. and Vibert, J.-F., *Effect of delay on the boundary of the basin of attraction in a self-excited single neuron,* Neural Computation **9** (1997), 319–336.

[51] Pakdaman, K., Grotta-Ragazzo, C., Malta, C. P., Arino, O. and Vibert, J.-F., *Effect of delay on the boundary of the basin of attraction in a system of two neurons,* preprint.

[52] Rinow, W., *Lehrbuch der Topologie,* VEB Deutscher Verlag der Wissenschaften, Berlin, 1975.

[53] Schoenfliess, A., *Die Entwicklung der Lehre von den Punktmannigfaltigkeiten. Bericht erstattet der Deutschen Mathematiker-Vereinigung, Teil II,* J.-Ber. Deutsch. Math.-Verein., Ergänzungsband II, 1908.

[54] Smith, H. L., *Monotone semiflows generated by functional differential equations,* J. Differential Equations **66** (1987), 420–442.

[55] Smith, H. L., *Oscillation and multiple steady states in a cyclic gene model with repression,* J. Math. Biol. **25** (1987), 169–190.

[56] Smith, H. L., *Monotone Dynamical Systems: An Introduction to the Theory of Competitive and Cooperative Systems,* Amer. Math. Soc., Providence, RI, 1995.

[57] Smith, H. L. and Thieme, H. R., *Strongly order preserving semiflows generated by functional differential equations,* J. Differential Equations **93** (1991), 332–363.

[58] Tank, D. and Hopfield, J. J., *Simple neural optimization networks: an A/D converter, single decision circuit and a linear programming circuit,* IEEE Trans. Circ. Syst. **33** 1986, 533–541.

[59] Vanderbauwhede, A. and van Gils, S. A., *Center manifolds and contractions on a scale of Banach spaces,* J. Functional Analysis **71** (1987), 209–224.

[60] Vanderbauwhede, A. and Iooss, G., *Center manifold theory in infinite dimensions,* in Dynamics Reported (New Series), Vol. 1 (C.K.R.T. Jones, U. Kirchgraber and H.-O. Walther, eds.), Springer-Verlag, New York, 1992.

[61] Walther, H.-O., *Über Ejektivität und periodische Lösungen bei Funktionaldifferentialgleichungeng mit verteilter Verzögerung,* Habilitationsschrift, Universität München, 1977.

[62] Walther, H.-O., *On instability, ω-limit sets and periodic solutions of nonlinear autonomous differential delay equations,* in *Functional Differential Equations and Approximation of Fixed Points* (H.-O. Peitgen, H.-O. Walther eds.), Lecture Notes in Math., Vol. 730, pp. 489–503, Springer-Verlag, New York, 1979.

[63] Walther, H.-O., *Homoclinic solutions and chaos in $\dot{x}(t) = f(x(t-1))$,* Nonlinear Anal. **5** (1981), 775–788.

[64] Walther, H.-O., *Bifurcation from periodic solutions in functional differential equations,* Math. Z. **182** (1983), 269–289.

[65] Walther, H.-O., *Hyperbolic periodic solutions, heteroclinic connections and transversal homoclinic points in autonomous differential delay equations,* Memoirs of the Amer. Math. Soc., Vol. 402, Amer. Math. Soc., Providence, RI, 1989.

[66] Walther, H.-O., *A differential delay equation with a planar attractor,* in *Proc. of the Int. Conf. on Differential Equations*, Université Cadi Ayyad, Marrakech, 1991.

[67] Walther, H.-O., *An invariant manifold of slowly oscillating solutions for $x'(t) = -\mu x(t) + f(x(t-1))$,* J. reine angew. Math. **414** (1991), 67–112.

[68] Walther, H.-O., *On Floquet multipliers of periodic solutions of delay equations with monotone nonlinearities,* in *Int. Symp. Functional Differential Equations (Kyoto, 1990)*, pp. 349–356, World Scientific, Singapore, 1991.

[69] Walther, H.-O., *Unstable manifolds of periodic orbits of a differential delay equation,* Contemp. Math. **129** (1992), 177–239.

[70] Walther, H.-O., *The 2-dimensional attractor of $\dot{x}(t) = -\mu x(t) + f(x(t-1))$,* Memoirs of the Amer. Math. Soc., Vol. 544, Amer. Math. Soc., Providence, RI, 1995.

[71] Walther, H.-O. and Yebdri, M., *Smoothness of the attractor of almost all solutions of a delay differential equation,* Dissertationes Mathematicae **368**, 1997.

[72] Wu, J., *Symmetric functional differential equations and neural networks with memory,* Trans. Amer. Math. Soc., to appear.

Index